Intelligent Systems Reference Library

Volume 156

Series Editors

Janusz Kacprzyk, Polish Academy of Sciences, Warsaw, Poland

Lakhmi C. Jain, Faculty of Engineering and Information Technology, Centre for
Artificial Intelligence, University of Technology, Sydney, NSW, Australia;
Faculty of Science, Technology and Mathematics, University of Canberra,
Canberra, ACT, Australia;
KES International, Shoreham-by-Sea, UK;
Liverpool Hope University, Liverpool, UK

The aim of this series is to publish a Reference Library, including novel advances and developments in all aspects of Intelligent Systems in an easily accessible and well structured form. The series includes reference works, handbooks, compendia, textbooks, well-structured monographs, dictionaries, and encyclopedias. It contains well integrated knowledge and current information in the field of Intelligent Systems. The series covers the theory, applications, and design methods of Intelligent Systems. Virtually all disciplines such as engineering, computer science, avionics, business, e-commerce, environment, healthcare, physics and life science are included. The list of topics spans all the areas of modern intelligent systems such as: Ambient intelligence, Computational intelligence, Social intelligence, Computational neuroscience, Artificial life, Virtual society, Cognitive systems, DNA and immunity-based systems, e-Learning and teaching, Human-centred computing and Machine ethics, Intelligent control, Intelligent data analysis, Knowledge-based paradigms, Knowledge management, Intelligent agents, Intelligent decision making, Intelligent network security, Interactive entertainment, Learning paradigms, Recommender systems, Robotics and Mechatronics including human-machine teaming, Self-organizing and adaptive systems, Soft computing including Neural systems, Fuzzy systems, Evolutionary computing and the Fusion of these paradigms, Perception and Vision, Web intelligence and Multimedia.
** Indexing: The books of this series are submitted to ISI Web of Science, SCOPUS, DBLP and Springerlink.

More information about this series at http://www.springer.com/series/8578

Fawaz Alsolami · Mohammad Azad ·
Igor Chikalov · Mikhail Moshkov

Decision and Inhibitory Trees and Rules for Decision Tables with Many-valued Decisions

 Springer

Fawaz Alsolami
Computer, Electrical and Mathematical
Sciences and Engineering Division
King Abdullah University of Science
and Technology
Thuwal, Saudi Arabia

Mohammad Azad
Computer, Electrical and Mathematical
Sciences and Engineering Division
King Abdullah University of Science
and Technology
Thuwal, Saudi Arabia

Igor Chikalov
Computer, Electrical and Mathematical
Sciences and Engineering Division
King Abdullah University of Science
and Technology
Thuwal, Saudi Arabia

Mikhail Moshkov
Computer, Electrical and Mathematical
Sciences and Engineering Division
King Abdullah University of Science
and Technology
Thuwal, Saudi Arabia

ISSN 1868-4394   ISSN 1868-4408   (electronic)
Intelligent Systems Reference Library
ISBN 978-3-030-12856-2   ISBN 978-3-030-12854-8   (eBook)
https://doi.org/10.1007/978-3-030-12854-8

Library of Congress Control Number: 2019930960

This Springer imprint is published by the registered company Springer Nature Switzerland AG
The registered company address is: Gewerbestrasse 11, 6330 Cham, Switzerland

To our families

Preface

This book is devoted to the study of decision and inhibitory trees and rules for decision tables with many-valued decisions. In conventional decision tables, a single decision is associated with each row. However, situations in which a set of decisions is associated with each row are often possible. For some decision tables, inhibitory trees and rules can represent more information than decision trees and rules.

We consider various examples of problems and decision tables with many-valued decisions and discuss the difference between decision and inhibitory trees and rules for decision tables with many-valued decisions. We mention without proofs some relatively simple results obtained earlier for decision trees, tests, rules, and rule systems for binary decision tables with many-valued decisions. We generalize these results to the inhibitory trees, tests, rules, and rule systems.

We extend the multi-stage and bi-criteria optimization approaches to the case of decision trees and rules for decision tables with many-valued decisions and then generalize them to the case of inhibitory trees and rules. The applications of these techniques include the study of totally optimal (optimal relative to a number of criteria simultaneously) decision and inhibitory trees and rules, the comparison of greedy heuristics for tree and rule construction as single-criterion and bi-criteria optimization algorithms, the development of the restricted multi-pruning approach used in classification and knowledge representation, etc.

We also study the time complexity of decision and inhibitory trees and rule systems over arbitrary sets of attributes represented by information systems.

The results presented in this book can be useful for researchers in data mining, knowledge discovery, and machine learning, especially those who work in rough set theory, test theory, and logical analysis of data. The book can be used for the creation of courses for graduate students.

Thuwal, Saudi Arabia
November 2018

Fawaz Alsolami
Mohammad Azad
Igor Chikalov
Mikhail Moshkov

Acknowledgements

We are greatly indebted to King Abdullah University of Science and Technology for the immense support.

We are grateful to our coauthors in papers devoted to the creation of extensions of dynamic programming for decision and inhibitory trees, rules, and rule systems: Hassan AbouEisha, Mohammed Al Farhan, Abdulaziz Alkhalid, Saad Alrawaf, Talha Amin, Monther Busbait, Shahid Hussain, and Beata Zielosko.

We are thankful to Prof. Andrzej Skowron for stimulating discussions.

We extend an expression of gratitude to Prof. Janusz Kacprzyk, to Dr. Thomas Ditzinger, and to the Series Intelligent Systems Reference Library staff at Springer for their support in making this book possible.

Contents

Chapter 1
Introduction

This book is devoted to the study of decision and inhibitory trees and rules for decision tables with many-valued decisions. In conventional decision tables (decision tables with single-valued decisions), a single decision is associated with each object (row). However, situations in which a set of decisions is associated with each row are often possible. For some decision tables, inhibitory trees and rules can represent more information than decision trees and rules.

We begin with the discussion of decision and inhibitory interpretations of decision tables with many-valued decisions, then consider results obtained in the frameworks of the three main directions of the study, describe the contents of the book, and finally add some words about its use.

1.1 Decision and Inhibitory Interpretations of Decision Tables with Many-valued Decisions

In each decision table with many-valued decisions T, columns are labeled with conditional attributes, rows are pairwise different, and each row r is labeled with a nonempty finite set of decisions $D(r)$. Let D be the union of the sets $D(r)$ among all rows r of the table T. We consider two interpretations of the decision table T: decision and inhibitory.

Decision interpretation: for a given row r of the table T, the objective is to find an arbitrary decision from the set of decisions $D(r)$.

Inhibitory interpretation: for a given row r of the table T, the objective is to find an arbitrary decision from the set $D \setminus D(r)$. When we consider the inhibitory interpretation, we require that $D(r) \neq D$ for every row r of T.

© Springer Nature Switzerland AG 2020
F. Alsolami et al., *Decision and Inhibitory Trees and Rules for Decision Tables with Many-valued Decisions*, Intelligent Systems Reference Library 156,
https://doi.org/10.1007/978-3-030-12854-8_1

1.1.1 Decision Interpretation

In the frameworks of the decision interpretation, we use decision trees and complete systems of decision rules for the table T to find a decision from the set $D(r)$ attached to a given row r. A complete system S of decision rules for T is a finite set of decision rules such that, for each row from T, there is at least one applicable rule from S, and each row from T satisfies all applicable rules from S. We also consider the notion of a decision test for T—a set of conditional attributes of T such that the values of these attributes are sufficient to find, for an arbitrary row r, a decision from the set $D(r)$.

Decision trees and rule systems are well known as means of knowledge representation, as classifiers to predict decisions for new objects, and as algorithms to solve various problems of combinatorial optimization, fault diagnosis, and computational geometry [14, 15, 22, 34, 44]. Decision tests and reducts (minimal tests) are one of the main objects of study in test theory and rough set theory [16, 41].

There are at least three sources of decision tables with many-valued decisions for which the decision interpretation can make sense.

The first source is decision tables for exactly formulated problems with many-valued decisions for which the set of inputs (objects) and the sets of decisions for these objects are completely known. In Chap. 2, we consider examples of exactly formulated problems and corresponding decision tables with many-valued decisions from the following areas: computational geometry, combinatorial optimization, and diagnosis of faults in combinatorial circuits. For each of the considered decision tables, decisions from the set attached to an arbitrary row are equally appropriate for this row and it is enough to find only one such decision.

The first examples of decision tables with many-valued decisions for which the decision interpretation makes sense were considered in 1981 [27] (see also a short book [32]) and were related mainly to the optimization problems. Let us have a finite set of linear forms over n-dimensional real space. For a given point of this space, the goal is to determine which form takes the minimum value (or absolute minimum value) at this point. We can also consider the problem of sorting of these forms. For each of the considered problems, there are inputs (n-tuples of real numbers) for which we have more than one possible decision (solution). These decisions are equally appropriate, and it is enough to find only one decision. We proved that, for each of the considered problems, there is a unique minimal set of attributes of the kind sign(l), where l is a linear form over n-dimensional real space in which the first coefficient is positive, such that the values of these attributes are sufficient to solve this problem. To obtain these results, we studied decision tables with many-valued decisions from some class and proved the uniqueness of dead-end tests (reducts) for these tables.

The second source is decision tables for problems with many-valued decisions based on experimental data in which objects can have, in a natural way, two or more decisions: a text, a music, an architecture can belong to multiple genres; a disease may belong to multiple categories; genes may have multiple functions; an image can belong to different semantic classes [13]. For such decision tables, the set of rows can be only a part of the set of all possible rows and the sets of decisions for rows can be incomplete. Such decision tables are known also as multi-label decision tables. They are studied intensively last decades [10, 43, 50, 51]. In many cases, the aim of the study is to construct a classifier which, for a given new object, will predict all decisions appropriate for this object. However, in some of the considered decision tables, decisions from the sets attached to rows are equally appropriate for objects corresponding to rows, and it is enough to find one decision from the set. An example is a decision table in which each row describes symptoms of some disease and the set of decisions attached to this row contains the most appropriate medicines for this disease.

The third source is inconsistent decision tables with single-valued decisions based on experimental data in which equal rows (objects with equal values of conditional attributes) can be labeled with different decisions. The causes of inconsistency are the following: either (i) we have no enough number of conditional attributes to distinguish these objects, or (ii) equal rows correspond to the same object but with different appropriate for this object decisions. We can keep one row from the group of equal rows and label this row with the set of decisions attached to all rows from the group. As a result, we obtain a decision table with many-valued decisions.

For the case (i), the decision interpretation of the obtained decision table makes some sense: for a given object, we recognize a decision attached to this object or to an object which is not distinguishable from the initial one by the existing set of conditional attributes.

In the case (ii), we deal with a decision table with many-valued decisions from the second source for which the decision interpretation can make sense. It makes sense, for example, if each decision is a possible behavior in the situation described by values of conditional attributes for the row, quality of behaviors is the same or does not differ much, and we would like to create a strategy of behavior in the form of a decision tree.

In Chap. 3, we compare the following three approaches to handle inconsistent decision tables: (a) to find, for a given group of equal rows, a decision attached to a row from the group (our approach), (b) to find, for a given group of equal rows, a decision which is the most common among all decisions attached to these rows, and (c) to find, for a given group of equal rows, all decisions attached to these rows (generalized decision approach widely used in rough set theory [41]). The experimental results show that the approach (a) is in some sense better than the approach (b), and the latter approach is better than the approach (c). "Better" here means that, for the most part of the considered decision tables, the decision trees constructed in the framework of one approach are simpler and have less misclassification error than the decision trees constructed in the framework of another approach. So, if our approach is appropriate to handle a given inconsistent decision table, then we can use it instead of the other two approaches.

1.1.2 Inhibitory Interpretation

In the frameworks of the inhibitory interpretation, we use inhibitory trees and complete systems of inhibitory rules for the table T to find a decision from the set $D \setminus D(r)$ for a given row r. Instead of decisions, in the terminal nodes of inhibitory trees and in the right-hand sides of inhibitory rules, there are expressions of the kind "$\neq$ decision". A complete system S of inhibitory rules for T is a finite set of inhibitory rules such that, for each row from T, there is at least one applicable rule from S, and each row from T satisfies all applicable rules from S. We consider also the notion of an inhibitory test for T—a set of conditional attributes of T such that the values of these attributes are sufficient to find, for an arbitrary row r, a decision from the set $D \setminus D(r)$.

Inhibitory trees, rules, rule systems, and tests are less known objects. The notion of an inhibitory rule was introduced in 2007 [20]. The first stage of the inhibitory rule study was devoted to the investigation of expressive abilities of inhibitory association rules, structure of the set of inhibitory association rules, greedy algorithms for inhibitory rule construction, and classifiers based on inhibitory rules. The obtained results were summarized in the book [21]. The second stage of the study is devoted mainly to the generalization of dynamic programming algorithms for decision rule optimization to the case of inhibitory rules [4–7]. The obtained results are summarized in the present book. The notion of an inhibitory tree was introduced in 2017 [8]. In this book, we extend many results obtained earlier for the decision trees to the case of inhibitory trees.

Two examples considered in Chap. 2 show that the use of inhibitory trees and rules gives us additional possibilities in comparison with decision trees and rules. These examples clarify why the inhibitory interpretation of decision tables makes sense.

The first example is related to the classification problem: for a new row that does not belong to a decision table, we should predict decisions corresponding to this row. The considered example shows that the inhibitory trees and rules can give us more information about possible decisions for the new row than the decision trees and rules.

In the book [21], different kinds of classifiers using decision and inhibitory rules were considered. The obtained results show that classifiers based on inhibitory rules are often better than the classifiers based on decision rules.

The second example is related to the problem of knowledge representation. We consider an information system I represented as a table in which columns are labeled with attributes and rows are pairwise different. We can fix sequentially each attribute of the information system as the decision attribute and consider decision and inhibitory trees and rules for the obtained decision tables as decision and inhibitory association trees and rules for the initial information system. The question under consideration: is it possible to describe the set of rows of I using only true and realizable trees and rules? We show that both decision association trees and rules cannot describe the set of rows of I. However, both inhibitory association trees and rules can describe this set.

The existence of information systems for which the information derived from the system by decision association rules is incomplete was discovered in [47, 49]. The fact that inhibitory association rules give us complete information for arbitrary information system was proved in [40]. At the end of Chap. 2, we show that the inhibitory association trees also give us complete information for arbitrary information system.

1.2   Main Directions of Study

There are three main directions of study in this book.

1.2.1   Explaining Examples and Preliminary Results

The first direction (Part I of this book) is in some sense preliminary. We consider various examples of problems and decision tables with many-valued decisions, discuss the difference between decision and inhibitory interpretations of decision tables with many-valued decisions, and compare three ways to handle inconsistent decision tables: our approach (to find, for a given group of equal rows, a decision attached to a row from the group), the approach based on the most common decisions, and the approach based on the generalized decisions. The aims are to show that (i) the decision interpretation of decision tables with many-valued decisions makes sense in many cases, (ii) the inhibitory trees and rules can give us more information about decision tables than the decision trees and rules, i.e., the inhibitory interpretation of decision tables makes sense, and (iii) if our approach to handle inconsistent decision tables is applicable, then it is better to use it than the two other approaches based on the most common and generalized decisions.

In [29], lower bounds on time complexity of decision trees for decision tables with many-valued decisions were studied. Upper bounds on complexity and algorithms for the construction of decision trees for such tables were considered in [35]. Both papers use the decision interpretation of decision tables with many-valued decisions. The book [39] contains some of these results and many others related to decision trees, tests, rules, and rule systems for binary decision tables with many-valued decisions. We mention these results without proofs and generalize them to the case of inhibitory trees, tests, rules, and rule systems. The obtained results include relationships among inhibitory trees, rules, and tests, bounds on their complexity, greedy algorithms for construction of inhibitory trees, rules, and tests, and dynamic programming algorithms for minimization of tree depth and rule length.

1.2.2 Extensions of Dynamic Programming for Decision and Inhibitory Trees, Rules, and Systems of Rules

The second direction (Parts II and III of this book) is related to the extensions of dynamic programming which includes multi-stage and bi-criteria optimization.

We work with a directed acyclic graph (DAG) which describes the structure of subtables of the initial decision table given by conditions of the kind "attribute = value". This DAG represents the sets of decision trees and rules under consideration. The multi-stage optimization means that we apply to this graph (to represented sets of trees or rules) sequentially the procedures of optimization relative to different cost functions. The bi-criteria optimization means that, based on the DAG, we construct the set of Pareto optimal points for a bi-criteria optimization problem for either decision trees or decision rules.

The conventional dynamic programming approach for decision tree optimization was created earlier [9, 23, 26, 45]. We began to work in this direction in 2000 [36] and published the first paper related to the multi-stage optimization of decision trees in 2002 [37] (see also an extended version of this paper [38]). Some results related to the bi-criteria optimization of decision trees can be found in [17, 18, 24]). The book [1] summarizes results on multi-stage and bi-criteria optimization of decision trees and rules for the decision tables with single-valued decisions.

We extend these results to the case of decision trees and rules for decision tables with many-valued decisions. Then we generalize them to the case of inhibitory trees and rules. To this end, we show that, instead of the study of inhibitory trees and rules for a decision table T, we can investigate decision trees and rules for the decision table T^C complementary to T. The table T^C has the same sets of attributes and rows as the table T. A row r in T^C is labeled with the set of decision $D \setminus D(r)$ where $D(r)$ is the set of decisions attached to the row r in the table T, and D is the union of sets $D(r)$ among all rows r of the table T.

The second direction contains not only theoretical results but also some applications based on computer experiments with the software system Dagger (from DAG) developed by our group in KAUST [1–3].

For the decision and inhibitory trees, the applications of the multi-stage optimization approach include the study of the minimum depth, minimum average depth, and minimum number of nodes in decision trees for sorting $n = 2, \ldots, 7$ elements among which there are, possibly, equal elements, and the study of totally optimal (simultaneously optimal relative to a number of cost functions) decision and inhibitory trees for decision tables with many-valued decisions obtained from decision tables from the UCI ML Repository [25] by the removal of some conditional attributes. The applications of the bi-criteria optimization approach include the comparison of 12 greedy heuristics for the construction of decision and inhibitory trees for decision tables with many-valued decisions as algorithms for single-criterion and bi-criteria optimization, the study of two relationships for decision trees related to the knowledge representation—number of nodes versus depth and number of nodes versus average depth, and the study of a new technique called restricted multi-pruning of decision trees which is used for data mining, knowledge representation, and classification.

For the decision and inhibitory rules and systems of rules, the applications of the multi-stage optimization approach include the study of decision and inhibitory rules that are totally optimal relative to the length and coverage (have minimum length and maximum coverage simultaneously), and the investigation of a simulation of a greedy algorithm for the construction of relatively small sets of decision rules that cover almost all rows in decision tables with many-valued decisions. The applications of the bi-criteria optimization approach include the comparison of 13 greedy heuristics for the construction of decision rules from the point of view of single-criterion optimization (relative to the length or coverage) and bi-criteria optimization (relative to the length and coverage).

1.2.3 Study of Decision and Inhibitory Trees and Rule Systems Over Arbitrary Information Systems

The third direction (Part IV of this book) is devoted to the study of time complexity of decision and inhibitory trees and rule systems over arbitrary sets of attributes represented by information systems. As the time complexity of a tree, we consider its depth—the maximum length of a path from the root to a terminal node, and as the time complexity of a rule system, we consider its length—the maximum length of a rule from the system.

An information system $\mathcal{U} = (A, B, F)$ consists of a set A, a finite set B, and a set F of attributes each of which is a function from A to B. This information system is called finite if F is a finite set and infinite otherwise. We define the notion of a problem with many-valued decisions over $\mathcal{U}$. Each such problem is described by a finite number of attributes from F which divide the set A into domains and a mapping that corresponds to each domain a finite set of decisions.

We consider two approaches to the study of decision and inhibitory trees and rule systems for problems with many-valued decisions over $\mathcal{U}$: local and global. Local approach assumes that trees and rules can use only attributes from the problem description. Global approach allows us to use arbitrary attributes from the set F in trees and rules.

For each kind of information systems (finite and infinite) and for each approach (local and global), we study the behavior of four Shannon functions which characterize the growth in the worst case of (i) minimum depth of decision trees, (ii) minimum length of decision rule systems, (iii) minimum depth of inhibitory trees, and (iv) minimum length of inhibitory rule systems with the growth of the number of attributes in problem description.

For the study of the most part of cases, it is enough to consider results published years ago in [28, 30, 31, 33, 34, 39]. The only exception is the case of global approach and finite information systems which is considered in details in Chap. 15.

Note that this is the only direction in which we go beyond the consideration of finite sets of attributes for the construction of trees and rules.

1.3   Contents of Book

This book includes Introduction and four parts consisting of 14 chapters.

1.3.1   Part I. Explaining Examples and Preliminary Results

In Part I, we consider preliminary definitions, examples, and results. It consists of three chapters.

In Chap. 2, we discuss notions of a problem with many-valued decisions, decision table with many-valued decisions, decision and inhibitory tree, rule, and rule system for a problem (in details) and for a decision table (in brief). We consider examples of problems in which objects can have two or more equally appropriate decisions in a natural way, and an example of inconsistent decision table that can be transformed into a decision table with many-valued decisions. At the end, we discuss two examples which explain why we study not only decision but also inhibitory trees and rules.

In Chap. 3, we discuss three approaches to handle inconsistency in decision tables: decision tables with many-valued decisions, decision tables with generalized decisions, and decision tables with most common decisions. We compare the complexity and classification accuracy of decision trees constructed by greedy heuristics in the frameworks of these approaches.

In Chap. 4, we consider in more details definitions related to binary decision tables with many-valued decisions, decision and inhibitory trees, tests, rules, and rule systems for these tables. We mention without proofs some relatively simple results obtained earlier for binary decision tables with many-valued decisions: relationships among decision trees, rules and tests, bounds on their complexity, greedy algorithms for construction of decision trees, rules and tests, and dynamic programming algorithms for minimization of tree depth and rule length. We extend these results to the inhibitory trees, tests, rules, and rule systems for binary decision tables with many-valued decisions.

1.3.2   Part II. Extensions of Dynamic Programming for Decision and Inhibitory Trees

Part II is devoted to the development of extensions of dynamic programming for decision and inhibitory trees. It consists of six chapters.

In Chap. 5, we consider some notions related to decision tables with many-valued decisions (the notions of table, directed acyclic graph for this table, uncertainty and completeness measures, and restricted information system) and discuss tools (statements and algorithms) for the work with Pareto optimal points. The definitions and results from this chapter are used both in Part II and Part III. Note that many results considered in this chapter were obtained jointly with Hassan AbouEisha, Talha Amin, and Shahid Hussain and already published in the book [1].

In Chap. 6, we define various types of decision and inhibitory trees. We discuss the notion of a cost function for trees, the notion of decision tree uncertainty, and the notion of inhibitory tree completeness. We also design an algorithm for counting the number of decision trees represented by a directed acyclic graph.

In Chap. 7, we consider multi-stage optimization of decision and inhibitory trees relative to a sequence of cost functions, and two applications of this technique: study of decision trees for sorting problem and study of totally optimal (simultaneously optimal relative to a number of cost functions) decision and inhibitory trees for modified decision tables from the UCI ML Repository.

In Chap. 8, we study bi-criteria optimization problem cost versus cost for decision and inhibitory trees. We design an algorithm which constructs the set of Pareto optimal points for bi-criteria optimization problem for decision trees, and show how the constructed set can be transformed into the graphs of functions that describe the relationships between the studied cost functions. We extend the obtained results to the case of inhibitory trees. As applications of bi-criteria optimization for two cost functions, we compare 12 greedy heuristics for construction of decision and inhibitory trees as single-criterion and bi-criteria optimization algorithms, and study two relationships for decision trees related to knowledge representation—number of nodes versus depth and number of nodes versus average depth.

In Chap. 9, we consider bi-criteria optimization problems cost versus uncertainty for decision trees and cost versus completeness for inhibitory trees, and discuss illustrative examples. The created tools allow us to understand complexity versus accuracy trade-off for decision and inhibitory trees and to choose appropriate trees.

In Chap. 10, we consider so-called multi-pruning approach based on dynamic programming algorithms for bi-criteria optimization of CART-like decision trees relative to the number of nodes and the number of misclassifications. This approach allows us to construct the set of all Pareto optimal points and to derive, for each such point, decision trees with parameters corresponding to that point. Experiments with decision tables from the UCI ML Repository show that, very often, we can find a suitable Pareto optimal point and derive a decision tree with a small number of nodes at the expense of a small increment in the number of misclassifications. Such decision trees can be used for knowledge representation. Multi-pruning approach includes a procedure which constructs decision trees that, as classifiers, often outperform decision trees constructed by CART [14]. We also consider a modification of multi-pruning approach (restricted multi-pruning) that requires less memory and time but usually keeps the quality of constructed trees as classifiers or as a way for knowledge representation. We extend the considered approaches to the case of decision tables with many-valued decisions.

1.3.3 Part III. Extensions of Dynamic Programming for Decision and Inhibitory Rules and Systems of Rules

Part III is devoted to the development of extensions of dynamic programming for decision and inhibitory rules and rule systems for decision tables with many-valued decisions. It consists of four chapters.

In Chap. 11, we define various types of decision and inhibitory rules and systems of rules. We discuss the notion of cost function for rules, the notion of decision rule uncertainty, and the notion of inhibitory rule completeness. Similar notions are introduced for systems of decision and inhibitory rules.

In Chap. 12, we consider the optimization of decision and inhibitory rules including multi-stage optimization relative to a sequence of cost functions. We discuss an algorithm for counting the number of optimal rules and consider simulation of a greedy algorithm for construction of decision rule sets. We also discuss results of computer experiments with decision and inhibitory rules: existence of small systems of sufficiently accurate decision rules that cover almost all rows, and existence of totally optimal decision and inhibitory rules that have minimum length and maximum coverage simultaneously.

In Chap. 13, we consider algorithms which construct the sets of Pareto optimal points for bi-criteria optimization problems for decision rules and rule systems relative to two cost functions. We show how the constructed set of Pareto optimal points can be transformed into the graphs of functions which describe the relationships between the considered cost functions. We compare 13 greedy heuristics for the construction of decision rules from the point of view of single-criterion optimization (relative to the length or coverage) and bi-criteria optimization (relative to the length and coverage). At the end of the chapter, we generalize the obtained results to the case of inhibitory rules and systems of inhibitory rules.

In Chap. 14, we consider algorithms which construct the sets of Pareto optimal points for bi-criteria optimization problems for decision (inhibitory) rules and rule systems relative to a cost function and an uncertainty (completeness) measure. We show how the constructed set of Pareto optimal points can be transformed into the graphs of functions which describe the relationships between the considered cost function and uncertainty (completeness) measure. Computer experiments provide us with examples of trade-off between complexity and accuracy for decision and inhibitory rule systems.

1.3.4 Part IV. Study of Decision and Inhibitory Trees and Rule Systems Over Arbitrary Information Systems

Part IV is devoted to the study of time complexity of decision and inhibitory trees and rule systems over arbitrary sets of attributes represented by information systems. It consists of Chap. 15.

In Chap. 15, we define the notions of an information system and a problem with many-valued decisions over this system. We consider local and global approaches to the study of decision and inhibitory trees and rule systems for problems with many-valued decisions over an information system. The local approach assumes that trees and rules can use only attributes from the problem description. The global approach allows us to use arbitrary attributes from the information system in trees and rules.

For each kind of information systems (finite and infinite) and for each approach (local and global), we study the behavior of four Shannon functions which characterize the growth in the worst case of (i) minimum depth of decision trees, (ii) minimum length of decision rule systems, (iii) minimum depth of inhibitory trees, and (iv) minimum length of inhibitory rule systems with the growth of the number of attributes in problem description.

1.4 Use of Book

Current versions of the designed tools have research goals: they can only work with medium-sized decision tables.

The obtained theoretical results and designed tools can be useful in rough set theory and its applications [41, 42, 46] where decision rules and decision trees are widely used, in logical analysis of data [11, 12, 19] which is based mainly on the use of patterns (decision rules), and in test theory [16, 48, 52] which study both decision trees and decision rules.

The book can also be used for the creation of courses for graduate students.

References

1. AbouEisha, H., Amin, T., Chikalov, I., Hussain, S., Moshkov, M.: Extensions of Dynamic Programming for Combinatorial Optimization and Data Mining, Intelligent Systems Reference Library, vol. 146. Springer, Berlin (2019)
2. Alkhalid, A., Amin, T., Chikalov, I., Hussain, S., Moshkov, M., Zielosko, B.: Dagger: a tool for analysis and optimization of decision trees and rules. In: Ficarra, F.V.C., Kratky, A., Veltman, K.H., Ficarra, M.C., Nicol, E., Brie, M. (eds.) Computational Informatics. Social Factors and New Information Technologies: Hypermedia Perspectives and Avant-Garde Experiencies in the Era of Communicability Expansion, pp. 29–39. Blue Herons (2011)
3. Alkhalid, A., Amin, T., Chikalov, I., Hussain, S., Moshkov, M., Zielosko, B.: Optimization and analysis of decision trees and rules: dynamic programming approach. Int. J. Gen. Syst. **42**(6), 614–634 (2013)
4. Alsolami, F., Chikalov, I., Moshkov, M., Zielosko, B.: Length and coverage of inhibitory decision rules. In: Nguyen, N.T., Hoang, K., Jedrzejowicz, P. (eds.) Computational Collective Intelligence. Technologies and Applications – 4th International Conference, ICCCI 2012, Ho Chi Minh City, Vietnam, 28–30 Nov 2012, Part II. Lecture Notes in Computer Science, vol. 7654, pp. 325–334. Springer (2012)
5. Alsolami, F., Chikalov, I., Moshkov, M., Zielosko, B.: Optimization of inhibitory decision rules relative to length and coverage. In: Li, T., Nguyen, H.S., Wang, G., Grzymala-Busse,

J.W., Janicki, R., Hassanien, A.E., Yu, H. (eds.) Rough Sets and Knowledge Technology – 7th International Conference, RSKT 2012, Chengdu, China, 17–20 Aug 2012. Lecture Notes in Computer Science, vol. 7414, pp. 149–154. Springer (2012)

6. Alsolami, F., Chikalov, I., Moshkov, M.: Sequential optimization of approximate inhibitory rules relative to the length, coverage and number of misclassifications. In: Lingras, P., Wolski, M., Cornelis, C., Mitra, S., Wasilewski, P. (eds.) Rough Sets and Knowledge Technology – 8th International Conference, RSKT 2013, Halifax, NS, Canada, 11–14 Oct 2013. Lecture Notes in Computer Science, vol. 8171, pp. 154–165. Springer, Berlin (2013)

7. Alsolami, F., Chikalov, I., Moshkov, M., Zielosko, B.: Optimization of approximate inhibitory rules relative to number of misclassifications. In: Watada, J., Jain, L.C., Howlett, R.J., Mukai, N., Asakura, K. (eds.) 17th International Conference in Knowledge Based and Intelligent Information and Engineering Systems, KES 2013, Kitakyushu, Japan, 9–11 Sept 2013. Procedia Computer Science, vol. 22, pp. 295–302. Elsevier (2013)

8. Azad, M., Moshkov, M.: Multi-stage optimization of decision and inhibitory trees for decision tables with many-valued decisions. Eur. J. Oper. Res. **263**(3), 910–921 (2017)

9. Bayes, A.J.: A dynamic programming algorithm to optimise decision table code. Aust. Comput. J. **5**(2), 77–79 (1973)

10. Blockeel, H., Schietgat, L., Struyf, J., Dzeroski, S., Clare, A.: Decision trees for hierarchical multilabel classification: A case study in functional genomics. In: Fürnkranz, J., Scheffer, T., Spiliopoulou, M. (eds.) Knowledge Discovery in Databases: 10th European Conference on Principles and Practice of Knowledge Discovery in Databases, PKDD 2006, Berlin, Germany, 18–22 Sept 2006. Lecture Notes in Computer Science, vol. 4213, pp. 18–29. Springer (2006)

11. Boros, E., Hammer, P.L., Ibaraki, T., Kogan, A.: Logical analysis of numerical data. Math. Program. **79**, 163–190 (1997)

12. Boros, E., Hammer, P.L., Ibaraki, T., Kogan, A., Mayoraz, E., Muchnik, I.: An implementation of logical analysis of data. IEEE Trans. Knowl. Data Eng. **12**, 292–306 (2000)

13. Boutell, M.R., Luo, J., Shen, X., Brown, C.M.: Learning multi-label scene classification. Pattern Recognit. **37**(9), 1757–1771 (2004)

14. Breiman, L., Friedman, J.H., Olshen, R.A., Stone, C.J.: Classification and Regression Trees. Wadsworth and Brooks, Monterey, CA (1984)

15. Carbonell, J.G., Michalski, R.S., Mitchell, T.M.: An overview of machine learning. In: Michalski, R.S., Carbonell, J.G., Mitchell, T.M. (eds.) Machine Learning. An Artificial Intelligence Approach, pp. 1–23. Tioga Publishing, Palo Alto (1983)

16. Chegis, I.A., Yablonskii, S.V.: Logical methods of control of work of electric schemes. Trudy Mat. Inst. Steklov (in Russian) **51**, 270–360 (1958)

17. Chikalov, I., Hussain, S., Moshkov, M.: Relationships between depth and number of misclassifications for decision trees. In: Kuznetsov, S.O., Slezak, D., Hepting, D.H., Mirkin, B.G. (eds.) 13th International Conference Rough Sets, Fuzzy Sets, Data Mining and Granular Computing, RSFDGrC 2011, Moscow, Russia, 25–27 June 2011. Lecture Notes in Computer Science, vol. 6743, pp. 286–292. Springer (2011)

18. Chikalov, I., Hussain, S., Moshkov, M.: Relationships between number of nodes and number of misclassifications for decision trees. In: Yao, J., Yang, Y., Slowinski, R., Greco, S., Li, H., Mitra, S., Polkowski, L. (eds.) Rough Sets and Current Trends in Computing – 8th International Conference, RSCTC 2012, Chengdu, China, 17–20 Aug 2012. Lecture Notes in Computer Science, vol. 7413, pp. 212–218. Springer (2012)

19. Crama, Y., Hammer, P.L., Ibaraki, T.: Cause-effect relationships and partially defined Boolean functions. Ann. Oper. Res. **16**, 299–326 (1988)

20. Delimata, P., Moshkov, M., Skowron, A., Suraj, Z.: Two families of classification algorithms. In: An, A., Stefanowski, J., Ramanna, S., Butz, C.J., Pedrycz, W., Wang, G. (eds.) Rough Sets, Fuzzy Sets, Data Mining and Granular Computing, 11th International Conference, RSFDGrC 2007, Toronto, Canada, 14–16 May 2007. Lecture Notes in Computer Science, vol. 4482, pp. 297–304. Springer (2007)

21. Delimata, P., Moshkov, M., Skowron, A., Suraj, Z.: Inhibitory Rules in Data Analysis: A Rough Set Approach. Studies in Computational Intelligence, vol. 163. Springer, Berlin (2009)

22. Fayyad, U., Piatetsky-Shapiro, G., Smyth, P.: From data mining to knowledge discovery in databases. AI Mag. **17**, 37–54 (1996)
23. Garey, M.R.: Optimal binary identification procedures. SIAM J. Appl. Math. **23**, 173–186 (1972)
24. Hussain, S.: Relationships among various parameters for decision tree optimization. In: Faucher, C., Jain, L.C. (eds.) Innovations in Intelligent Machines-4 – Recent Advances in Knowledge Engineering. Studies in Computational Intelligence, vol. 514, pp. 393–410. Springer, Berlin (2014)
25. Lichman, M.: UCI Machine Learning Repository. University of California, Irvine, School of Information and Computer Sciences (2013). http://archive.ics.uci.edu/ml
26. Martelli, A., Montanari, U.: Optimizing decision trees through heuristically guided search. Commun. ACM **21**(12), 1025–1039 (1978)
27. Moshkov, M.: On the uniqueness of dead-end tests for recognition problems with linear decision rules. In: Markov, A.A. (ed.) Combinatorial-Algebraic Methods in Applied Mathematics (in Russian), pp. 97–109. Gorky University Press, Gorky (1981)
28. Moshkov, M.: Decision Trees. Theory and Applications (in Russian). Nizhny Novgorod University Publishers, Nizhny Novgorod (1994)
29. Moshkov, M.: Lower bounds for the time complexity of deterministic conditional tests. Diskret. Mat. (in Russian) **8**(3), 98–110 (1996). https://doi.org/10.4213/dm538
30. Moshkov, M.: Unimprovable upper bounds on complexity of decision trees over information systems. Found. Comput. Decis. Sci. **21**, 219–231 (1996)
31. Moshkov, M.: On time complexity of decision trees. In: Polkowski, L., Skowron, A. (eds.) Rough Sets in Knowledge Discovery 1: Methodology and Applications. Studies in Fuzziness and Soft Computing, vol. 18, pp. 160–191. Physica-Verlag, Heidelberg (1998)
32. Moshkov, M.: Elements of Mathematical Theory of Tests with Applications to Problems of Discrete Optimization. Nizhni Novgorod University Publishers, Nizhni Novgorod (2001). (in Russian)
33. Moshkov, M.: Classification of infinite information systems depending on complexity of decision trees and decision rule systems. Fundam. Inform. **54**, 345–368 (2003)
34. Moshkov, M.: Time complexity of decision trees. In: Peters, J.F., Skowron, A. (eds.) Trans. Rough Sets III. Lecture Notes in Computer Science, vol. 3400, pp. 244–459. Springer (2005)
35. Moshkov, M.: Bounds on complexity and algorithms for construction of deterministic conditional tests. In: Mathematical Problems of Cybernetics (in Russian), vol. 16, pp. 79–124. Fizmatlit, Moscow (2007). http://library.keldysh.ru/mvk.asp?id=2007-79
36. Moshkov, M., Chikalov, I.: On algorithm for constructing of decision trees with minimal depth. Fundam. Inform. **41**(3), 295–299 (2000)
37. Moshkov, M., Chikalov, I.: Sequential optimization of decision trees relatively different complexity measures. In: 6th International Conference Soft Computing and Distributed Processing, Rzeszòw, Poland, 24–25 June, pp. 53–56 (2002)
38. Moshkov, M., Chikalov, I.: Consecutive optimization of decision trees concerning various complexity measures. Fundam. Inform. **61**(2), 87–96 (2004)
39. Moshkov, M., Zielosko, B.: Combinatorial Machine Learning - A Rough Set Approach. Studies in Computational Intelligence, vol. 360. Springer, Heidelberg (2011)
40. Moshkov, M., Skowron, A., Suraj, Z.: Maximal consistent extensions of information systems relative to their theories. Inf. Sci. **178**(12), 2600–2620 (2008)
41. Pawlak, Z.: Rough Sets - Theoretical Aspect of Reasoning About Data. Kluwer Academic Publishers, Dordrecht (1991)
42. Pawlak, Z., Skowron, A.: Rudiments of rough sets. Inf. Sci. **177**(1), 3–27 (2007)
43. Read, J., Martino, L., Olmos, P.M., Luengo, D.: Scalable multi-output label prediction: from classifier chains to classifier trellises. Pattern Recognit. **48**(6), 2096–2109 (2015)
44. Rokach, L., Maimon, O.: Data Mining with Decision Trees: Theory and Applications. World Scientific Publishing, River Edge (2008)
45. Schumacher, H., Sevcik, K.C.: The synthetic approach to decision table conversion. Commun. ACM **19**(6), 343–351 (1976)

46. Skowron, A., Rauszer, C.: The discernibility matrices and functions in information systems. In: Słowiński, R. (ed.) Intelligent Decision Support: Handbook of Applications and Advances of the Rough Sets Theory, pp. 331–362. Kluwer Academic Publishers, Dordrecht (1992)

47. Skowron, A., Suraj, Z.: Rough sets and concurrency. Bull. Pol. Acad. Sci. **41**(3), 237–254 (1993)

48. Soloviev, N.A.: Tests (Theory, Construction, Applications). Nauka, Novosibirsk (1978). (in Russian)

49. Suraj, Z.: Some remarks on extensions and restrictions of information systems. In: Ziarko, W., Yao, Y.Y. (eds.) Rough Sets and Current Trends in Computing, Second International Conference, RSCTC 2000, Banff, Canada, 16–19 Oct 2000, Revised Papers. Lecture Notes in Computer Science, vol. 2005, pp. 204–211. Springer (2001)

50. Wieczorkowska, A., Synak, P., Lewis, R.A., Ras, Z.W.: Extracting emotions from music data. In: Hacid, M., Murray, N.V., Ras, Z.W., Tsumoto, S. (eds.) Foundations of Intelligent Systems, 15th International Symposium, ISMIS 2005, Saratoga Springs, NY, USA, 25–28 May 2005. Lecture Notes in Computer Science, vol. 3488, pp. 456–465. Springer (2005)

51. Zhou, Z.H., Jiang, K., Li, M.: Multi-instance learning based web mining. Appl. Intell. **22**(2), 135–147 (2005)

52. Zhuravlev, J.I.: On a class of partial Boolean functions. Diskret. Analiz (in Russian) **2**, 23–27 (1964)

Part I
Explaining Examples and Preliminary Results

In this part, we consider preliminary definitions, examples, and results. It consists of three chapters.

In Chap. 2, we discuss notions of a problem with many-valued decisions, decision table with many-valued decisions, decision and inhibitory tree, rule, and rule system for a problem (in details) and for a decision table (in brief). We consider examples of problems in which objects can have two or more equally appropriate decisions in a natural way, and an example of inconsistent decision table that can be transformed into a decision table with many-valued decisions. At the end, we discuss two examples which explain why we study not only decision but also inhibitory trees and rules.

In Chap. 3, we discuss three approaches to handle inconsistency in decision tables: decision tables with many-valued decisions, decision tables with generalized decisions, and decision tables with most common decisions. We compare the complexity and classification accuracy of decision trees constructed by greedy heuristics in the frameworks of these approaches.

In Chap. 4, we consider in more details definitions related to binary decision tables with many-valued decisions, decision and inhibitory trees, tests, rules, and rule systems for these tables. We mention without proofs some relatively simple results obtained earlier for binary decision tables with many-valued decisions: relationships among decision trees, rules and tests, bounds on their complexity, greedy algorithms for construction of decision trees, rules and tests, and dynamic programming algorithms for minimization of tree depth and rule length. We extend these results to inhibitory trees, tests, rules, and rule systems for binary decision tables with many-valued decisions.

Chapter 2
Explaining Examples

In this chapter, we discuss definitions of a problem with many-valued decisions, corresponding decision table with many-valued decisions, decision and inhibitory trees, rules, and rule systems for a problem (in details) and for a decision table (in brief). We consider only depth of trees, and length of rules and systems of rules.

After that, we concentrate on the consideration of simple examples of problems with many-valued decisions from different areas of applications: fault diagnosis, computational geometry, and combinatorial optimization. In these problems, objects, in a natural way, can have two or more decisions which are equally appropriate. Separately, we consider an example of inconsistent decision table with single-valued decisions. Such tables can be transformed into decision tables with many-value decisions. However, the rationality of this transformation depends on content of the problem under consideration. We consider this question in more details in Chap. 3.

At the end of the chapter, we discuss two examples which explain why we study not only decision but also inhibitory rules and trees.

2.1 Problems and Decision Tables

In this section, we discuss the notions of problems, decision tables, and decision and inhibitory trees, rules and rule systems.

2.1.1 Problems with Many-valued Decisions

We begin with simple model of a *problem*. Let A be a set (elements from A will be called sometimes *objects*). It is possible that A is an infinite set. Let $f_1, \ldots, f_n$ be attributes, each of which is a function from A to B where B is a nonempty finite set. Attributes $f_1, \ldots, f_n$ divide the set A into a number of domains in each of which

© Springer Nature Switzerland AG 2020
F. Alsolami et al., *Decision and Inhibitory Trees and Rules for Decision
Tables with Many-valued Decisions*, Intelligent Systems Reference Library 156,
https://doi.org/10.1007/978-3-030-12854-8_2

values of these attributes are constant. These domains are labeled with nonempty
finite subsets of the set $\omega = \{0, 1, 2, \ldots\}$ of nonnegative integers. We will interpret
these subsets as sets of decisions.

More formally, a *problem* is a tuple $z = (\nu, f_1, \ldots, f_n)$ where ν is a mapping from
B^n to the set of all nonempty finite subsets of the set ω. Each domain corresponds
to the nonempty set of solutions over A of a system of equations of the kind

$$\{f_1(x) = \delta_1, \ldots, f_n(x) = \delta_n\}$$

where $\delta_1, \ldots, \delta_n \in B$. Denote $D(z) = \bigcup \nu(\delta_1, \ldots, \delta_n)$ where union is considered
over all tuples $(\delta_1, \ldots, \delta_n) \in B^n$ for which the system of equations

$$\{f_1(x) = \delta_1, \ldots, f_n(x) = \delta_n\}$$

has a solution over A. For a given $a \in A$, denote $z(a) = \nu(f_1(a), \ldots, f_n(a))$.

We will consider two interpretations of the problem z: decision and inhibitory.

Decision interpretation: for a given $a \in A$, we should find a number from the
set $z(a)$. This interpretation means that the decisions from the set $z(a)$ are equally
appropriate, for example, all decisions from $z(a)$ are optimal for the object a relative
to some criteria, and it is enough to find one decision from the set $z(a)$.

Inhibitory interpretation: for a given $a \in A$, we should find a number from the set
$D(z) \setminus z(a)$ (we will assume here that $D(z) \neq z(a)$ for any $a \in A$). This interpretation means that the decisions from the set $D(z) \setminus z(a)$ are equally inappropriate, for
example, all decisions from $D(z) \setminus z(a)$ are not optimal for the object a relative to
the considered criteria, and it is enough to find one decision from the set $D(z) \setminus z(a)$.

In the case of decision interpretation, we study decision trees and decision rules.

A *decision tree over z* is a finite directed tree with the root in which each terminal
node is labeled with a number from ω (decision), each nonterminal node (such nodes
will be called *working* nodes) is labeled with an attribute from the set $\{f_1, \ldots, f_n\}$.
Edges starting in a working node are labeled with pairwise different elements from
B.

Let Γ be a decision tree over z. For a given element $a \in A$, the work of the tree
starts in the root. If the current node v is terminal then Γ finishes its work in this
node. Let the current node v be a working node labeled with an attribute f_i. If there
is an edge starting in v and labeled with $f_i(a)$ then the computation passes along this
edge, etc. Otherwise, Γ finishes its work in v.

We will say that Γ *solves* the problem z in decision interpretation if, for any
$a \in A$, Γ finishes its work in a terminal node labeled with a number belonging to
$z(a)$.

As time complexity of Γ we will consider the *depth* $h(\Gamma)$ of Γ which is the
maximum length of a path from the root to a terminal node of Γ. We denote by $h(z)$
the minimum depth of a decision tree over z which solves the problem z in decision
interpretation.

A *decision rule* ρ over z is an expression of the kind

$$(f_{i_1} = b_1) \wedge \ldots \wedge (f_{i_m} = b_m) \to t$$

where $f_{i_1}, \ldots, f_{i_m} \in \{f_1, \ldots, f_n\}$, $b_1, \ldots, b_m \in B$, and $t \in \omega$. The number m is called the *length* of the rule ρ. This rule is called *realizable* for an element $a \in A$ if

$$f_{i_1}(a) = b_1, \ldots, f_{i_m}(a) = b_m .$$

The rule ρ is called *true* for z if, for any $a \in A$ such that ρ is realizable for a, $t \in z(a)$.

A *decision rule system* S over z is a nonempty finite set of decision rules over z. A system S is called a *complete* decision rule system for z if each rule from S is true for z and, for every $a \in A$, there exists a rule from S which is realizable for a.

We denote by $l(S)$ the maximum length of a rule from S (the *length* of S), and by $l(z)$ we denote the minimum value of $l(S)$ among all complete decision rule systems S for z.

In the case of inhibitory interpretation, we study inhibitory trees and rules.

An *inhibitory tree* over z is a finite directed tree with the root in which each terminal node is labeled with an expression $\neq t$, $t \in \omega$, each nonterminal node (such nodes will be called *working* nodes) is labeled with an attribute from the set $\{f_1, \ldots, f_n\}$. Edges starting in a working node are labeled with pairwise different elements from B.

Let Γ be an inhibitory tree over z. For a given element $a \in A$, the work of the tree starts in the root. If the current node v is terminal then Γ finishes its work in this node. Let the current node v be a working node labeled with an attribute f_i. If there is an edge starting in v and labeled with $f_i(a)$ then the computation passes along this edge, etc. Otherwise, Γ finishes its work in v.

We will say that Γ *solves* the problem z in inhibitory interpretation if, for any $a \in A$, Γ finishes its work in a terminal node labeled with an expression $\neq t$ such that $t \in D(z) \setminus z(a)$.

As time complexity of Γ we will consider the *depth* $h(\Gamma)$ of Γ which is the maximum length of a path from the root to a terminal node of Γ. We denote by $ih(z)$ the minimum depth of an inhibitory tree over z which solves the problem z in inhibitory interpretation.

An *inhibitory rule* ρ over z is an expression of the kind

$$(f_{i_1} = b_1) \wedge \ldots \wedge (f_{i_m} = b_m) \to \neq t$$

where $f_{i_1}, \ldots, f_{i_m} \in \{f_1, \ldots, f_n\}$, $b_1, \ldots, b_m \in B$, and $t \in \omega$. The number m is called the *length* of the rule ρ. This rule is called *realizable* for an element $a \in A$ if

$$f_{i_1}(a) = b_1, \ldots, f_{i_m}(a) = b_m .$$

The rule ρ is called *true* for z if, for any $a \in A$ such that ρ is realizable for a, $t \in D(z) \setminus z(a)$.

An *inhibitory rule system* S over z is a nonempty finite set of inhibitory rules over z. A system S is called a *complete* inhibitory rule system for z if each rule from S is true for z and, for every $a \in A$, there exists a rule from S which is realizable for a.

We denote by $l(S)$ the maximum length of a rule from S (the *length* of S), and by $il(z)$ we denote the minimum value of $l(S)$ among all complete inhibitory rule systems S for z.

2.1.2 Decision Tables Corresponding to Problems

We associate a *decision table* $T = T(z)$ with the considered problem z. This table is a rectangular table with n columns labeled with attributes $f_1, \ldots, f_n$. A tuple $(\delta_1, \ldots, \delta_n) \in B^n$ is a row of T if and only if the system of equations

$$\{f_1(x) = \delta_1, \ldots, f_n(x) = \delta_n\}$$

is consistent (has a solution over the set A). This row is labeled with the set $v(\delta_1, \ldots, \delta_n)$. Let $Row(T)$ be the set of rows of T. Denote

$$D(T) = \bigcup_{(\delta_1, \ldots, \delta_n) \in Row(T)} v(\delta_1, \ldots, \delta_n) .$$

It is clear that $D(T) = D(z)$.

We will consider two interpretations of the decision table T: decision and inhibitory.

Decision interpretation: for a given row $(\delta_1, \ldots, \delta_n)$ of T, we should find a decision from the set of decisions $v(\delta_1, \ldots, \delta_n)$ attached to this row. This interpretation means that the decisions from the set $v(\delta_1, \ldots, \delta_n)$ are equally appropriate and it is enough to find one decision from the set $v(\delta_1, \ldots, \delta_n)$.

Inhibitory interpretation: for a given row $(\delta_1, \ldots, \delta_n)$ of T, we should find a decision from the set $D(T) \setminus v(\delta_1, \ldots, \delta_n)$ (when we study inhibitory interpretation of T, we assume that $v(\delta_1, \ldots, \delta_n) \neq D(T)$ for any row $(\delta_1, \ldots, \delta_n)$ of T). This interpretation means that the decisions from the set $D(T) \setminus v(\delta_1, \ldots, \delta_n)$ are equally inappropriate and it is enough to find one decision from the set $D(T) \setminus v(\delta_1, \ldots, \delta_n)$.

In the case of decision interpretation, we study decision trees and decision rules.

We can formulate the notion of a decision tree over T, and describe the work of a decision tree over T on a row of T in a natural way. We will say that a decision tree Γ over T is a *decision tree for* T if, for any row $(\delta_1, \ldots, \delta_n)$ of T, Γ finishes its work in a terminal node labeled with a decision t such that $t \in v(\delta_1, \ldots, \delta_n)$. It is not difficult to show that the set of decision trees for T coincides with the set of decision trees over z solving the problem z in decision interpretation. We denote by $h(T)$ the minimum depth of a decision tree for the table $T = T(z)$. Then $h(z) = h(T(z))$.

We can formulate the notion of a decision rule over T, the notion of a decision rule realizable for a row of T, and the notion of a decision rule true for T in a natural way. We will say that a system S of decision rules over T is a *complete* decision rule system for T if each rule from S is true for T and, for every row of T, there exists a rule from S which is realizable for this row. We denote by $l(T)$ the minimum value of $l(S)$ among all complete decision rule systems S for T. One can show that a decision rule system S over z is complete for z if and only if S is complete for $T = T(z)$. So $l(z) = l(T(z))$.

In the case of inhibitory interpretation, we study inhibitory trees and inhibitory rules.

We can formulate the notion of inhibitory tree over T, and describe the work of an inhibitory tree over T on a row of T in a natural way. We will say that an inhibitory tree Γ over T is an *inhibitory tree for* T if, for any row $(\delta_1, \ldots, \delta_n)$ of T, Γ finishes its work in a terminal node labeled with an expression $\neq t$ such that $t \in D(T) \setminus \nu(\delta_1, \ldots, \delta_n)$. It is not difficult to show that the set of inhibitory trees for T coincides with the set of inhibitory trees over z solving the problem z in inhibitory interpretation. We denote by $ih(T)$ the minimum depth of inhibitory tree for the table $T = T(z)$. Then $ih(z) = ih(T(z))$.

We can formulate the notion of an inhibitory rule over T, the notion of an inhibitory rule realizable for a row of T, and the notion of an inhibitory rule true for T in a natural way. We will say that a system S of inhibitory rules over T is a *complete* inhibitory rule system for T if each rule from S is true for T and, for every row of T, there exists a rule from S which is realizable for this row. We denote by $il(T)$ the minimum value of $l(S)$ among all complete inhibitory rule systems S for T. One can show that an inhibitory rule system S over z is complete for z if and only if S is complete for $T = T(z)$. So $il(z) = il(T(z))$.

As a result, instead of the problem z in decision or inhibitory interpretation we can study the decision or inhibitory interpretation of the table $T(z)$.

2.2 Examples of Decision Tables with Many-valued Decisions

There are at least three sources of decision tables with many-valued decisions for which the decision interpretation or both the decision and inhibitory interpretations make sense.

The first source is classes of exactly formulated problems with many-valued decisions for which objects can have, in a natural way, two or more decisions. For such problems, the set A contains all possible objects and sets of decisions for these objects are completely known. We consider examples of exactly formulated problems and corresponding decision tables with many-valued decisions from the following areas: computational geometry, combinatorial optimization, and diagnosis of faults in combinatorial circuits. For each of the considered decision tables, the decisions from the sets attached to rows are equally appropriate for these rows (objects corresponding to rows), and decisions that do not belong to the sets attached to rows are equally inappropriate for these rows (objects corresponding to rows).

The second source is problems with many-valued decisions based on experimental data in which objects can have, in a natural way, two or more decisions: a text, a music, an architecture can belong to multiple genres; a disease may belong to multiple categories; genes may have multiple functions; an image can belong to different semantic classes [1]. For such problems, the set A can be only a part of the set of all possible objects and the sets of decisions for objects from A can be incomplete. In some of the obtained decision tables, decisions from the sets attached to rows are equally appropriate for objects corresponding to rows, and it is enough to find one decision from the set. An example is a decision table in which each row describes symptoms of some disease and the set of decisions contains the most appropriate drugs for this disease.

The third source is inconsistent decision tables with single-valued decisions based on experimental data in which equal rows (objects with equal values of conditional attributes) can be labeled with different decisions. The causes of inconsistency are the following: either (i) the set of conditional attributes is not sufficient to distinguish these objects, or (ii) equal rows correspond to the same object but with different appropriate for this object decisions. We can keep one row from the group of equal rows and label this row with the set of decisions attached to all rows from the group. As a result, we obtain a decision table with many-valued decisions. We consider an example related to inconsistent decision tables.

For the case (i), the decision interpretation of the decision table considered in this book (for a given row, we should find a decision from the set of decisions attached to this row) makes some sense: for a given object, we recognize a decision attached to this object or to an object which is not distinguishable from the initial one by existing set of conditional attributes. The inhibitory interpretation (for a given row, we should find a decision which does not belong to the set of decisions attached to this row) also makes some sense: we recognize a decision which cannot be attached to an object that is not distinguishable from the initial one by the set of conditional attributes.

For the case (ii), the considered decision interpretation can make sense. It makes sense, for example, if each decision is a possible behavior in the situation described by values of conditional attributes for the row, quality of behaviors does not differ much, and we would like to create a strategy of behavior in the form of a decision tree.

There are different approaches to handle inconsistent decision tables. We compare the considered one with two other approaches in the next chapter.

Fig. 2.1 Problem of three post-offices

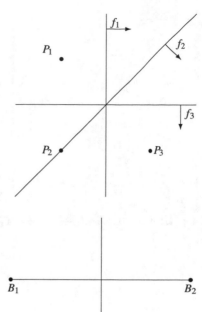

Fig. 2.2 Distance to points B_1 and B_2

2.2.1 Problem of Three Post-offices

Figure 2.1 from [4] shows a hypothetical topographic map where positions of three post-offices are marked as P_1, P_2, and P_3. The objective is given an arbitrary point indicating the client location, assign the client to a nearest post-office.

Let we have two points B_1 and B_2. We connect these points with a line segment, then draw a perpendicular line through the midpoint of the segment (see Fig. 2.2 from [4]).

One can see all points located left to the perpendicular are closer to B_1, all points located on the perpendicular are equidistant to B_1 and to B_2, and all points located right to the perpendicular are closer to B_2. This reasoning allows us to construct attributes for the problem of three post-offices which will be denoted z_1.

We connect all pairs of post-offices P_1, P_2, P_3 by line segments (these segments are invisible in Fig. 2.1) and draw perpendiculars through midpoints of these segments. These perpendiculars (lines) correspond to the three attributes f_1, f_2, f_3. Each such attribute takes value -1 from the left of the considered line, takes value 0 on the line, and takes value $+1$ from the right of the considered line (arrow points to the right).

The attributes f_1, f_2, and f_3 divide the plane into 13 regions, such that for all points within a region, all attribute values and the nearest post-offices are the same. We assign to each region the tuple of attribute values and indices of the nearest post-

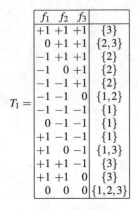

$$T_1 = \begin{array}{ccc|c} f_1 & f_2 & f_3 & \\ \hline +1 & +1 & +1 & \{3\} \\ 0 & +1 & +1 & \{2,3\} \\ -1 & +1 & +1 & \{2\} \\ -1 & 0 & +1 & \{2\} \\ -1 & -1 & +1 & \{2\} \\ -1 & -1 & 0 & \{1,2\} \\ -1 & -1 & -1 & \{1\} \\ 0 & -1 & -1 & \{1\} \\ +1 & -1 & -1 & \{1\} \\ +1 & 0 & -1 & \{1,3\} \\ +1 & +1 & -1 & \{3\} \\ +1 & +1 & 0 & \{3\} \\ 0 & 0 & 0 & \{1,2,3\} \end{array}$$

Fig. 2.3 Decision table T_1 for problem of three post-offices

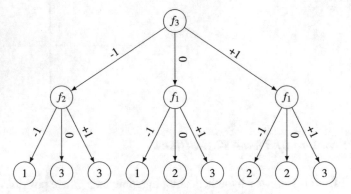

Fig. 2.4 Decision tree with minimum depth for decision table T_1

offices, and thus construct the decision table T_1 for the considered problem (see Fig. 2.3).

One can show that

$$\{(f_3 = -1) \wedge (f_2 = -1) \to 1, (f_3 = +1) \wedge (f_1 = -1) \to 2,$$
$$(f_2 = +1) \wedge (f_1 = +1) \to 3, (f_1 = 0) \wedge (f_3 = +1) \to 2,$$
$$(f_3 = 0) \wedge (f_1 = -1) \to 1, (f_2 = 0) \wedge (f_1 = +1) \to 1, (f_1 = 0) \wedge (f_2 = 0) \to 1\}$$

is a complete decision rule system with minimum length for the table T_1, and the tree depicted in Fig. 2.4 is a decision tree with minimum depth for the table T_1.

For the consideration of inhibitory rules and trees, we have to remove the central point which is the intersection of lines corresponding to attributes f_1, f_2, f_3 (see Fig. 2.1) since the whole set of decisions $\{1, 2, 3\}$ corresponds to this point. We denote the obtained problem z_1'. We also have to remove from the table T_1 the row

$(0, 0, 0)$ labeled with the whole set of decisions $\{1, 2, 3\}$. We denote the obtained decision table T_1'.

One can show that

$$\{(f_3 = +1) \to \neq 1, (f_3 = -1) \to \neq 2, (f_1 = +1) \to \neq 2, (f_1 = -1) \to \neq 3\}$$

is a complete inhibitory rule system with minimum length for the table T_1', and the tree depicted in Fig. 2.5 is an inhibitory tree with minimum depth for the table T_1'.

The considered problem is an example of problems studied in *computational geometry*.

2.2.2 Traveling Salesman Problem with Four Cities

Let we have a complete undirected graph with four nodes in which each edge is labeled with a real number – the length of this edge (see Fig. 2.6 from [4]).

Fig. 2.5 Inhibitory tree with minimum depth for decision table T_1'

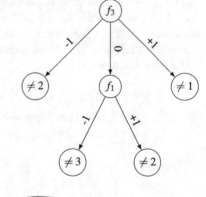

Fig. 2.6 Traveling salesman problem with four cities

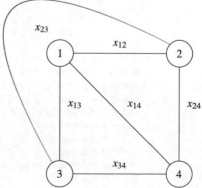

A Hamiltonian circuit is a closed path which passes through each node exactly once. The objective is to find a minimum length Hamiltonian circuit. The considered graph contains three Hamiltonian circuits:

H_1: 12341 or, which is the same, 14321,

H_2: 12431 or 13421,

H_3: 13241 or 14231.

For $i = 1, 2, 3$, we denote by L_i the length of H_i. Then

$$L_1 = x_{12} + x_{23} + x_{34} + x_{14} = \overset{\alpha}{(x_{12} + x_{34})} + \overset{\beta}{(x_{23} + x_{14})},$$
$$L_2 = x_{12} + x_{24} + x_{34} + x_{13} = \overset{\alpha}{(x_{12} + x_{34})} + \overset{\gamma}{(x_{24} + x_{13})},$$
$$L_3 = x_{13} + x_{23} + x_{24} + x_{14} = \overset{\gamma}{(x_{24} + x_{13})} + \overset{\beta}{(x_{23} + x_{14})}.$$

In the capacity of attributes we will use three functions $f_1 = \text{sign}(L_1 - L_2)$, $f_2 = \text{sign}(L_1 - L_3)$, and $f_3 = \text{sign}(L_2 - L_3)$ where $\text{sign}(x) = -1$ if $x < 0$, $\text{sign}(x) = 0$ if $x = 0$, and $\text{sign}(x) = +1$ if $x > 0$.

Values L_1, L_2 and L_3 are linearly ordered. It is clear that values of α, β and γ can be chosen independently. Based on this fact, we can show that any order of L_1, L_2, and L_3 is possible. As a result, we can construct the decision table T_2 for the considered problem (see Fig. 2.7). In this table, the set of decisions for each row consists of indices of Hamiltonian circuits which have minimum length.

One can show that

$$\{(f_1 = -1) \wedge (f_2 = -1) \rightarrow 1, (f_1 = +1) \wedge (f_3 = -1) \rightarrow 2,$$
$$(f_2 = +1) \wedge (f_3 = +1) \rightarrow 3, (f_1 = 0) \wedge (f_2 = 0) \rightarrow 1,$$
$$(f_1 = 0) \wedge (f_2 = -1) \rightarrow 1, (f_1 = -1) \wedge (f_2 = 0) \rightarrow 1,$$
$$(f_2 = +1) \wedge (f_3 = 0) \rightarrow 2\}$$

	f_1	f_2	f_3	
If $\alpha < \beta < \gamma$ then $L_1 < L_2 < L_3$	−1	−1	−1	{1}
If $\alpha = \beta < \gamma$ then $L_1 < L_2 = L_3$	−1	−1	0	{1}
If $\alpha < \beta = \gamma$ then $L_1 = L_2 < L_3$	0	−1	−1	{1,2}
If $\alpha = \beta = \gamma$ then $L_1 = L_2 = L_3$	0	0	0	{1,2,3}
If $\alpha < \gamma < \beta$ then $L_2 < L_1 < L_3$	+1	−1	−1	{2}
If $\alpha = \gamma < \beta$ then $L_2 < L_1 = L_3$	+1	0	−1	{2}
If $\beta < \alpha < \gamma$ then $L_1 < L_3 < L_2$	−1	−1	+1	{1}
If $\beta < \alpha = \gamma$ then $L_1 = L_3 < L_2$	−1	0	+1	{1,3}
If $\beta < \gamma < \alpha$ then $L_3 < L_1 < L_2$	−1	+1	+1	{3}
If $\beta = \gamma < \alpha$ then $L_3 < L_1 = L_2$	0	+1	+1	{3}
If $\gamma < \alpha < \beta$ then $L_2 < L_3 < L_1$	+1	+1	−1	{2}
If $\gamma < \alpha = \beta$ then $L_2 = L_3 < L_1$	+1	+1	0	{2,3}
If $\gamma < \beta < \alpha$ then $L_3 < L_2 < L_1$	+1	+1	+1	{3}

$= T_2$

Fig. 2.7 Decision table T_2 for traveling salesman problem with four cities

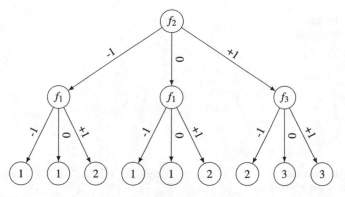

Fig. 2.8 Decision tree with minimum depth for decision table T_2

Fig. 2.9 Inhibitory tree with minimum depth for decision table T_2'

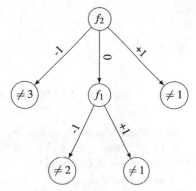

is a complete decision rule system with minimum length for T_2, and the tree depicted in Fig. 2.8 is a decision tree with minimum depth for the table T_2.

For the consideration of inhibitory rules and trees, we have to remove from the table T_2 the row $(0, 0, 0)$ labeled with the whole set of decisions $\{1, 2, 3\}$. We denote the obtained decision table T_2'.

One can show that

$$\{(f_2 = -1) \to \neq 3, (f_2 = +1) \to \neq 1, (f_3 = -1) \to \neq 3, (f_3 = +1) \to \neq 2\}$$

is a complete inhibitory rule system with minimum length for T_2', and the tree depicted in Fig. 2.9 is an inhibitory tree with minimum depth for the table T_2'.

It was an example of *combinatorial optimization* problem.

Fig. 2.10 One-gate
combinatorial circuit

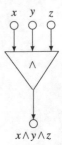

Fig. 2.11 Decision table T_3
for problem of diagnosis of
one-gate circuit

$$T_3 = \begin{array}{c|cccccccc|c} & 000 & 001 & 010 & 011 & 100 & 101 & 110 & 111 & \\ \hline 1 & 1 & 1 & 1 & 1 & 1 & 1 & 1 & 1 & \{x,y,z\} \\ x & 0 & 0 & 0 & 0 & 1 & 1 & 1 & 1 & \{y,z\} \\ y & 0 & 0 & 1 & 1 & 0 & 0 & 1 & 1 & \{x,z\} \\ z & 0 & 1 & 0 & 1 & 0 & 1 & 0 & 1 & \{x,y\} \\ xy & 0 & 0 & 0 & 0 & 0 & 0 & 1 & 1 & \{z\} \\ xz & 0 & 0 & 0 & 0 & 0 & 1 & 0 & 1 & \{y\} \\ yz & 0 & 0 & 0 & 1 & 0 & 0 & 0 & 1 & \{x\} \end{array}$$

2.2.3 Diagnosis of One-Gate Circuit

This example was discussed in [4]. We add here only results about a complete system of inhibitory rules and an inhibitory tree. Consider the combinatorial circuit S depicted in Fig. 2.10 from [4].

Each input of S can work correctly or can have constant fault 1. Let us assume that at least one such fault exists in S.

The objective is to find a faulty input. We will use attributes corresponding to tuples from the set $\{0, 1\}^3$. For a given circuit with faults and attribute, the attribute value is the circuit output value, when the corresponding tuple is passed to the circuit inputs.

It is clear that the circuit S with at least one fault 1 on an input implements a function from the set $\{1, x, y, z, xy, xz, yz\}$ (we write xy instead of $x \wedge y$). Corresponding decision table T_3 is represented in Fig. 2.11 from [4]. We do not encode decisions x, y, z by numbers 1, 2, 3. In this table, the set of decisions for each row consists of all inputs with faults.

One can show that

$$\{(011 = 1) \rightarrow x, (101 = 1) \rightarrow y, (110 = 1) \rightarrow z\}$$

is a complete decision rule system with minimum length for the table T_3, and the tree depicted in Fig. 2.12 is a decision tree with minimum depth for the table T_3.

For the consideration of inhibitory rules and trees, we have to remove from the table T_3 the first row labeled with the whole set of decisions $\{x, y, x\}$. We denote the obtained decision table T_3'.

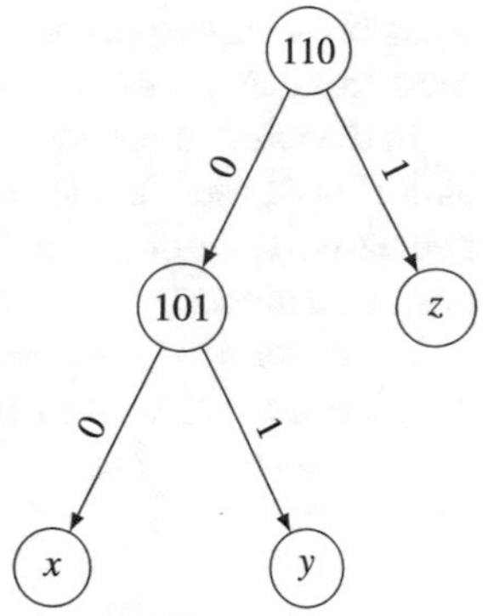

Fig. 2.12 Decision tree with minimum depth for decision table T_3

Fig. 2.13 Inhibitory tree with minimum depth for decision table T_3'

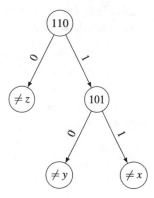

One can show that

$$\{(110 = 0) \to \neq z,\ (101 = 0) \to \neq y,\ (011 = 0) \to \neq x\}$$

is a complete inhibitory rule system with minimum length for the table T_3', and the tree depicted in Fig. 2.13 is an inhibitory tree with minimum depth for the table T_3'.

It was an example of *fault diagnosis* problem.

2.2.4 Example of Inconsistent Decision Table

It is possible that, in a decision table with single-valued decisions, there are equal rows with different decisions, i.e., the table is inconsistent. We keep one row from each group of equal rows and label this row with the set of all decisions attached to rows from the group. As a result, we obtain a decision table with many-valued decisions.

Note that there are other possibilities to form a set of decisions attached to a row. For example, we can include to this set all the most common decisions among the decisions attached to rows from the group, or the first k most frequent decisions for rows from the group, etc. We will not consider these possibilities in the book.

In the next chapter, we compare three approaches to handle inconsistent decision tables: one based on the creation of decision tables with many-valued decisions and two based on the creation of decision tables with single-valued decisions (generalized or most common).

Let we have the inconsistent decision table D depicted in Fig. 2.14. Corresponding decision table T_4 with many-valued decisions is depicted in Fig. 2.15.

One can show that

$$\{(f_3 = 0) \to 1, (f_3 = 1) \to 2\}$$

is a complete decision rule system with minimum length for the table T_4, and the tree depicted in Fig. 2.16 is a decision tree with minimum depth for the table T_4.

One can show that

$$\{(f_3 = 0) \to \neq 2, (f_3 = 1) \to \neq 1\}$$

Fig. 2.14 Inconsistent decision table D

$$D = \begin{array}{|ccc|c|}\hline f_1 & f_2 & f_3 & \\\hline 0 & 0 & 1 & 2 \\ 0 & 0 & 1 & 3 \\ 0 & 0 & 1 & 3 \\ 1 & 0 & 0 & 1 \\ 1 & 0 & 0 & 1 \\ 1 & 0 & 0 & 3 \\ 1 & 0 & 1 & 2 \\ 1 & 0 & 1 & 3 \\ 0 & 1 & 1 & 2 \\ 1 & 1 & 0 & 1 \\ 1 & 1 & 0 & 1 \\\hline\end{array}$$

Fig. 2.15 Decision table T_4 with many-valued decisions corresponding to decision table D

$$T_4 = \begin{array}{|ccc|c|}\hline f_1 & f_2 & f_3 & \\\hline 0 & 0 & 1 & \{2,3\} \\ 1 & 0 & 0 & \{1,3\} \\ 1 & 0 & 1 & \{2,3\} \\ 0 & 1 & 1 & \{2\} \\ 1 & 1 & 0 & \{1\} \\\hline\end{array}$$

Fig. 2.16 Decision tree with minimum depth for decision table T_4

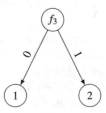

Fig. 2.17 Inhibitory tree
with minimum depth for
decision table T_4

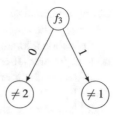

is a complete inhibitory rule system with minimum length for the table T_4, and the
tree depicted in Fig. 2.17 is an inhibitory tree with minimum depth for the table T_4.

2.3 Difference Between Decision and Inhibitory Rules and Trees

We consider two examples which show that the use of inhibitory rules and trees can
give us additional possibilities in comparison with decisions rules and trees.

2.3.1 Prediction Problem

Let us consider the decision table T_5 with many-valued decisions depicted in Fig.
2.18.

We will study only true for T_5 and realizable for at least one row from T_5 decision
and inhibitory rules. It is easy to see that decision rules can be simulated in some
sense by inhibitory rules. Let us consider, for example, the pair of decision rules
$(f_1 = 2) \rightarrow 3$ and $(f_1 = 2) \rightarrow 4$. This pair can be simulated by the following pair
of inhibitory rules: $(f_1 = 2) \rightarrow \neq 1$ and $(f_1 = 2) \rightarrow \neq 2$.

However, not each inhibitory rule can be simulated by decision rules. In particular,
there are two inhibitory rules which are realizable for the row $(0, 0)$ that does not
belong to T_5: $(f_1 = 0) \rightarrow \neq 3$ and $(f_2 = 0) \rightarrow \neq 1$, but there are no decision rules
which are realizable for this row.

Let us discuss the problem of prediction of decision value for a new object given
by values of condition attributes $f_1 = 0$ and $f_2 = 0$. Decision rules will not give

Fig. 2.18 Decision table T_5
with many-valued decisions

$$T_5 = \begin{array}{|cc|c|} \hline f_1 & f_2 & \\ \hline 0 & 1 & \{1\} \\ 0 & 2 & \{2,4\} \\ 1 & 0 & \{2\} \\ 2 & 0 & \{3,4\} \\ \hline \end{array}$$

us any information about decisions corresponding to this new object. However, the
inhibitory rules will restrict the set of possible decisions to $\{2, 4\}$.

Let us consider now decision and inhibitory trees for T_5 such that, for each terminal
node of the tree, there is a row of T_5 for which the tree work finishes in this node.

It is clear that there is no a decision tree for T_5 the work of which, for a new object
given by values of condition attributes $f_1 = 0$ and $f_2 = 0$, finishes in a terminal
node. It means that the decision trees will not give us any information about decisions
corresponding to this new object. However, the inhibitory trees Γ_1 and Γ_2 for the
decision table T_5 (see Figs. 2.19 and 2.20) will restrict the set of possible decisions
to $\{2, 4\}$.

Note that, in the book [2], different kinds of classifiers using decision and
inhibitory rules were considered. The obtained results show that classifiers based
on inhibitory rules are often better than the classifiers based on decision rules.

2.3.2 Knowledge Representation Problem

Let us consider the data table (information system) I depicted in Fig. 2.21.

We will compare now knowledge that can be derived from I by inhibitory and
decision association rules which are true for I and realizable for at least one row of I.
Decision association rules have on the right-hand side expressions of the kind $f_i = a$
where $a \in V_I(f_i)$ and $V_I(f_i)$ is the set of values of the attribute f_i in the information

Fig. 2.19 Inhibitory tree Γ_1
for decision table T_5

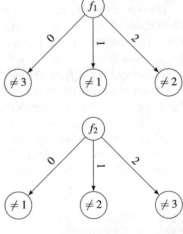

Fig. 2.20 Inhibitory tree Γ_2
for decision table T_5

Fig. 2.21 Data table
(information system) I

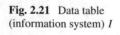

$$I = \begin{array}{|cc|} \hline f_1 & f_2 \\ \hline 0 & 1 \\ 1 & 0 \\ 0 & 2 \\ 2 & 0 \\ \hline \end{array}$$

system I. *Inhibitory association* rules have on the right-hand side expressions of the kind $f_i \neq a$ where $a \in V_I(f_i)$.

For the information system I, $V_I(f_1) = V_I(f_2) = \{0, 1, 2\}$. We will consider the set

$$V_I(f_1) \times V_I(f_2) = \{(0, 0), (0, 1), (0, 2), (1, 0), (1, 1), (1, 2), (2, 0), (2, 1), (2, 2)\}$$

which contains all 2-tuples from I and some additional 2-tuples, for example, 2-tuple $(0, 0)$.

The question under consideration is about possibility to remove from the set $V_I(f_1) \times V_I(f_2)$ all 2-tuples that do not belong to I using decision (inhibitory) association rules that are true for I and realizable for at least one 2-tuple (row) from I.

One can show that the set of decision association rules that are true for I and realizable for at least one 2-tuple from I is the following:

$$\begin{aligned} \{(f_1 = 1) &\to (f_2 = 0), (f_1 = 2) \to (f_2 = 0), \\ (f_2 = 1) &\to (f_1 = 0), (f_2 = 2) \to (f_1 = 0)\} \,. \end{aligned} \tag{2.1}$$

Based on these rules we can remove all 2-tuples from $V_I(f_1) \times V_I(f_2)$ that do not belong to I with the exception of $(0, 0)$. For example, the tuple $(2, 2)$ can be removed by the rule $(f_2 = 2) \to (f_1 = 0)$. We cannot remove the 2-tuple $(0, 0)$ since no rule is realizable for this 2-tuple. So, we have that the information derived from I by decision association rules is in some sense incomplete. The existence of such information systems was discovered in [5, 6].

One can show that the set of inhibitory association rules that are true for I and realizable for at least one 2-tuple from I is the following:

$$\begin{aligned} \{(f_1 = 0) &\to (f_2 \neq 0), (f_2 = 1) \to (f_1 \neq 1), \\ (f_1 = 1) &\to (f_2 \neq 1), (f_2 = 1) \to (f_1 \neq 2), \\ (f_1 = 1) &\to (f_2 \neq 2), (f_2 = 0) \to (f_1 \neq 0), \\ (f_1 = 2) &\to (f_2 \neq 1), (f_2 = 2) \to (f_1 \neq 1), \\ (f_1 = 2) &\to (f_2 \neq 2), (f_2 = 2) \to (f_1 \neq 2)\} \,. \end{aligned} \tag{2.2}$$

Based on these rules we can remove all 2-tuples from $V_I(f_1) \times V_I(f_2)$ that do not belong to I. For example, the 2-tuple $(2, 2)$ can be removed by the rule $(f_1 = 2) \to (f_2 \neq 2)$, and the 2-tuple $(0, 0)$ can be removed by the rule $(f_1 = 0) \to (f_2 \neq 0)$.

So, we have that the information derived from I by inhibitory association rules is in some sense complete. This fact was proved for all information systems in [3].

We can fix sequentially each attribute of the information system as the decision attribute and consider decision and inhibitory trees for the obtained decision tables as decision and inhibitory association trees for the initial information system. We study only realizable decision and inhibitory trees. In such trees, for each terminal node of

Fig. 2.22 Decision
association tree G_1 for
information system I

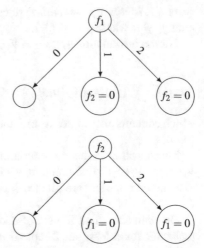

Fig. 2.23 Decision
association tree G_2 for
information system I

the tree, there is a row of the decision table for which the tree work finishes in this
node. In some cases, these trees can have empty terminal nodes. Let the obtained
decision table contain two equal rows r that are labeled with different decisions,
Γ be a decision tree for this table, and the work of Γ for the row r be finished in
the terminal node v. Then v is empty since there is no definite value of the decision
attribute for the row r.

Let the obtained decision table contain a group of equal rows r such that the set
of decisions attached to rows from the group is equal to the whole set of decisions in
the considered decision table. Let Γ be a inhibitory tree for this table, and the work
of Γ for the row r be finished in the terminal node v. Then v is empty since each
value of the decision attribute can be attached to the row r.

For the considered information system I, there are two decision association trees
G_1 and G_2 depicted in Figs. 2.22 and 2.23, respectively. The tree G_1 contains an
empty terminal node corresponding to the case when $f_1 = 0$. In this case, the attribute
f_2 can have two values 1 and 2, so there is no definite value of f_2. The same situation
is with the tree G_2. Note that the set of decision association rules corresponding to
paths in G_1 and G_2 from the roots to nonempty terminal nodes is equal to the set
(2.1).

So, we have that the information derived from I by the decision association trees
is incomplete.

For the considered information system I, all eight inhibitory association trees
can be represented as trees $\Gamma_1(a, b)$ and $\Gamma_2(a, b)$ depicted in Figs. 2.24 and 2.25,
respectively, where $a, b \in \{1, 2\}$. Note that the set of inhibitory association rules
corresponding to paths in $\Gamma_1(a, b)$ and $\Gamma_2(a, b)$, $a, b \in \{1, 2\}$, from the roots to
nonempty terminal nodes is equal to the set (2.2).

As a result, we have that the information derived from I by the inhibitory association trees is complete. We can prove that it is true for all information systems.

Fig. 2.24 Inhibitory
association trees $\Gamma_1(a, b)$,
$a, b \in \{1, 2\}$, for information
system I

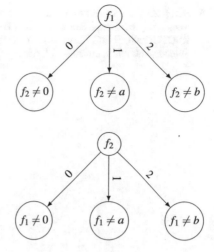

Fig. 2.25 Inhibitory
association trees $\Gamma_2(a, b)$,
$a, b \in \{1, 2\}$, for information
system I

Let J be an information system. We denote by $Irr(J)$ the set of inhibitory association rules which are true for J, realizable for at least one row of J, and satisfy the following condition: if we remove an arbitrary descriptor $f_i = a$ from the left-hand side of the considered rule, then the obtained rule is not true for the information system J. We call such rules irreducible. Using the results obtained in [3], one can show that the information derived from J by the rules from $Irr(J)$ is complete.

Let $\rho \in Irr(J)$. It is not difficult to construct an inhibitory association tree $\Gamma(\rho)$ for J such that (i) for each terminal node of the tree $\Gamma(\rho)$, there is a row of J for which the work of $\Gamma(\rho)$ finishes in this node (i.e., $\Gamma(\rho)$ is realizable), and (ii) there is a path in $\Gamma(\rho)$ from the root to a nonempty terminal node for which the corresponding rule coincides with the rule ρ.

The set of inhibitory association rules corresponding to paths in the trees $\Gamma(\rho)$, $\rho \in Irr(J)$, from the roots to nonempty terminal nodes contains the set $Irr(J)$. From here it follows that the information derived from the system J by the inhibitory association trees is complete.

References

1. Boutell, M.R., Luo, J., Shen, X., Brown, C.M.: Learning multi-label scene classification. Pattern Recognit. **37**(9), 1757–1771 (2004)
2. Delimata, P., Moshkov, M., Skowron, A., Suraj, Z.: Inhibitory Rules in Data Analysis: A Rough Set Approach. Studies in Computational Intelligence, vol. 163. Springer, Berlin (2009)
3. Moshkov, M., Skowron, A., Suraj, Z.: Maximal consistent extensions of information systems relative to their theories. Inf. Sci. **178**(12), 2600–2620 (2008)
4. Moshkov, M., Zielosko, B.: Combinatorial Machine Learning - A Rough Set Approach. Studies in Computational Intelligence, vol. 360. Springer, Heidelberg (2011)
5. Skowron, A., Suraj, Z.: Rough sets and concurrency. Bull. Pol. Acad. Sci. **41**(3), 237–254 (1993)

6. Suraj, Z.: Some remarks on extensions and restrictions of information systems. In: Ziarko, W., Yao, Y.Y. (eds.) Rough Sets and Current Trends in Computing, Second International Conference, RSCTC 2000, Banff, Canada, 16–19 Oct 2000, Revised Papers. Lecture Notes in Computer Science, vol. 2005, pp. 204–211. Springer, Berlin (2001)

Chapter 3
Three Approaches to Handle Inconsistency in Decision Tables

In this chapter, we discuss main notions related to inconsistent decision tables (decision tables with single-valued decisions that have equal rows labeled with different decisions), and three approaches to handle inconsistency in decision tables: decision tables with many-valued decisions (*MVD* approach), decision tables with generalized decisions (*GD* approach), and decision tables with most common decisions (*MCD* approach). After that, we compare complexity and classification accuracy of decision trees constructed by greedy heuristics in the frameworks of these approaches.

The obtained results show that the *MVD* approach is in some sense better than the *MCD* approach, and the latter approach is better than the *GD* approach. "Better" here means that, for the most part of the considered cases, the decision trees constructed in the framework of one approach are simpler and have less classification error rate than the decision trees constructed in the framework of another approach. So, if the *MVD* approach is appropriate to handle a given inconsistent decision table, then we can use it instead of the other two approaches. Note that this chapter contains some results from [2, 3].

3.1 Inconsistent Decision Tables and Three Approaches to Handle Them

It is pretty common in real life problems to deal with inconsistent decision tables with single-valued decisions in which there are groups of equal rows (objects with equal values of conditional attributes) labeled with different decisions (values of the decision attribute). The main reason is that we do not have enough number of conditional attributes to separate such objects (rows). Another possible reason is that equal rows correspond to the same object but with different appropriate for this object decisions. In Fig. 3.1, we can see an example of inconsistent decision table T^0. In this section, we discuss three approaches to handle such tables.

© Springer Nature Switzerland AG 2020
F. Alsolami et al., *Decision and Inhibitory Trees and Rules for Decision Tables with Many-valued Decisions*, Intelligent Systems Reference Library 156,
https://doi.org/10.1007/978-3-030-12854-8_3

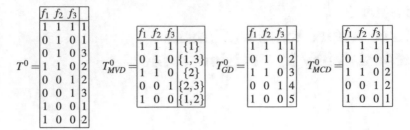

Fig. 3.1 Transformation of inconsistent decision table T^0 into decision tables T_{MVD}^0, T_{GD}^0, and T_{MCD}^0

In the book, we study decision tables with many-valued decisions. In this approach (*MVD*), a single row is considered from a group of equal rows of an inconsistent decision table T, and this row is labeled with a set of decisions that consists of different decisions attached to rows from the group. As a result, the inconsistent decision table T is transformed into a decision table T_{MVD} with many-valued decisions (see table T_{MVD}^0 in Fig. 3.1).

In rough set theory [5], generalized decisions are used to handle inconsistency. In this approach (*GD*), each set of decisions in the table T_{MVD} is encoded by a number (decision) such that equal sets are encoded by equal numbers and different sets – by different numbers. As a result, we obtain a consistent decision table T_{GD} with single-valued decisions called a decision table with generalized decisions (see table T_{GD}^0 in Fig. 3.1).

We use also another approach (*MCD*) based on the most common decisions. Instead of a group of equal rows, we consider one row from the group and label this row with the most common decision for rows from the group. If there are two or more such decisions, then we chose the minimum one. As a result, the initial table T is transformed into a consistent decision table T_{MCD} with single-valued decisions called a decision table with most common decisions (see table T_{MCD}^0 in Fig. 3.1).

Let T be an inconsistent decision table with single-valued decisions and S be a group of equal rows of T. If we need to find all decisions attached to rows from the group S, then we should consider the decision table with generalized decisions T_{GD} corresponding to T. If we should find the most common decision among decisions attached to rows from the group, then we should consider the decision table with most common decisions T_{MCD} corresponding to T. If all decisions attached to rows from the group S are equally appropriate and it is enough to find one of these decisions, we can consider the decision table with many-valued decisions T_{MVD} corresponding to T.

We interpret decision tables with single-valued decisions, i.e., T_{GD} and T_{MCD}, as a special case of decision tables with many-valued decisions where each row is labeled with a set of decisions that has only a single decision. Hence, we can apply the same algorithms for all three cases.

The aim of this chapter is to compare complexity and classification accuracy of decision trees constructed by greedy heuristics in the frameworks of the three approaches, i.e., compare decision trees for tables T_{MVD}, T_{GD}, and T_{MCD}.

3.2 Decision Tables Used in Experiments

We consider a number of decision tables from the UCI ML Repository [4] and from the KEEL-Dataset Repository [1].

Some conditional attributes are removed from the UCI ML Repository decision tables to convert them into inconsistent decision tables. We also performed other preprocessing steps, i.e., we removed a conditional attribute if it has a unique value for each row, and imputed the missing value for a conditional attribute with the most common value for this attribute. Each of the obtained inconsistent decision tables T was further transformed into tables T_{MVD}, T_{GD}, and T_{MCD} by the procedures described in Sect. 3.1. These modified decision tables were renamed. The new name includes the name of the initial table and an index equal to the number of removed conditional attributes.

Table 3.1 Characteristics of decision tables related to the UCI ML repository

Decision table T	Rows	Attr	Spectrum						Removed attributes
			#1	#2	#3	#4	#5	#6	
BALANCE-SCALE-1	125	3	45	50	30				1
BREAST-CANCER-1	193	8	169	24					3
BREAST-CANCER-5	98	4	58	40					4, 5, 6, 8, 9
CARS-1	432	5	258	161	13				1
FLAGS-5	171	21	159	12					1, 2, 3, 5, 19
HAYES-ROTH-DATA-1	39	3	22	13	4				4
KR-VS- KP-5	1987	31	1564	423					15, 16, 21, 27, 33
KR-VS- KP-4	2061	32	1652	409					15, 16, 21, 33
LYMPHOGRAPHY-5	122	13	113	9					1, 13, 14, 15, 18
MUSHROOM-5	4078	17	4048	30					5, 8, 11, 13, 22
NURSERY-4	240	4	97	96	47				1, 5, 6, 7
NURSERY-1	4320	7	2858	1460	2				1
SPECT-TEST-1	164	21	161	3					3
TEETH-1	22	7	12	10					1
TEETH-5	14	3	6	3	0	5	0	2	2, 3, 4, 5, 8
TIC-TAC-TOE-4	231	5	102	129					3, 5, 7, 9
TIC-TAC-TOE-3	449	6	300	149					5, 7, 9
ZOO-DATA-5	42	11	36	6					2, 9, 10, 13, 14

Table 3.2 Characteristics of decision tables from the KEEL-Dataset Repository

Decision table T	Rows	Attr	Spectrum									
			#1	#2	#3	#4	#5	#6	#7	#8	#9	#10
BIBTEX*	7355	1836	2791	1825	1302	669	399	179	87	46	18	7
COREL5K	4998	499	3	376	1559	3013	17	0	1	0	0	0
DELICIOUS*	15862	944	95	207	292	340	422	536	714	930	1108	1460
ENRON*	1561	1001	179	238	441	337	200	91	51	15	3	3
GENBASE-1	662	1186	27	560	58	31	8	2	3	0	0	0
MEDICAL	967	1449	741	212	14	0	0	0	0	0	0	0

Each decision table T from the KEEL-Dataset Repository is already in the format of a decision table with many-valued decisions, i.e., $T = T_{MVD}$. This table is further transformed into the format T_{MCD} (in this case, the first decision is selected from the set of decisions attached to a row as the most common decision for this row) and into the format T_{GD} (in this case, we use the procedure described in Sect. 3.1).

The information about the considered decision tables is shown in Tables 3.1 and 3.2. These two tables contain the name of a decision table T, the number of rows (column "Rows"), the number of conditional attributes (column "Attr"), and the spectrum of the table T_{MVD} (column "Spectrum"). The spectrum of a decision table with many-valued decisions is a sequence #1, #2,..., where #i, $i = 1, 2, \ldots$, is the number of rows labeled with sets of decisions with the cardinality equal to i. For some tables (marked with * in Table 3.2), the spectrum is too long to fit in the page. Therefore it is shown up to the element #10. Table 3.1 contains also indexes of conditional attributes removed from the initial decision table (column "Removed attributes").

3.3   Comparison of Complexity of Decision Trees

In this section, we compare the depth, average depth, and number of nodes of decision trees constructed by a greedy heuristic for decision tables T_{MVD}, T_{GD}, and T_{MCD}. We use ws_me greedy heuristic (see Sect. 8.3) which is based on weighted sum type of impurity function ws and misclassification error uncertainty measure me.

The depth of a decision tree is already defined. It is not necessary to explain what means the number of nodes in a decision tree. However, we need to define the notion of average depth. Let T be a nonempty decision table and Γ be a decision tree for T. For each row of T, there is exactly one path in Γ from the root to a terminal node which accepts this row. We consider the sum of lengths of such paths for all rows of T and divide it by the number of rows in T. As a result, we obtain the average depth of the decision tree Γ for the table T.

Table 3.3 Depth, average depth, and number of nodes for decision trees constructed for tables T_{MVD}, T_{MCD}, and T_{GD}, UCI ML Repository

Decision table T	Depth			Average depth			Number of nodes		
	MVD	MCD	GD	MVD	MCD	GD	MVD	MCD	GD
BALANCE-SCALE-1	2	3	3	2.00	2.52	3.00	31	96	156
BREAST-CANCER-1	6	6	7	3.68	3.74	4.10	152	160	217
BREAST-CANCER-5	3	4	4	1.82	2.18	2.60	46	77	102
CARS-1	5	5	5	1.96	2.58	3.81	43	101	280
FLAGS-5	6	6	6	3.75	3.80	3.84	210	216	223
HAYES-ROTH-DATA-1	2	3	3	1.74	1.97	2.31	17	26	39
KR-VS-KP-5	13	14	15	7.80	8.33	9.50	543	873	1539
KR-VS-KP-4	14	14	14	7.87	8.50	9.48	555	915	1635
LYMPHOGRAPHY-5	7	7	7	3.79	4.12	4.31	77	94	112
MUSHROOM-5	7	7	8	2.77	2.78	2.80	246	253	265
NURSERY-1	7	7	7	2.17	3.47	4.18	198	832	1433
NURSERY-4	2	4	4	1.33	2.28	2.08	9	53	54
SPECT-TEST-1	6	10	10	3.13	3.27	3.54	35	43	53
TEETH-1	4	4	4	2.82	2.82	2.82	35	35	35
TEETH-5	3	3	3	2.21	2.21	2.21	20	20	20
TIC-TAC-TOE-4	5	5	5	2.97	4.14	4.29	76	182	216
TIC-TAC-TOE-3	6	6	6	4.15	4.78	5.21	199	362	490
ZOO-DATA-5	4	7	7	3.21	3.71	4.12	19	25	41
Average	5.67	6.39	6.56	3.29	3.73	4.12	139.5	242.39	383.89

Table 3.4 Depth, average depth, and number of nodes for decision trees constructed for tables T_{MVD}, T_{MCD}, and T_{GD}, KEEL-Dataset Repository

Decision table T	Depth			Average depth			Number of nodes		
	MVD	MCD	GD	MVD	MCD	GD	MVD	MCD	GD
BIBTEX	39	42	43	11.52	12.24	12.97	9357	10583	13521
COREL5K	156	156	157	36.10	36.41	36.29	6899	8235	9823
DELICIOUS	79	92	92	13.74	15.90	16.72	6455	18463	31531
ENRON	28	26	41	9.18	9.62	11.18	743	1071	2667
MEDICAL	16	16	16	8.42	8.42	8.42	747	747	747
Average	63.6	66.4	69.8	15.79	16.52	17.12	4840.2	7819.8	11657.8

Tables 3.3 and 3.4 contain the depth, average depth, and number of nodes for decision trees constructed by the greedy heuristic *ws_me* for decision tables T_{MVD}, T_{MCD}, and T_{GD}.

For each cost function and for each decision table (with the exception of the average depth for NURSERY-4 and the depth for ENRON), the cost of decision trees constructed in the framework of *MVD* approach is at most the cost of decision trees constructed in the framework of *MCD* approach, and the cost of latter trees is at most the cost of trees constructed in the framework of *GD* approach.

For each decision table (with the exception of NURSERY-4, TEETH-1, TEETH-5, ENRON, and MEDICAL), the decision trees constructed in the framework of *MVD* approach are simpler than the decision trees constructed in the framework of *MCD* approach, and the decision trees constructed in the framework of *MCD* approach are simpler than the decision trees constructed in the framework of *GD* approach.

Hence, if we are interested in the reducing the complexity of the constructed decision trees, and the *MVD* approach is appropriate for a given inconsistent decision table, then we can use it instead of the approaches *MCD* and *GD*.

3.4   Comparison of Accuracy of Classifiers

In this section, we study 12 greedy heuristics for construction of decision trees. We use four uncertainty measures (*me*, *abs*, *entS*, and *entM*) and three types of impurity function (*ws*, *wm*, and *M_ws*). Each uncertainty measure and each type of impurity function define a greedy heuristic for decision tree construction (see details in Sect. 8.3). We consider the decision trees constructed by these heuristics as classifiers, and compare their accuracy in the frameworks of *MVD*, *MCD*, and *GD* approaches.

Let T be one of the considered decision tables and H be one of the considered greedy heuristics. We use 3-fold cross validation in which the decision table T is divided into 3 folds. At ith ($i = 1, 2, 3$) iteration, ith fold is used as the test subtable, and the rest of table is partitioned randomly into train (70%) and validation (30%) subtables. We apply the heuristic H to the train subtable and, as a result, we obtain the trained decision tree Γ_1. We successively prune the nodes of the trained decision tree Γ_1, calculate the error rates of obtained classifiers on the validation subtable, and return the pruned decision tree Γ_2 with minimum error rate. After that, we calculate the error rate of the decision tree Γ_2 on the test subtable. We repeat the experiment 5 times and take the average of the obtained 15 error rates. As a result, we obtain the classification error rate $E(T, H)$.

Let us consider one of the approaches *MVD*, *MCD*, or *GD*. We have 12 greedy heuristics $H_1, \ldots, H_{12}$ for decision tree construction and 10 decision tables $T_1, \ldots, T_{10}$ (see Table 3.6). For each decision table T_i, $i = 1, \ldots, 10$, we rank the heuristics $H_1, \ldots, H_{12}$ on T_i based on the classification error rates $E(T_i, H_1), \ldots, E(T_i, H_{12})$. We assign the rank 1 to the best heuristic (heuristic with minimum classification error rate), the rank 2 to the second best heuristic, etc. In case of ties, average ranks are assigned. Let r_i^j be the rank of the heuristic H_j on the decision table T_i. For $j = 1, \ldots, 12$, we correspond to the heuristic H_j the average rank $R_j = \frac{1}{10} \cdot \sum_{i=1}^{10} r_i^j$.

Table 3.5 shows the three best ranked heuristics (with minimum average ranks) for the approaches *MVD*, *MCD*, and *GD*. It is interesting that, for the considered three

Table 3.5 Average ranks of the three best ranked greedy heuristics

Name of heuristic	Average rank		
	MVD	MCD	GD
$\mathcal{H}_1$ (M_ws_abs)	3.45	3.65	4.05
$\mathcal{H}_2$ (M_ws_entS)	3.90	3.90	4.55
$\mathcal{H}_3$ (M_ws_entM)	3.60	3.80	2.90

Table 3.6 Classification error rate (in %)

Decision table T	MVD			MCD			GD		
	$\mathcal{H}_1$	$\mathcal{H}_2$	$\mathcal{H}_3$	$\mathcal{H}_1$	$\mathcal{H}_2$	$\mathcal{H}_3$	$\mathcal{H}_1$	$\mathcal{H}_2$	$\mathcal{H}_3$
BIBTEX	57.38	59.96	57.72	54.61	65.37	63.95	90.97	90.97	83.07
CARS-1	3.66	3.80	4.44	6.67	6.67	6.67	16.16	17.04	17.08
COREL5K	74.39	76.52	77.55	79.54	82.14	81.96	98.70	98.70	97.46
ENRON	35.93	27.05	30.29	32.30	30.72	32.22	85.93	85.93	86.48
FLAGS-5	61.75	63.86	63.51	60.82	63.39	63.98	68.19	72.05	65.15
GENBASE-1	4.46	5.34	4.85	6.60	6.80	6.90	13.41	13.41	16.70
LYMPHOGRAPHY-5	25.20	24.87	24.54	29.34	29.67	29.01	29.34	40.50	30.00
MEDICAL	20.00	23.43	21.72	23.27	26.00	25.07	36.53	36.53	33.71
NURSERY-1	2.05	2.70	2.71	5.40	5.25	5.35	8.25	6.69	8.39
ZOO-DATA-5	34.76	32.38	30.48	35.24	34.76	32.86	48.10	48.10	50.00
Average	31.96	31.99	31.78	33.38	35.08	34.80	49.55	50.99	48.80

approaches, we have the same three best heuristics but ranked in different order. We show the classification error rate for 10 decision tables, for the three best ranked heuristics, and for the three approaches MVD, MCD, and GD in Table 3.6.

For each decision table and for each heuristic (with the exception of BIBTEX and $\mathcal{H}_1$, ENRON and $\mathcal{H}_1$, FLAGS-5 and $\mathcal{H}_1$, $\mathcal{H}_2$, and LYMPHOGRAPHY-5 and $\mathcal{H}_1$), the MVD approach gives better classification results compared to the MCD approach, and the MCD approach gives better classification results than the GD approach.

Hence, if we are interested in the reducing of the classification error rate for the constructed decision trees, and the MVD approach is appropriate for a given inconsistent decision table, then we can use it instead of the approaches MCD and GD.

References

1. Alcalá-Fdez, J., Fernandez, A., Luengo, J., Derrac, J., García, S., Herrera, F.: KEEL data-mining software tool: data set repository, integration of algorithms and experimental analysis framework. J. Mult. Valued Log. Soft Comput. **17**, 255–287 (2011)

2. Azad, M., Moshkov, M.: 'Misclassification error' greedy heuristic to construct decision trees for inconsistent decision tables. In: Fred, A.L.N., Filipe, J. (eds.) International Conference on Knowledge Discovery and Information Retrieval, KDIR 2014, Rome, Italy, 21–24 Oct 2014, pp. 184–191. SciTePress (2014)
3. Azad, M., Moshkov, M.: Classification for inconsistent decision tables. In: Flores, V., Gomide, F., Janusz, A., Meneses, C., Miao, D., Peters, G., Slezak, D., Wang, G., Weber, R., Yao, Y. (eds.) Rough Sets – International Joint Conference, IJCRS 2016, Santiago de Chile, Chile, 7–11 Oct 2016. Lecture Notes in Computer Science, vol. 9920, pp. 525–534. Springer (2016)
4. Lichman, M.: UCI Machine Learning Repository. University of California, Irvine, School of Information and Computer Sciences (2013). http://archive.ics.uci.edu/ml
5. Pawlak, Z.: Rough Sets - Theoretical Aspect of Reasoning About Data. Kluwer Academic Publishers, Dordrecht (1991)

Chapter 4
Preliminary Results for Decision and Inhibitory Trees, Tests, Rules, and Rule Systems

In the book [2], some relatively simple results were considered for binary decision tables with many-valued decisions: relationships among decision trees, rules and tests, bounds on their complexity, greedy algorithms for construction of decision trees, rules and tests, and dynamic programming algorithms for minimization of tree depth and rule length.

In this chapter, we consider in more details definitions related to binary decision tables with many-valued decisions, decision and inhibitory trees, tests, rules, and rule systems for these tables. After that, we mention without proofs results from the book [2] related to decision trees, tests, rules, and rule systems for binary decision tables with many-valued decisions and extend them to inhibitory trees, tests, rules, and rule systems.

4.1 Main Notions

We now consider formal definitions of notions corresponding to binary decision tables with many-valued decisions including decision and inhibitory trees, tests, rules, and rule systems, and complementary decision tables.

4.1.1 Binary Decision Tables with Many-valued Decisions

A *binary decision table with many-valued decisions* is a rectangular table T filled by numbers from the set $\{0, 1\}$. Columns of this table are labeled with attributes $f_1, \ldots, f_n$. Rows of the table are pairwise different, and each row r is labeled with a nonempty finite set $D(r) = D_T(r)$ of numbers from $\omega = \{0, 1, 2, \ldots\}$ which is

© Springer Nature Switzerland AG 2020
F. Alsolami et al., *Decision and Inhibitory Trees and Rules for Decision Tables with Many-valued Decisions*, Intelligent Systems Reference Library 156, https://doi.org/10.1007/978-3-030-12854-8_4

interpreted as the set of decisions. Let $Row(T)$ be the set of rows of T and $D(T) = \bigcup_{r \in Row(T)} D(r)$. We denote by $N(T)$ the number of rows in the table T.

A table obtained by removal of some rows of T is called a *subtable of* T. Let $f_{i_1}, \ldots, f_{i_m} \in \{f_1, \ldots, f_n\}$ and $\delta_1, \ldots, \delta_m \in \{0, 1\}$. We denote by

$$T(f_{i_1}, \delta_1) \ldots (f_{i_m}, \delta_m)$$

a subtable of the table T which contains only rows that at the intersection with columns $f_{i_1}, \ldots, f_{i_m}$ have numbers $\delta_1, \ldots, \delta_m$, respectively. We denote by $SEP(T)$ the set of all nonempty subtables of the table T including the table T which can be represented in the form $T(f_{i_1}, \delta_1) \ldots (f_{i_m}, \delta_m)$. We will call such tables *separable subtables* of the table T.

We will consider two interpretations of the decision table T: decision and inhibitory. In the case of *decision interpretation*, for a given row r of T, we should find a decision from the set $D(r)$. In the case of *inhibitory interpretation*, for a given row r of T, we should find a decision from the set $D(T) \setminus D(r)$. When we consider inhibitory interpretation and study inhibitory trees, rules and tests, we assume that $D(r) \neq D(T)$ for any row $r \in Row(T)$.

4.1.2   Decision Trees, Rule Systems, and Tests

We begin from the consideration of *decision interpretation* of the decision table T.

A *decision tree over* T is a finite tree with root in which each terminal node is labeled with a decision (a number from ω), each nonterminal node (such nodes will be called *working*) is labeled with an attribute from the set $\{f_1, \ldots, f_n\}$. Two edges start in each working node. These edges are labeled with 0 and 1, respectively.

Let Γ be a decision tree over T. For a given row r of T, this tree works in the following way. We begin the work in the root of Γ. If the considered node is terminal then the result of Γ work is the number attached to this node. Let the considered node be working node which is labeled with an attribute f_i. If the value of f_i in the considered row is 0 then we pass along the edge which is labeled with 0. Otherwise, we pass along the edge which is labeled with 1, etc.

We will say that a decision tree Γ over the decision table T is a *decision tree for* T if, for any row r of T, the work of Γ finishes in a terminal node which is labeled with a number from the set $D(r)$ attached to the row r.

We denote by $h(\Gamma)$ the *depth* of Γ which is the maximum length of a path from the root to a terminal node. We denote by $h(T)$ the minimum depth of a decision tree for the table T.

A *decision rule over* T is an expression of the kind

$$(f_{i_1} = b_1) \wedge \cdots \wedge (f_{i_m} = b_m) \rightarrow t$$

where $f_{i_1}, \ldots, f_{i_m} \in \{f_1, \ldots, f_n\}$, $b_1, \ldots, b_m \in \{0, 1\}$, and $t \in \omega$. We denote this rule ρ. The number m is called the *length* of the rule ρ and is denoted $l(\rho)$. The decision t is called the *right-hand side of the rule ρ*. This rule is called *realizable* for a row $r = (\delta_1, \ldots, \delta_n)$ if

$$\delta_{i_1} = b_1, \ldots, \delta_{i_m} = b_m .$$

The rule ρ is called *true* for T if, for any row r of T such that the rule ρ is realizable for row r, $t \in D(r)$. We denote by $l(T, r)$ the minimum length of a decision rule over T which is true for T and realizable for r. We will say that the considered rule is a *rule for T and r* if this rule is true for T and realizable for r.

A nonempty finite set S of decision rules over T is called a *complete decision rule system for T* if each rule from S is true for T and, for every row of T, there exists a rule from S which is realizable for this row. We denote by $l(S)$ the maximum length of a rule from S (we will call $l(S)$ the *length* of S), and by $l(T)$ we denote the minimum value of $l(S)$ among all complete decision rule systems S for T.

We will say that T is a *degenerate* table if either T has no rows, or the intersection of sets of decisions attached to rows of T is nonempty (in this case, we will say that T has a *common* decision).

A *decision test for the table T* is a subset of columns $\{f_{i_1}, \ldots, f_{i_m}\}$ such that, for any numbers $\delta_1, \ldots, \delta_m \in \{0, 1\}$, the subtable $T(f_{i_1}, \delta_1) \ldots T(f_{i_m}, \delta_m)$ is a degenerate table. Empty set is a decision test for T if and only if T is a degenerate table.

A *decision reduct* for the table T is a decision test for T for which each proper subset is not a decision test. It is clear that each decision test has a decision reduct as a subset. We denote by $R(T)$ the minimum cardinality of a decision reduct for T.

4.1.3 Inhibitory Trees, Rule Systems, and Tests

We consider now the *inhibitory interpretation* of the decision table T. We will assume that $D(r) \neq D(T)$ for any row $r \in Row(T)$.

An *inhibitory tree over T* is a finite tree with root in which each terminal node is labeled with an expression $\neq t$ where t is a decision (a number from ω), each nonterminal node (such nodes will be called *working*) is labeled with an attribute from the set $\{f_1, \ldots, f_n\}$. Two edges start in each working node. These edges are labeled with 0 and 1, respectively.

Let Γ be an inhibitory tree over T. For a given row r of T, this tree works in the following way. We begin the work in the root of Γ. If the considered node is terminal then the result of Γ work is the expression attached to this node. Let the considered node be working node which is labeled with an attribute f_i. If the value of f_i in the considered row is 0 then we pass along the edge which is labeled with 0. Otherwise, we pass along the edge which is labeled with 1, etc.

We will say that an inhibitory tree Γ over the decision table T is an *inhibitory tree for T* if, for any row r of T, the work of Γ finishes in a terminal node which is labeled with an expression $\neq t$ where $t \in D(T) \setminus D(r)$.

We denote by $h(\Gamma)$ the *depth* of Γ which is the maximum length of a path from the root to a terminal node. We denote by $ih(T)$ the minimum depth of an inhibitory tree for the table T.

An *inhibitory rule over T* is an expression of the kind

$$(f_{i_1} = b_1) \wedge \cdots \wedge (f_{i_m} = b_m) \to \neq t$$

where $f_{i_1}, \ldots, f_{i_m} \in \{f_1, \ldots, f_n\}$, $b_1, \ldots, b_m \in \{0, 1\}$, and $t \in \omega$. We denote this rule ρ. The number m is called the *length* of the rule ρ and is denoted by $l(\rho)$. The expression $\neq t$ is called the *right-hand side of the rule ρ*. This rule is called *realizable* for a row $r = (\delta_1, \ldots, \delta_n)$ if

$$\delta_{i_1} = b_1, \ldots, \delta_{i_m} = b_m .$$

The rule ρ is called *true* for T if, for any row r of T such that the rule ρ is realizable for row r, $t \in D(T) \setminus D(r)$. We denote by $il(T, r)$ the minimum length of an inhibitory rule over T which is true for T and realizable for r. We will say that the considered rule is a *rule for T and r* if this rule is true for T and realizable for r.

A nonempty finite set S of inhibitory rules over T is called a *complete inhibitory rule system for T* if each rule from S is true for T and, for every row of T, there exists a rule from S which is realizable for this row. We denote by $l(S)$ the maximum length of a rule from S (we will call $l(S)$ the *length* of S), and by $il(T)$ we denote the minimum value of $l(S)$ among all complete inhibitory rule systems S for T.

Let Θ be a subtable of T. We will say that Θ is *incomplete* relative to T if $D(\Theta) \subset D(T)$.

An *inhibitory test for the table T* is a subset of columns $\{f_{i_1}, \ldots, f_{i_m}\}$ such that, for any numbers $\delta_1, \ldots, \delta_m \in \{0, 1\}$, the subtable $T(f_{i_1}, \delta_1) \ldots T(f_{i_m}, \delta_m)$ is incomplete relative to T.

An *inhibitory reduct* for the table T is an inhibitory test for T for which each proper subset is not an inhibitory test. It is clear that each inhibitory test has an inhibitory reduct as a subset. We denote by $iR(T)$ the minimum cardinality of an inhibitory reduct for T.

4.1.4 Complementary Decision Table

Let T be a nondegenerate binary decision table with many-valued decisions such that $D_T(r) \neq D(T)$ for any $r \in Row(T)$. We denote by T^C *complementary* to T decision table obtained from the table T by changing, for each row $r \in Row(T)$, the set $D_T(r)$ with the set $D(T) \setminus D_T(r)$, i.e., $D_{T^C}(r) = D(T) \setminus D_T(r)$ for any $r \in Row(T)$. It is clear that $Row(T) = Row(T^C)$, $D(T) = D(T^C)$, and $D_T(r) = D(T) \setminus D_{T^C}(r)$ for any row $r \in Row(T)$. In particular, we have $T^{CC} = T$.

Let Γ_1 be a decision tree over T^C and Γ_2 be an inhibitory tree over T. We denote by Γ_1^- an inhibitory tree over T obtained from Γ_1 by changing expressions attached to terminal nodes: if a terminal node in Γ_1 is labeled with t then the corresponding node in Γ_1^- is labeled with $\neq t$. We denote by Γ_2^+ a decision tree over T^C obtained from Γ_2 by changing expressions attached to terminal nodes: if a terminal node in Γ_2 is labeled with $\neq t$ then the corresponding node in Γ_2^+ is labeled with t.

Let ρ_1 be a decision rule over T^C and ρ_2 be an inhibitory rule over T. We denote by ρ_1^- an inhibitory rule over T obtained from ρ_1 by changing the right-hand side of ρ_1: if the right-hand side of ρ_1 is t then the right-hand side of ρ_1^- is $\neq t$. We denote by ρ_2^+ a decision rule over T^C obtained from ρ_2 by changing the right-hand side of ρ_2: if the right-hand side of ρ_2 is $\neq t$ then the right-hand side of ρ_2^+ is t.

Let S_1 be a nonempty finite set of decision rules over T^C and S_2 be a nonempty finite system of inhibitory rules over T. We denote $S_1^- = \{\rho^- : \rho \in S_1\}$ and $S_2^+ = \{\rho^+ : \rho \in S_2\}$.

Proposition 4.1 *Let T be a nondegenerate binary decision table with many-valued decisions containing n columns labeled with attributes $f_1, \ldots, f_n$, and T^C be complementary to T decision table. Then*

1. *A decision tree Γ over T^C is a decision tree for T^C if and only if Γ^- is an inhibitory tree for T;*
2. *A decision rule ρ over T^C is true for T^C if and only if the inhibitory rule ρ^- is true for T;*
3. *A nonempty finite set S of decision rules over T^C is a complete system of decision rules for T^C if and only if S^- is a complete system of inhibitory rules for T;*
4. *A subset $\{f_{i_1}, \ldots, f_{i_m}\}$ of the set $\{f_1, \ldots, f_n\}$ is a decision test for T^C if and only if the subset $\{f_{i_1}, \ldots, f_{i_m}\}$ is an inhibitory test for T.*

Proof 1. Let Γ be a decision tree over T^C. Let us assume that Γ is a decision tree for T^C. Then, for any row r of T^C, the work of Γ finishes in a terminal node which is labeled with a number $t \in D_{T^C}(r)$. From here it follows that, for any row r of T, the work of Γ^- finishes in a terminal node which is labeled with an expression $\neq t$ where $t \in D_{T^C}(r) = D(T) \setminus D_T(r)$. Therefore Γ^- is an inhibitory tree for T.

Let us assume now that Γ^- is an inhibitory tree for T. Then, for any row r of T, the work of Γ^- finishes in a terminal node which is labeled with an expression $\neq t$ where $t \in D(T) \setminus D_T(r)$. From here it follows that, for any row r of T^C, the work of Γ finishes in a terminal node which is labeled with a number t where $t \in D(T) \setminus D_T(r) = D_{T^C}(r)$. Therefore Γ is a decision tree for T^C.

2. Let ρ be a decision rule over T^C and ρ be equal to $(f_{i_1} = b_1) \wedge \cdots \wedge (f_{i_m} = b_m) \to t$. Then ρ^- is equal to $(f_{i_1} = b_1) \wedge \cdots \wedge (f_{i_m} = b_m) \to \neq t$. Denote

$$T_*^C = T^C(f_{i_1}, b_1) \ldots (f_{i_m}, b_m)$$

and $T_* = T(f_{i_1}, b_1) \ldots (f_{i_m}, b_m)$. One can show that ρ is true for T^C if and only if $t \in D_{T^C}(r)$ for any $r \in Row(T_*^C)$, and ρ^- is true for T if and only if $t \in D(T) \setminus D_T(r)$ for any $r \in Row(T_*)$. We know that $Row(T_*^C) = Row(T_*)$ and $D_{T^C}(r) = D(T) \setminus D_T(r)$ for any $r \in Row(T_*)$. From here it follows that ρ is true for T^C if and only if ρ^- is true for T.

3. Let S be a nonempty finite set of decision rules over T^C. From statement 2 it follows that each rule ρ from S is true for T^C if and only if each rule ρ^- from S^- is true for T. It is clear that a rule ρ from S is realizable for a row r from $Row(T^C) = Row(T)$ if and only if ρ^- is realizable for r. From here it follows that S is a complete system of decision rules for T^C if and only if S^- is a complete system of inhibitory rules for T.

4. Let $\{f_{i_1}, \ldots, f_{i_m}\}$ be a subset of the set $\{f_1, \ldots, f_n\}$ and $\delta_1, \ldots, \delta_m \in \{0, 1\}$. One can show that $T^C(f_{i_1}, \delta_1) \ldots (f_{i_m}, \delta_m)$ is degenerate if and only if

$$T(f_{i_1}, \delta_1) \ldots (f_{i_m}, \delta_m)$$

is incomplete relative to T. Therefore the subset $\{f_{i_1}, \ldots, f_{i_m}\}$ is a decision test for T^C if and only the subset $\{f_{i_1}, \ldots, f_{i_m}\}$ is an inhibitory test for T.   □

Corollary 4.1 *Let T be a nondegenerate binary decision table with many-valued decisions, T^C be complementary to T decision table, and r be a row of T. Then $ih(T) = h(T^C)$, $il(T, r) = l(T^C, r)$, $il(T) = l(T^C)$, and $iR(T) = R(T^C)$.*

Proof 1. Let Γ_1 be a decision tree for T^C such that $h(\Gamma_1) = h(T^C)$. Then, by Proposition 4.1, Γ_1^- is an inhibitory tree for T. It is clear that $h(\Gamma_1) = h(\Gamma_1^-)$. Therefore $ih(T) \leq h(T^C)$. Let Γ_2 be an inhibitory tree for T such that $h(\Gamma_2) = ih(T)$. We now consider a decision tree Γ_2^+ over T^C. Evidently, $(\Gamma_2^+)^- = \Gamma_2$. By Proposition 4.1, Γ_2^+ is a decision tree for T^C. It is clear that $h(\Gamma_2) = h(\Gamma_2^+)$. Therefore $h(T^C) \leq ih(T)$ and $h(T^C) = ih(T)$.

2. Let ρ_1 be a decision rule for T^C and r such that $l(\rho_1) = l(T^C, r)$. It is clear that ρ_1 is realizable for r if and only if the rule ρ_1^- is realizable for r. Then, by Proposition 4.1, ρ_1^- is an inhibitory rule for T and r. It is clear that $l(\rho_1) = l(\rho_1^-)$. Therefore $il(T, r) \leq l(T^C, r)$. Let ρ_2 be an inhibitory rule for T and r such that $l(\rho_2) = il(T, r)$. We now consider a decision rule ρ_2^+ over T^C. Evidently, $(\rho_2^+)^- = \rho_2$, and ρ_2 is realizable for r if and only if the rule ρ_2^+ is realizable for r. By Proposition 4.1, ρ_2^+ is a decision rule for T^C. It is clear that $l(\rho_2) = l(\rho_2^+)$. Therefore $l(T^C, r) \leq il(T, r)$ and $l(T^C, r) = il(T, r)$.

3. Let S_1 be a complete system of decision rules for T^C such that $l(S_1) = l(T^C)$. Then, by Proposition 4.1, S_1^- is a complete system of inhibitory rules for T. It is clear that $l(S_1) = l(S_1^-)$. Therefore $il(T) \leq l(T^C)$. Let S_2 be a complete system of inhibitory rules for T such that $l(S_2) = il(T)$. We now consider a set of decision rules S_2^+ over T^C. Evidently, $(S_2^+)^- = S_2$. By Proposition 4.1, S_2^+ is a complete system of decision rules for T^C. It is clear that $l(S_2) = l(S_2^+)$. Therefore $l(T^C) \leq il(T)$ and $l(T^C) = il(T)$.

4. The equality $iR(T) = R(T^C)$ follows directly from Proposition 4.1.   □

4.2   Relationships Among Trees, Rule Systems, and Tests

In this section, we consider relationships among decision (inhibitory) trees, rule systems, and tests.

4.2.1   Decision Interpretation

We now consider relationships among decision trees, rule systems, and tests.

Theorem 4.1 ([2]) *Let T be a binary decision table with many-valued decisions.*

1. *If Γ is a decision tree for T then the set of attributes attached to working nodes of Γ is a decision test for the table T.*
2. *Let $\{f_{i_1}, \ldots, f_{i_m}\}$ be a decision test for T. Then there exists a decision tree Γ for T which uses only attributes from $\{f_{i_1}, \ldots, f_{i_m}\}$ and for which $h(\Gamma) = m$.*

Corollary 4.2 ([2]) *Let T be a binary decision table with many-valued decisions. Then*

$$h(T) \leq R(T) .$$

Theorem 4.2 ([2]) *Let T be a binary decision table with many-valued decisions containing n columns labeled with attributes $f_1, \ldots, f_n$.*

1. *If S is a complete system of decision rules for T then the set of attributes from rules in S is a decision test for T.*
2. *If $F = \{f_{i_1}, \ldots, f_{i_m}\}$ is a decision test for T then there exists a complete system S of decision rules for T which uses only attributes from F and for which $l(S) = m$.*

Corollary 4.3 ([2]) *Let T be a binary decision table with many-valued decisions. Then*

$$l(T) \leq R(T) .$$

Let Γ be a decision tree for T and τ be a path in Γ from the root to a terminal node in which working nodes are labeled with attributes $f_{i_1}, \ldots, f_{i_m}$, edges are labeled with numbers $b_1, \ldots, b_m$, and the terminal node of τ is labeled with the decision t. We correspond to τ the decision rule

$$(f_{i_1} = b_1) \wedge \cdots \wedge (f_{i_m} = b_m) \to t .$$

Theorem 4.3 ([2]) *Let T be a binary decision table with many-valued decisions, Γ be a decision tree for T, and S be the set of decision rules corresponding to paths in Γ from the root to terminal nodes. Then S is a complete system of decision rules for T, and $l(S) = h(\Gamma)$.*

Corollary 4.4 ([2]) *Let T be a binary decision table with many-valued decisions. Then*

$$l(T) \leq h(T) .$$

4.2.2 Inhibitory Interpretation

We now consider relationships among inhibitory trees, rule systems, and tests. Here and later, under the consideration of inhibitory trees, rules or tests for a decision table T, we will assume that $D_T(r) \neq D(T)$ for any row r of T.

Theorem 4.4 *Let T be a nondegenerate binary decision table with many-valued decisions.*

1. *If Γ is an inhibitory tree for T then the set of attributes attached to working nodes of Γ is an inhibitory test for the table T.*
2. *Let $\{f_{i_1}, \ldots, f_{i_m}\}$ be an inhibitory test for T. Then there exists an inhibitory tree for T which uses only attributes from $\{f_{i_1}, \ldots, f_{i_m}\}$ and which depth is equal to m.*

Proof 1. Let Γ be an inhibitory tree for T and $F(\Gamma)$ be the set of attributes attached to working nodes of Γ. We now consider a decision tree Γ^+ over T^C. It is clear that $F(\Gamma)$ is the set of attributes attached to working nodes of Γ^+. By Proposition 4.1, Γ^+ is a decision tree for T^C. Using Theorem 4.1 we obtain that $F(\Gamma)$ is a decision test for T^C. By Proposition 4.1, $F(\Gamma)$ is an inhibitory test for T.

2. Let $\{f_{i_1}, \ldots, f_{i_m}\}$ be an inhibitory test for T. By Proposition 4.1, $\{f_{i_1}, \ldots, f_{i_m}\}$ is a decision test for T^C. Then, by Theorem 4.1, there exists a decision tree Γ for T^C which uses only attributes from $\{f_{i_1}, \ldots, f_{i_m}\}$ and for which $h(\Gamma) = m$. By Proposition 4.1, Γ^- is an inhibitory tree for T. It is clear that Γ^- uses only attributes from $\{f_{i_1}, \ldots, f_{i_m}\}$, and $h(\Gamma^-) = m$. $\square$

Corollary 4.5 *Let T be a nondegenerate binary decision table with many-valued decisions. Then*

$$ih(T) \leq iR(T) .$$

Theorem 4.5 *Let T be a nondegenerate binary decision table with many-valued decisions.*

1. *If S is a complete system of inhibitory rules for T then the set of attributes from rules in S is an inhibitory test for T.*
2. *If F is an inhibitory test for T then there exists a complete system of inhibitory rules for T which uses only attributes from F and for which the length is equal to $|F|$.*

Proof 1. Let S be a complete system of inhibitory rules for T and $F(S)$ be the set of attributes from rules in S. We now consider a set of decision rules S^+ over T^C. It is clear that $F(S)$ is the set of attributes from rules in S^+. By Proposition 4.1, S^+ is a complete system of decision rules for T^C. Using Theorem 4.2 we obtain that $F(S)$ is a decision test for T^C. By Proposition 4.1, $F(S)$ is an inhibitory test for T.

2. Let F be an inhibitory test for T and $|F| = m$. By Proposition 4.1, F is a decision test for T^C. Then, by Theorem 4.2, there exists a complete system G of decision rules for T which uses only attributes from F and for which $l(G) = m$. By Proposition 4.1, G^- is a complete system of inhibitory rules for T. It is clear that G^- uses only attributes from F, and $l(G^-) = m$. □

Corollary 4.6 *Let T be a nondegenerate binary decision table with many-valued decisions. Then*

$$il(T) \leq iR(T) .$$

Let Γ be an inhibitory tree for T and τ be a path in Γ from the root to a terminal node in which working nodes are labeled with attributes $f_{i_1}, \ldots, f_{i_m}$, edges are labeled with numbers $b_1, \ldots, b_m$, and the terminal node of τ is labeled with the expression $\neq t$. We correspond to τ the inhibitory rule

$$(f_{i_1} = b_1) \wedge \cdots \wedge (f_{i_m} = b_m) \to \neq t .$$

Theorem 4.6 *Let T be a nondegenerate binary decision table with many-valued decisions, Γ be an inhibitory tree for T, and S be the set of inhibitory rules corresponding to paths in Γ from the root to terminal nodes. Then S is a complete system of inhibitory rules for T, and $l(S) = h(\Gamma)$.*

Proof Let Γ be an inhibitory tree for T, and S be the set of inhibitory rules corresponding to paths in Γ from the root to terminal nodes. Let us consider a decision tree Γ^+ over T^C. By Proposition 4.1, Γ^+ is a decision tree for T^C. It is clear that S^+ is the set of decision rules corresponding to paths in Γ^+ from the root to terminal nodes. Using Theorem 4.3 we obtain that S^+ is a complete system of decision rules for T^C. By Proposition 4.1, $(S^+)^- = S$ is a complete system of inhibitory rules for T. □

Corollary 4.7 *Let T be a nondegenerate binary decision table with many-valued decisions. Then*

$$il(T) \leq ih(T) .$$

4.3 Lower Bounds on Complexity of Trees, Rules, Rule Systems, and Tests

In this section, we consider lower bounds on complexity of decision and inhibitory trees, rules, rule systems, and tests.

4.3.1 Decision Interpretation

From Corollaries 4.2 and 4.4 it follows that $l(T) \leq h(T) \leq R(T)$. So each lower bound on $l(T)$ is also a lower bound on $h(T)$ and $R(T)$, and each lower bound on $h(T)$ is also a lower bound on $R(T)$.

Let T be a nonempty binary decision table with many-valued decisions. A nonempty subset B of the set $D(T)$ is called a *system of representatives for the table T* if, for each row r of T, $B \cap D_T(r) \neq \emptyset$. We denote by $S(T)$ the minimum cardinality of a system of representatives for the table T.

Theorem 4.7 ([2]) *Let T be a nonempty binary decision table with many-valued decisions. Then*

$$h(T) \geq \log_2 S(T) \, .$$

Theorem 4.8 ([2]) *Let T be a binary decision table with many-valued decisions. Then*

$$h(T) \geq \log_2 (R(T) + 1).$$

Let T be a binary decision table with many-valued decisions which has n columns labeled with attributes $f_1, \ldots, f_n$, and $\bar{\delta} = (\delta_1, \ldots, \delta_n) \in \{0, 1\}^n$. We define the parameters $M(T, \bar{\delta})$ and $M(T)$. If T is a degenerate table then $M(T, \bar{\delta}) = 0$ and $M(T) = 0$. Let now T be a nondegenerate table. Then $M(T, \bar{\delta})$ is the minimum natural m such that there exist attributes $f_{i_1}, \ldots, f_{i_m} \in \{f_1, \ldots, f_n\}$ for which $T(f_{i_1}, \delta_{i_1}) \ldots (f_{i_m}, \delta_{i_m})$ is a degenerate table. We denote $M(T) = \max\{M(T, \bar{\delta}) : \bar{\delta} \in \{0, 1\}^n\}$.

Theorem 4.9 ([2]) *Let T be a binary decision table with many-valued decisions. Then*

$$h(T) \geq M(T) \, .$$

Let T be a binary decision table with many-valued decisions which has n columns labeled with attributes $\{f_1, \ldots, f_n\}$, m be a natural number, and $m \leq n$.

A (T, m)-*proof-tree* is a finite directed tree G with the root in which the length of each path from the root to a terminal node is equal to $m - 1$. Nodes of this tree are not labeled. In each nonterminal node, exactly n edges start. These edges are labeled with pairs of the kind $(f_1, \delta_1), \ldots, (f_n, \delta_n)$, respectively, where $\delta_1, \ldots, \delta_n \in \{0, 1\}$.

Let v be an arbitrary terminal node of G and $(f_{i_1}, \delta_1), \ldots, (f_{i_{m-1}}, \delta_{m-1})$ be pairs attached to edges in the path from the root of G to the terminal node v. Denote $T(v) = T(f_{i_1}, \delta_1) \ldots (f_{i_{m-1}}, \delta_{m-1})$.

We will say that G is a *proof-tree for the bound $h(T) \geq m$* if, for any terminal node v, the subtable $T(v)$ is not degenerate. The following statement allows us to use proof-trees to obtain lower bounds on $h(T)$.

Theorem 4.10 ([2]) *Let T be a nondegenerate binary decision table with many-valued decisions having n columns, and m be a natural number such that $m \leq n$. Then* *a proof-tree for the bound* $h(T) \geq m$ *exists if and only if the inequality* $h(T) \geq m$ *holds.*

Theorem 4.11 ([2]) *Let* T *be a binary decision table with many-valued decisions and* $Row(T)$ *be the set of rows of* T. *Then* $l(T, \bar{\delta}) = M(T, \bar{\delta})$ *for any row* $\bar{\delta} \in Row(T)$ *and* $l(T) = \max\{M(T, \bar{\delta}) : \bar{\delta} \in Row(T)\}$.

4.3.2 Inhibitory Interpretation

From Corollaries 4.5 and 4.7 it follows that $il(T) \leq ih(T) \leq iR(T)$. So each lower bound on $il(T)$ is also a lower bound on $ih(T)$ and $iR(T)$, and each lower bound on $ih(T)$ is also a lower bound on $iR(T)$.

Let T be a nondegenerate binary decision table with many-valued decisions. A nonempty subset B of the set $D(T)$ is called a *system of non-representatives for the table* T if, for each row r of T, $B \cap (D(T) \setminus D_T(r)) \neq \emptyset$. We denote by $iS(T)$ the minimum cardinality of a system of non-representatives for the table T.

Theorem 4.12 *Let* T *be a nondegenerate binary decision table with many-valued decisions. Then*

$$ih(T) \geq \log_2(iS(T)) .$$

Proof Let T^C be complementary to T decision table. Since T is a nondegenerate table, $D(T) = D(T^C)$. One can show that a nonempty subset B of the set $D(T)$ is a system of representatives for the table T^C if and only if B is a system of non-representatives for the table T. Therefore $iS(T) = S(T^C)$. From Corollary 4.1 and from Theorem 4.7 it follows that $ih(T) = h(T^C)$ and $h(T^C) \geq \log_2 S(T^C)$. Therefore $ih(T) \geq \log_2(iS(T))$. □

Theorem 4.13 *Let* T *be a nondegenerate binary decision table with many-valued decisions. Then*

$$ih(T) \geq \log_2(iR(T) + 1) .$$

Proof Let T^C be complementary to T decision table. From Corollary 4.1 it follows that $ih(T) = h(T^C)$ and $iR(T) = R(T^C)$. By Theorem 4.8, $h(T^C) \geq \log_2(R(T^C) + 1)$. Therefore $ih(T) \geq \log_2(iR(T) + 1)$. □

Let T be a nondegenerate binary decision table with many-valued decisions which has n columns labeled with attributes $f_1, \ldots, f_n$ and $\bar{\delta} = (\delta_1, \ldots, \delta_n) \in \{0, 1\}^n$. We define now the parameters $iM(T, \bar{\delta})$ and $iM(T)$. Denote $iM(T, \bar{\delta})$ the minimum natural m such that there exist attributes $f_{i_1}, \ldots, f_{i_m} \in \{f_1, \ldots, f_n\}$ for which subtable $T(f_{i_1}, \delta_{i_1}) \ldots (f_{i_m}, \delta_{i_m})$ is incomplete relative to T. We denote $iM(T) = \max\{iM(T, \bar{\delta}) : \bar{\delta} \in \{0, 1\}^n\}$.

Lemma 4.1 *Let* T *be a nondegenerate binary decision table with many-valued decisions and* T^C *be complementary to* T *decision table. Then*

$$iM(T) = M(T^C) .$$

Proof Let T have n columns labeled with attributes $f_1, \ldots, f_n$. One can show that, for any $f_{i_1}, \ldots, f_{i_m} \in \{f_1, \ldots, f_n\}$ and any $\delta_1, \ldots, \delta_m \in \{0, 1\}$, the subtable $T(f_{i_1}, \delta_1) \ldots (f_{i_m}, \delta_m)$ is incomplete relative to T if and only if the subtable

$$T^C(f_{i_1}, \delta_1) \ldots (f_{i_m}, \delta_m)$$

is degenerate. From here it follows that $iM(T) = M(T^C)$. □

Theorem 4.14 *Let T be a nondegenerate binary decision table with many-valued decisions. Then*

$$ih(T) \geq iM(T) .$$

Proof By Corollary 4.1, $ih(T) = h(T^C)$. From Lemma 4.1 it follows that $iM(T) = M(T^C)$. By Theorem 4.9, $h(T^C) \geq M(T^C)$. Therefore $ih(T) \geq iM(T)$. □

Let T be a binary decision table with many-valued decisions which has n columns labeled with attributes $\{f_1, \ldots, f_n\}$, m be a natural number, and $m \leq n$.

A (T, m)-*proof-tree* is a finite directed tree G with the root in which the length of each path from the root to a terminal node is equal to $m - 1$. Nodes of this tree are not labeled. In each nonterminal node, exactly n edges start. These edges are labeled with pairs of the kind $(f_1, \delta_1), \ldots, (f_n, \delta_n)$, respectively, where $\delta_1, \ldots, \delta_n \in \{0, 1\}$.

Let v be an arbitrary terminal node of G and $(f_{i_1}, \delta_1), \ldots, (f_{i_{m-1}}, \delta_{m-1})$ be pairs attached to edges in the path from the root of G to the terminal node v. Denote $T(v) = T(f_{i_1}, \delta_1) \ldots (f_{i_{m-1}}, \delta_{m-1})$.

We will say that G is a *proof-tree for the bound $ih(T) \geq m$* if, for any terminal node v, the subtable $T(v)$ is not incomplete relative to T. The following statement allows us to use proof-trees to obtain lower bounds on $ih(T)$.

Theorem 4.15 *Let T be a nondegenerate binary decision table with many-valued decisions having n columns, and m be a natural number such that $m \leq n$. Then a proof-tree for the bound $ih(T) \geq m$ exists if and only if the inequality $ih(T) \geq m$ holds.*

Proof Let T^C be complementary to T decision table, G be a (T, m)-*proof-tree*, and v be a terminal node of G. One can show that the subtable $T(v)$ is not incomplete relative to T if and only if the subtable $T^C(v)$ is not degenerate. From here it follows that G is a proof-tree for the bound $ih(T) \geq m$ if and only if G is a proof-tree for the bound $h(T^C) \geq m$. Using Theorem 4.10 we obtain that a proof-tree for the bound $ih(T) \geq m$ exists if and only if the inequality $h(T^C) \geq m$ holds. By Corollary 4.1, $h(T^C) \geq m$ if and only if $ih(T) \geq m$. □

Theorem 4.16 *Let T be a nondegenerate binary decision table with many-valued decisions and $Row(T)$ be the set of rows of T. Then $il(T, \bar{\delta}) = iM(T, \bar{\delta})$ for any row $\bar{\delta} \in Row(T)$, and $il(T) = \max\{iM(T, \bar{\delta}) : \bar{\delta} \in Row(T)\}$.*

Proof Let T have n columns labeled with attributes $f_1, \ldots, f_n$,

$$\bar{\delta} = (\delta_1, \ldots, \delta_n) \in Row(T) \,,$$

and T^C be complementary to T decision table. It is clear that $Row(T) = Row(T^C)$. One can show that, for any $f_{i_1}, \ldots, f_{i_m} \in \{f_1, \ldots, f_n\}$, the subtable

$$T(f_{i_1}, \delta_{i_1}) \ldots (f_{i_m}, \delta_{i_m})$$

is incomplete relative to T if and only if the subtable $T^C(f_{i_1}, \delta_{i_1}) \ldots (f_{i_m}, \delta_{i_m})$ is degenerate. From here it follows that $iM(T, \bar{\delta}) = M(T^C, \bar{\delta})$. By Corollary 4.1, $il(T, \bar{\delta}) = l(T^C, \bar{\delta})$ and $il(T) = l(T^C)$. By Theorem 4.11, $l(T^C, \bar{\delta}) = M(T^C, \bar{\delta})$ for any row $\bar{\delta} \in Row(T^C)$. Therefore $il(T, \bar{\delta}) = iM(T, \bar{\delta})$ for any row $\bar{\delta} \in Row(T)$. By Theorem 4.11, $l(T^C) = \max\{M(T^C, \bar{\delta}) : \bar{\delta} \in Row(T^C)\}$. Therefore $il(T) = \max\{iM(T, \bar{\delta}) : \bar{\delta} \in Row(T)\}$. $\qquad\square$

4.4   Upper Bounds on Complexity of Trees, Rule Systems, and Tests

In this section, we consider upper bounds on complexity of decision and inhibitory trees, rule systems, and tests.

4.4.1   Decision Interpretation

We know that $l(T) \leq h(T) \leq R(T)$. Therefore each upper bound on $R(T)$ is also an upper bound on $h(T)$ and $l(T)$, and each upper bound on $h(T)$ is also an upper bound on $l(T)$.

Theorem 4.17 ([2]) *Let T be a nonempty binary decision table with many-valued decisions. Then*

$$R(T) \leq N(T) - 1 \,.$$

Theorem 4.18 ([2]) *Let T be a nonempty binary decision table with many-valued decisions. Then*

$$h(T) \leq M(T) \log_2 N(T) \,.$$

4.4.2 Inhibitory Interpretation

We know that $il(T) \leq ih(T) \leq iR(T)$. Therefore each upper bound on $iR(T)$ is also an upper bound on $ih(T)$ and $il(T)$, and each upper bound on $ih(T)$ is also an upper bound on $il(T)$.

Theorem 4.19 *Let T be a nondegenerate binary decision table with many-valued decisions. Then*

$$iR(T) \leq N(T) - 1 .$$

Proof Let T^C be complementary to T decision table. From Corollary 4.1 it follows that $iR(T) = R(T^C)$. It is clear that $N(T) = N(T^C)$. By Theorem 4.17, $R(T^C) \leq N(T^C) - 1$. Therefore $iR(T) \leq N(T) - 1$.   □

Theorem 4.20 *Let T be a nondegenerate binary decision table with many-valued decisions. Then*

$$ih(T) \leq iM(T) \log_2 N(T) .$$

Proof Let T have n columns labeled with attributes $f_1, \ldots, f_n$, and T^C be complementary to T decision table. From Corollary 4.1 it follows that $ih(T) = h(T^C)$. It is clear that $N(T) = N(T^C)$. From Lemma 4.1 it follows that $iM(T) = M(T^C)$. By Theorem 4.18, $h(T^C) \leq M(T^C) \log_2 N(T^C)$. Therefore $ih(T) \leq iM(T) \log_2 N(T)$.   □

4.5 Approximate Algorithms for Optimization of Decision and Inhibitory Tests and Rules

In this section, we consider approximate polynomial algorithms for problems of minimization of decision and inhibitory test cardinality, problems of minimization of decision and inhibitory rule length, and problems of minimization of decision and inhibitory rule system length. First, we consider well known greedy algorithm for set cover problem which will be used later.

4.5.1 Greedy Algorithm for Set Cover Problem

Let A be a set containing $N > 0$ elements, and $F = \{S_1, \ldots, S_p\}$ be a family of subsets of the set A such that $A = \bigcup_{i=1}^{p} S_i$. A subfamily $\{S_{i_1}, \ldots, S_{i_t}\}$ of the family F is called a *cover* if $\bigcup_{j=1}^{t} S_{i_j} = A$. Set cover problem: it is required to find a cover with minimum cardinality t.

We now consider well known *greedy* algorithm for set cover problem. During each step this algorithm chooses a set from the family F which covers the maximum number of previously uncovered elements from A.

We denote by C_{greedy} the cardinality of the cover constructed by the greedy algorithm and by C_{min} – the minimum cardinality of a cover. The proof of the following theorem which was discovered by different authors [1, 3] can be found in [2].

Theorem 4.21 $C_{\text{greedy}} \le C_{\text{min}} \ln N + 1$.

4.5.2 Optimization of Decision Tests

Let T be a binary decision table with many-valued decisions. A subtable T' of T is called *boundary* subtable if T' is not degenerate but each proper subtable of T' is degenerate.

We denote by $B(T)$ the number of boundary subtables of the table T. We denote by $Tab(t)$, where t is a natural number, the set of decision tables with many-valued decisions such that each row in the table is labeled with a set of decisions which cardinality is at most t.

Proposition 4.2 ([2]) *Let T' be a boundary subtable with m rows. Then each row of T' is labeled with a set of decisions which cardinality is at least $m - 1$.*

Corollary 4.8 ([2]) *Each boundary subtable of a table $T \in Tab(t)$ has at most $t + 1$ rows.*

Therefore, for tables from $Tab(t)$, there exists a polynomial algorithm for the computation of the parameter $B(T)$ and for the construction of the set of boundary subtables of the table T.

Let T be a binary decision table with many-valued decisions. It is clear that T is a degenerate table if and only if $B(T) = 0$.

Let us consider an algorithm which, for a given binary decision table with many-valued decisions T, constructs a decision test for T. Let T contain n columns labeled with attributes $f_1, \ldots, f_n$. We construct a set cover problem $A(T)$, $F(T)$ corresponding to the table T, where $A(T)$ is the set of all boundary subtables of T, $F(T) = \{S_1, \ldots, S_n\}$ is a family of subsets of $A(T)$ and, for $i = 1, \ldots, n$, S_i is the set of boundary subtables from $A(T)$ in each of which there exists a pair of rows that are different in the column f_i. One can show that $\{f_{i_1}, \ldots, f_{i_m}\}$ is a decision test for T if and only if $\{S_{i_1}, \ldots, S_{i_m}\}$ is a cover for $A(T)$, $F(T)$. Let us apply the greedy algorithm for set cover problem to $A(T)$, $F(T)$. As a result, we obtain a cover corresponding to a decision test for T. This test is a result of the considered algorithm work. We denote by $R_{\text{greedy}}(T)$ the cardinality of the constructed decision test.

The next statement follows from Theorem 4.21.

Theorem 4.22 ([2]) *Let T be a binary decision table with many-valued decisions. Then*

$$R_{\text{greedy}}(T) \leq R(T) \ln B(T) + 1 \; .$$

For any natural t, for tables from the class $Tab(t)$, the considered algorithm has polynomial time complexity.

Proposition 4.3 ([2]) *The problem of minimization of decision test cardinality for nondegenerate binary decision tables with many-valued decisions is NP-hard.*

Theorem 4.23 ([2]) *If $NP \nsubseteq DTIME(n^{O(\log \log n)})$ then, for any ε, $0 < \varepsilon < 1$, there is no a polynomial algorithm which, for a given nondegenerate binary decision table T with many-valued decisions, constructs a decision test for T which cardinality is at most*

$$(1 - \varepsilon)R(T) \ln B(T) \; .$$

4.5.3 Optimization of Inhibitory Tests

Let T be a nondegenerate binary decision table with many-valued decisions. A subtable T' of T is called *i-boundary* subtable if T' is not incomplete relative to T but each proper subtable of T' is incomplete relative to T.

Let Θ be a subtable of T. We denote by $iB(T, \Theta)$ the number of i-boundary subtables of the table T which are subtables of Θ. We denote by $iTab(t)$, where t is a natural number, the set of binary decision tables T with many-valued decisions such that $|D(T)| \leq t$.

Proposition 4.4 *Let T' be an i-boundary subtable of the table T, and T' contain m rows. Then $|D(T)| \geq m$.*

Proof According to the definition of i-boundary subtable, each row r of T' should have in the set $D(r)$ a decision $d(r)$ such that

$$d(r) \notin \bigcup_{r' \in Row(T) \setminus \{r\}} D(r') \; .$$

From here it follows that $|D(T)| \geq m$.   □

Corollary 4.9 *Each i-boundary subtable of a table $T \in iTab(t)$ has at most t rows.*

Therefore, for tables from $iTab(t)$, there exists a polynomial algorithm for the computation of the parameter $iB(T)$ and for the construction of the set of i-boundary subtables of the table T.

Let T be a nondegenerate binary decision table with many-valued decisions and Θ be a subtable of T. It is clear that Θ is incomplete relative to T if and only if $iB(T, \Theta) = 0$.

Let us consider an algorithm which, for a given nondegenerate binary decision table with many-valued decisions T, constructs an inhibitory test for T. Let T contain n columns labeled with attributes $f_1, \ldots, f_n$. We construct a set cover problem $iA(T), iF(T)$ corresponding to the table T, where $iA(T)$ is the set of all i-boundary subtables of T, $iF(T) = \{iS_1, \ldots, iS_n\}$ is a family of subsets of $iA(T)$ and, for $j = 1, \ldots, n$, iS_j is the set of i-boundary subtables from $iA(T)$ in each of which there exists a pair of rows that are different in the column f_j. One can show that $\{f_{j_1}, \ldots, f_{j_m}\}$ is an inhibitory test for T if and only if $\{iS_{j_1}, \ldots, iS_{j_m}\}$ is a cover for $iA(T), iF(T)$.

Let us apply the greedy algorithm for set cover problem to $iA(T), iF(T)$. As a result, we obtain a cover corresponding to an inhibitory test for T. This test is a result of the considered algorithm work. We denote by $iR_{\text{greedy}}(T)$ the cardinality of the constructed inhibitory test.

The next statement follows from Theorem 4.21.

Theorem 4.24 *Let T be a nondegenerate binary decision table with many-valued decisions. Then*

$$iR_{\text{greedy}}(T) \leq iR(T) \ln(iB(T, T)) + 1 .$$

For any natural t, for tables from the class $iTab(t)$, the considered algorithm has polynomial time complexity.

Lemma 4.2 *Let T be a nondegenerate binary decision table with many-valued decisions, T^C be complementary to T decision table, T_1 be a subtable of T, T_2 be a subtable of T^C, and $Row(T_1) = Row(T_2)$. Then*

$$iB(T, T_1) = B(T_2) .$$

Proof Let Θ_1 be a subtable of T, Θ_2 be a subtable of T^C, and $Row(\Theta_1) = Row(\Theta_2)$. One can show that Θ_1 is an i-boundary subtable of T if and only if Θ_2 is a boundary subtable of T^C. From here it follows that $iB(T, T_1) = B(T_2)$. $\square$

Proposition 4.5 *The problem of minimization of inhibitory test cardinality for nondegenerate binary decision tables with many-valued decisions is NP-hard.*

Proof Let T be a nondegenerate decision table with n columns labeled with attributes $f_1, \ldots, f_n$. We denote $\Theta = T^C$. It is clear that Θ is nondegenerate and $\Theta^C = T$. From Proposition 4.1 it follows that a subset $\{f_{i_1}, \ldots, f_{i_m}\}$ of the set $\{f_1, \ldots, f_n\}$ is a decision test for $\Theta^C = T$ if and only if the subset $\{f_{i_1}, \ldots, f_{i_m}\}$ is an inhibitory test for $\Theta = T^C$. So we have a polynomial time reduction of the problem of minimization of decision test cardinality for nondegenerate binary decision table T with many-valued decisions to the problem of minimization of inhibitory test cardinality for nondegenerate binary decision table T^C with many-valued decisions. By Proposition 4.3, the problem of minimization of decision test cardinality for nondegenerate binary decision tables with many-valued decisions is NP-hard. Therefore the problem of minimization of inhibitory test cardinality for nondegenerate binary decision tables with many-valued decisions is NP-hard. $\square$

Theorem 4.25 *If $NP \nsubseteq DTIME(n^{O(\log \log n)})$ then, for any ε, $0 < \varepsilon < 1$, there is no a polynomial algorithm which, for a given nondegenerate binary decision table T with many-valued decisions, constructs an inhibitory test for T which cardinality is at most*

$$(1 - \varepsilon) i R(T) \ln(i B(T, T)) .$$

Proof Let us assume that, for some ε, $0 < \varepsilon < 1$, there exists a polynomial algorithm $\mathscr{A}$ which, for a given nondegenerate binary decision table T, constructs an inhibitory test for T which cardinality is at most $(1 - \varepsilon) i R(T) \ln(i B(T, T))$.

Let T be a nondegenerate binary decision table with many-valued decisions. We denote $\Theta = T^C$. It is clear that Θ is nondegenerate and $\Theta^C = T$. From Proposition 4.1 it follows that a subset $\{f_{i_1}, \ldots, f_{i_m}\}$ of the set $\{f_1, \ldots, f_n\}$ is a decision test for $\Theta^C = T$ if and only if the subset $\{f_{i_1}, \ldots, f_{i_m}\}$ is an inhibitory test for $\Theta = T^C$. We apply the algorithm $\mathscr{A}$ to the table $\Theta = T^C$. As a result, we obtain an inhibitory test for $\Theta = T^C$ and in the same time a decision test for $\Theta^C = T$ which cardinality is at most $(1 - \varepsilon) i R(\Theta) \ln(i B(\Theta, \Theta))$. We denote this test by F.

From Lemma 4.2 it follows that $i B(\Theta, \Theta) = B(\Theta^C) = B(T)$. From Corollary 4.1 it follows that $i R(\Theta) = R(\Theta^C) = R(T)$. Therefore the cardinality of the set F is at most $(1 - \varepsilon) R(T) \ln B(T)$. As a result, there is a polynomial algorithm that, for a given nondegenerate binary decision table T with many-valued decisions, constructs a decision test for T which cardinality is at most $(1 - \varepsilon) R(T) \ln B(T)$. But this is impossible if $NP \nsubseteq DTIME(n^{O(\log \log n)})$ (see Theorem 4.23). Thus, if $NP \nsubseteq DTIME(n^{O(\log \log n)})$ then, for any ε, $0 < \varepsilon < 1$, there is no a polynomial algorithm that, for a given nondegenerate binary decision table T with many-valued decisions, constructs an inhibitory test for T which cardinality is at most $(1 - \varepsilon) i R(T) \ln(i B(T, T))$. □

4.5.4 Optimization of Decision Rules

We can apply the greedy algorithm for set cover problem to construct decision rules for decision tables with many-valued decisions.

Let T be a nondegenerate binary decision table with many-valued decisions containing n columns labeled with attributes $f_1, \ldots, f_n$. Let $r = (b_1, \ldots, b_n)$ be a row of T, $D(r)$ be the set of decisions attached to r, and $d \in D(r)$.

We consider a set cover problem $A(T, r, d)$, $F(T, r, d) = \{S_1, \ldots, S_n\}$, where $A(T, r, d)$ is the set of all rows r' of T such that $d \notin D(r')$. For $i = 1, \ldots, n$, the set S_i coincides with the set of all rows from $A(T, r, d)$ which are different from r in the column f_i. One can show that the decision rule

$$(f_{i_1} = b_{i_1}) \wedge \cdots \wedge (f_{i_m} = b_{i_m}) \rightarrow d$$

is true for T (it is clear that this rule is realizable for r) if and only if the subfamily $\{S_{i_1}, \ldots, S_{i_m}\}$ is a cover for the set cover problem $A(T, r, d)$, $F(T, r, d)$.

We denote $P(T, r, d) = |A(T, r, d)|$ and $l(T, r, d)$ the minimum length of a decision rule over T which is true for T, realizable for r, and has d on the right-hand side. It is clear that, for the constructed set cover problem, $C_{min} = l(T, r, d)$ where C_{min} is the minimum cardinality of a cover.

Let us apply the greedy algorithm to set cover problem $A(T, r, d)$, $F(T, r, d)$. This algorithm constructs a cover which corresponds to a decision rule $rule(T, r, d)$ that is true for T, realizable for r, and has d on the right-hand side. We denote by $l_{greedy}(T, r, d)$ the length of $rule(T, r, d)$. From Theorem 4.21 it follows that

$$l_{greedy}(T, r, d) \leq l(T, r, d) \ln P(T, r, d) + 1 .$$

We denote by $l_{greedy}(T, r)$ the length of the rule constructed by the following polynomial algorithm (we will say about this algorithm as about modified greedy algorithm). For a given binary decision table T with many-valued decisions and row r of T, for each $d \in D(r)$, we construct the set cover problem $A(T, r, d)$, $F(T, r, d)$ and then apply to this problem the greedy algorithm. We transform the constructed cover to the rule $rule(T, r, d)$. Among the rules $rule(T, r, d)$, $d \in D(r)$, we choose a rule with minimum length. This rule is the output of the considered algorithm. We have

$$l_{greedy}(T, r) = \min\{l_{greedy}(T, r, d) : d \in D(r)\} .$$

It is clear that

$$l(T, r) = \min\{l(T, r, d) : d \in D(r)\} .$$

Let $K(T, r) = \max\{P(T, r, d) : d \in D(r)\}$. Then

$$l_{greedy}(T, r) \leq l(T, r) \ln K(T, r) + 1 .$$

So we have the following statement.

Theorem 4.26 ([2]) *Let T be a nondegenerate binary decision table with many-valued decisions and r be a row of T. Then*

$$l_{greedy}(T, r) \leq l(T, r) \ln K(T, r) + 1 .$$

We can use the considered modified greedy algorithm to construct a complete decision rule system for the decision table T with many-valued decisions. To this end, we apply this algorithm sequentially to the table T and to each row r of T. As a result, we obtain a system of rules S in which each rule is true for T and, for every row of T, there exists a rule from S which is realizable for this row.

We denote $l_{greedy}(T) = l(S)$ and

$$K(T) = \max\{K(T, r) : r \in Row(T)\} ,$$

where $Row(T)$ is the set of rows of T. It is clear that $l(T) = \max\{l(T, r) : r \in Row(T)\}$. Using Theorem 4.26 we obtain

Theorem 4.27 ([2]) *Let T be a nondegenerate binary decision table with many-valued decisions. Then*

$$l_{\text{greedy}}(T) \le l(T) \ln K(T) + 1 .$$

Proposition 4.6 ([2]) *The problem of minimization of decision rule length for binary decision tables with many-valued decisions is NP-hard.*

Theorem 4.28 ([2]) *If $NP \nsubseteq DTIME(n^{O(\log \log n)})$ then, for any ε, $0 < \varepsilon < 1$, there is no a polynomial algorithm that, for a given nondegenerate binary decision table T with many-valued decisions and row r of T, constructs a decision rule which is true for T, realizable for r, and which length is at most*

$$(1 - \varepsilon)l(T, r) \ln K(T, r) .$$

Proposition 4.7 ([2]) *The problem of minimization of complete decision rule system length for binary decision tables with many-valued decisions is NP-hard.*

Theorem 4.29 ([2]) *If $NP \nsubseteq DTIME(n^{O(\log \log n)})$ then, for any ε, $0 < \varepsilon < 1$, there is no a polynomial algorithm that, for a given nondegenerate binary decision table T with many-valued decisions, constructs a complete decision rule system S for T such that*

$$l(S) \le (1 - \varepsilon)l(T) \ln K(T) .$$

4.5.5 Optimization of Inhibitory Rules

We can apply the greedy algorithm for set cover problem to construct inhibitory rules for binary decision tables with many-valued decisions.

Let T be a nondegenerate binary decision table with many-valued decisions containing n columns labeled with attributes $f_1, \dots, f_n$. Let $r = (b_1, \dots, b_n)$ be a row of T, $D(r)$ be the set of decisions attached to r, and $d \in D(T) \setminus D(r)$.

We consider a set cover problem $iA(T, r, d), iF(T, r, d) = \{iS_1, \dots, iS_n\}$, where $iA(T, r, d)$ is the set of all rows r' of T such that $d \in D(r')$. For $j = 1, \dots, n$, the set iS_j coincides with the set of all rows from $iA(T, r, d)$ which are different from r in the column f_j. One can show that the inhibitory rule

$$(f_{j_1} = b_{j_1}) \wedge \cdots \wedge (f_{j_m} = b_{j_m}) \to \neq d$$

is true for T (it is clear that this rule is realizable for r) if and only if the subfamily $\{iS_{j_1}, \dots, iS_{j_m}\}$ is a cover for the set cover problem $iA(T, r, d), iF(T, r, d)$.

We denote $iP(T, r, d) = |iA(T, r, d)|$ and $il(T, r, d)$ the minimum length of an inhibitory rule over T which is true for T, realizable for r, and has $\neq d$ on the right-hand side. It is clear that, for the constructed set cover problem, $il(T, r, d)$ is the minimum cardinality of a cover.

Let us apply the greedy algorithm to the set cover problem $iA(T, r, d), iF(T, r, d)$. This algorithm constructs a cover which corresponds to an inhibitory rule

$$irule(T, r, d)$$

which is true for T, realizable for r, and has $\neq d$ on the right-hand side. We denote by $il_{\text{greedy}}(T, r, d)$ the length of $irule(T, r, d)$. From Theorem 4.21 it follows that

$$il_{\text{greedy}}(T, r, d) \leq il(T, r, d) \ln(iP(T, r, d)) + 1 .$$

We denote by $il_{\text{greedy}}(T, r)$ the length of the rule constructed by the following polynomial algorithm (we will say about this algorithm as about modified greedy algorithm). For a given nondegenerate binary decision table T with many-valued decisions and row r of T, for each $d \in D(T) \setminus D(r)$, we construct the set cover problem $iA(T, r, d), iF(T, r, d)$ and then apply to this problem greedy algorithm. We transform the constructed cover to the rule $irule(T, r, d)$. Among the rules $irule(T, r, d)$, $d \in D(T) \setminus D(r)$, we choose a rule with minimum length. This rule is the output of the considered algorithm. We have

$$il_{\text{greedy}}(T, r) = \min\{il_{\text{greedy}}(T, r, d) : d \in D(T) \setminus D(r)\} .$$

It is clear that
$$il(T, r) = \min\{il(T, r, d) : d \in D(T) \setminus D(r)\} .$$

Let $iK(T, r) = \max\{iP(T, r, d) : d \in D(T) \setminus D(r)\}$. Then

$$il_{\text{greedy}}(T, r) \leq il(T, r) \ln(iK(T, r)) + 1 .$$

So we have the following statement.

Theorem 4.30 *Let T be a nondegenerate binary decision table with many-valued decisions and r be a row of T. Then*

$$il_{\text{greedy}}(T, r) \leq il(T, r) \ln(iK(T, r)) + 1 .$$

We can use the considered modified greedy algorithm to construct a complete inhibitory rule system for the nondegenerate binary decision table T with many-valued decisions. To this end, we apply this algorithm sequentially to the table T and to each row r of T. As a result, we obtain a system of rules S in which each inhibitory rule is true for T and, for every row of T, there exists a rule from S which is realizable for this row.

We denote $il_{\text{greedy}}(T) = l(S)$ and

$$iK(T) = \max\{iK(T, r) : r \in Row(T)\} \,,$$

where $Row(T)$ is the set of rows of T. It is clear that $il(T) = \max\{il(T, r) : r \in Row(T)\}$. Using Theorem 4.30 we obtain the following statement.

Theorem 4.31 *Let T be a nondegenerate binary decision table with many-valued decisions. Then*

$$il_{\text{greedy}}(T) \leq il(T) \ln(iK(T)) + 1 \,.$$

Proposition 4.8 *The problem of minimization of inhibitory rule length for nondegenerate binary decision tables with many-valued decisions is NP-hard.*

Proof Let T be a nondegenerate binary decision table with many-valued decisions and r be a row of T. We denote $\Theta = T^C$. It is clear that Θ is nondegenerate and $\Theta^C = T$. From Proposition 4.1 it follows that a decision rule ρ over $\Theta^C = T$ is true for $\Theta^C = T$ if and only if the inhibitory rule ρ^- over $\Theta = T^C$ is true for $\Theta = T^C$. It is clear that ρ is realizable for r if and only if ρ^- is realizable for r. By Corollary 4.1, the minimum length of a decision rule for $\Theta^C = T$ and r is equal to the minimum length of an inhibitory rule for $\Theta = T^C$ and r. So we have a polynomial time reduction of the problem of minimization of decision rule length for nondegenerate binary decision table T with many-valued decisions and row r of T to the problem of minimization of inhibitory rule length for nondegenerate binary decision table T^C with many-valued decisions and row r of T^C. By Proposition 4.6, the problem of minimization of decision rule length for nondegenerate binary decision tables with many-valued decisions is NP-hard. Therefore the problem of minimization of inhibitory rule length for nondegenerate binary decision tables with many-valued decisions is NP-hard.   □

Theorem 4.32 *If $NP \not\subseteq DTIME(n^{O(\log \log n)})$ then, for any ε, $0 < \varepsilon < 1$, there is no a polynomial algorithm that, for a given nondegenerate binary decision table T with many-valued decisions and row r of T, constructs an inhibitory rule for T and r which length is at most*

$$(1 - \varepsilon) il(T, r) \ln(iK(T, r)) \,.$$

Proof Let us assume that, for some ε, $0 < \varepsilon < 1$, there exists a polynomial algorithm $\mathscr{A}$ that, for a given nondegenerate binary decision table T with many-valued decisions and row r of T, constructs an inhibitory rule for T and r which length is at most $(1 - \varepsilon) il(T, r) \ln(iK(T, r))$.

Let T be a nondegenerate binary decision table with many-valued decisions and r be a row of T. We denote $\Theta = T^C$. It is clear that Θ is nondegenerate and $\Theta^C = T$. We apply the algorithm $\mathscr{A}$ to the table $\Theta = T^C$ and row r. As a result, we obtain an inhibitory rule ρ for $\Theta = T^C$ and r which length is at most $(1 - \varepsilon) il(\Theta, r) \ln(iK(\Theta, r))$. Let us consider a decision rule ρ^+ over $\Theta^C = T$.

Using Proposition 4.1 one can show that ρ^+ is a decision rule for $\Theta^C = T$ and r. The length of ρ^+ is at most $(1 - \varepsilon)il(\Theta, r)\ln(iK(\Theta, r))$. By Corollary 4.1, $il(\Theta, r) = l(\Theta^C, r)$. One can show that $iK(\Theta, r) = K(\Theta^C, r)$. Therefore the length of rule ρ^+ is at most $(1 - \varepsilon)l(T, r)\ln K(T, r)$.

As a result, there is a polynomial algorithm that, for a given nondegenerate binary decision table T with many-valued decisions and row r of T, constructs a decision rule for T and r which length is at most $(1 - \varepsilon)l(T, r)\ln K(T, r)$. But this is impossible if $NP \nsubseteq DTIME(n^{O(\log \log n)})$ (see Theorem 4.28). Thus, if $NP \nsubseteq DTIME(n^{O(\log \log n)})$ then, for any $\varepsilon, 0 < \varepsilon < 1$, there is no a polynomial algorithm that, for a given nondegenerate binary decision table T with many-valued decisions and row r of T, constructs an inhibitory rule which is true for T, realizable for r, and which length is at most $(1 - \varepsilon)il(T, r)\ln(iK(T, r))$. □

Proposition 4.9 *The problem of minimization of complete inhibitory rule system length for nondegenerate binary decision tables with many-valued decisions is NP-hard.*

Proof Let T be a nondegenerate binary decision table with many-valued decisions. We denote $\Theta = T^C$. It is clear that Θ is nondegenerate and $\Theta^C = T$. From Proposition 4.1 it follows that a finite nonempty set S of decision rules over $\Theta^C = T$ is complete for $\Theta^C = T$ if and only if the set S^- of inhibitory rules over $\Theta = T^C$ is complete for $\Theta = T^C$. By Corollary 4.1, $il(\Theta) = l(\Theta^C)$. So we have a polynomial time reduction of the problem of minimization of complete decision rule system length for nondegenerate binary decision table T with many-valued decisions to the problem of minimization of complete inhibitory rule system length for nondegenerate binary decision table T^C with many-valued decisions. By Proposition 4.7, the problem of minimization of complete decision rule system length for nondegenerate binary decision tables with many-valued decisions is NP-hard. Therefore the problem of minimization of complete inhibitory rule system length for nondegenerate binary decision tables with many-valued decisions is NP-hard. □

Theorem 4.33 *If $NP \nsubseteq DTIME(n^{O(\log \log n)})$ then, for any $\varepsilon, 0 < \varepsilon < 1$, there is no a polynomial algorithm that, for a given nondegenerate binary decision table T with many-valued decisions, constructs a complete inhibitory rule system S for T such that*

$$l(S) \le (1 - \varepsilon)il(T)\ln(iK(T)) .$$

Proof Let us assume that, for some $\varepsilon, 0 < \varepsilon < 1$, there exists a polynomial algorithm $\mathscr{A}$ which, for a given nondegenerate binary decision table T with many-valued decisions, constructs a complete inhibitory rule system S for T such that $l(S) \le (1 - \varepsilon)il(T)\ln(iK(T))$.

Let T be a nondegenerate binary decision table with many-valued decisions. We denote $\Theta = T^C$. It is clear that Θ is nondegenerate and $\Theta^C = T$. We apply the algorithm $\mathscr{A}$ to the table $\Theta = T^C$. As a result, we obtain a complete inhibitory rule system S for Θ such that $l(S) \le (1 - \varepsilon)il(\Theta)\ln(iK(\Theta))$. Let us consider a system S^+ of decision rules over $\Theta^C = T$. From Proposition 4.1 it follows that S^+ is a complete

decision rule system for $\Theta^C = T$. We have $l(S^+) \leq (1 - \varepsilon)il(\Theta, r) \ln(iK(\Theta, r))$.
By Corollary 4.1, $il(\Theta) = l(\Theta^C)$. One can show that $iK(\Theta) = K(\Theta^C)$. Therefore
$l(S^+) \leq (1 - \varepsilon)l(T) \ln K(T)$.

As a result, there is a polynomial algorithm that, for a given nondegenerate binary
decision table T with many-valued decisions, constructs a complete decision rule
system S for T such that $l(S) \leq (1 - \varepsilon)l(T) \ln K(T)$ which is impossible if $NP \nsubseteq$
$DTIME(n^{O(\log \log n)})$. Thus, if $NP \nsubseteq DTIME(n^{O(\log \log n)})$ then, for any ε, $0 <$
$\varepsilon < 1$, there is no a polynomial algorithm that, for a given nondegenerate binary
decision table T with many-valued decisions, constructs a complete inhibitory rule
system S for T such that $l(S) \leq (1 - \varepsilon)il(T) \ln(iK(T))$.   □

4.6   Approximate Algorithms for Decision and Inhibitory Tree Optimization

In this section, we consider approximate polynomial algorithms for problems of
minimization of decision and inhibitory tree depth.

4.6.1   Optimization of Decision Trees

We now describe an algorithm U which, for a binary decision table T with many-
valued decisions, constructs a decision tree $U(T)$ for the table T. Let T have n
columns labeled with attributes $f_1, \ldots, f_n$.

Step 1: Construct a tree consisting of a single node labeled with the table T and
proceed to the second step.

Suppose $t \geq 1$ steps have been made already. The tree obtained at the step t will
be denoted by G.

Step $(t + 1)$: If no one node of the tree G is labeled with a table then we denote
by $U(T)$ the tree G. The work of the algorithm U is completed.

Otherwise, we choose certain node v in the tree G which is labeled with a subtable
of the table T. Let the node v be labeled with the table Θ. If Θ is a degenerate table
and d is the minimum decision such that $d \in D(r)$ for any row r of Θ, then instead
of Θ we mark the node v by the number d and proceed to the step $(t + 2)$. Let Θ be
a nondegenerate table. Then, for $i = 1, \ldots, n$, we compute the value

$$Q(f_i) = \max\{B(\Theta(f_i, 0)), B(\Theta(f_i, 1))\} .$$

We mark the node v by the attribute f_{i_0} where i_0 is the minimum i for which $Q(f_i)$
has minimum value. For each $\delta \in \{0, 1\}$, we add to the tree G the node $v(\delta)$, mark
this node by the table $\Theta(f_{i_0}, \delta)$, draw the edge from v to $v(\delta)$, and mark this edge by
δ. Proceed to the step $(t + 2)$.

Theorem 4.34 ([2]) *Let T be a nondegenerate binary decision table with many-valued decisions. Then*

$$h(U(T)) \leq M(T) \ln B(T) + 1 .$$

Corollary 4.10 ([2]) *For any nondegenerate binary decision table T with many-valued decisions,*
$$h(U(T)) \leq h(T) \ln B(T) + 1 .$$

Proposition 4.10 ([2]) *The problem of minimization of decision tree depth for non-degenerate binary decision tables with many-valued decisions is NP-hard.*

Theorem 4.35 ([2]) *If $NP \nsubseteq DTIME(n^{O(\log \log n)})$ then, for any $\varepsilon > 0$, there is no a polynomial algorithm which, for a given nondegenerate binary decision table T with many-valued decisions, constructs a decision tree for T which depth is at most*

$$(1 - \varepsilon) h(T) \ln B(T) .$$

4.6.2   Optimization of Inhibitory Trees

We now describe an algorithm iU which, for a nondegenerate binary decision table with many-valued decisions T, constructs an inhibitory tree $iU(T)$ for the table T. Let T have n columns labeled with attributes $f_1, \ldots, f_n$.

Step 1: Construct a tree consisting of a single node labeled with the table T and proceed to the second step.

Suppose $t \geq 1$ steps have been made already. The tree obtained at the step t will be denoted by G.

Step $(t + 1)$: If no one node of the tree G is labeled with a table then we denote by $iU(T)$ the tree G. The work of the algorithm iU is completed.

Otherwise, we choose certain node v in the tree G which is labeled with a subtable of the table T. Let the node v be labeled with the table Θ. If Θ is an incomplete subtable relative to T and d is the minimum decision such that $d \in D(T) \setminus D(\Theta)$, then instead of Θ we mark the node v by the expression $\neq d$ and proceed to the step $(t + 2)$. Let Θ be not incomplete relative to T. Then, for $j = 1, \ldots, n$, we compute the value

$$iQ(f_j) = \max\{iB(T, \Theta(f_j, 0)), iB(T, \Theta(f_j, 1))\} .$$

We mark the node v by the attribute f_{j_0} where j_0 is the minimum j for which $iQ(f_j)$ has minimum value. For each $\delta \in \{0, 1\}$, we add to the tree G the node $v(\delta)$, mark this node by the table $\Theta(f_{j_0}, \delta)$, draw the edge from v to $v(\delta)$, and mark this edge by δ. Proceed to the step $(t + 2)$.

Theorem 4.36 *Let T be a nondegenerate binary decision table with many-valued decisions. Then*

$$h(iU(T)) \le iM(T)\ln(iB(T,T)) + 1.$$

Proof Let T have n columns labeled with attributes $f_1, \ldots, f_n$,

$$f_{i_1}, \ldots, f_{i_m} \in \{f_1, \ldots, f_n\},$$

$\delta_1, \ldots, \delta_m \in \{0, 1\}$, and $\alpha = (f_{i_1}, \delta_1) \ldots (f_{i_m}, \delta_m)$. One can show that $T\alpha$ is incomplete relative to T if and only if $T^C\alpha$ is degenerate. By Lemma 4.2, $B(T^C\alpha) = iB(T, T\alpha)$. It is not difficult to show that $D(T) \setminus D(T_\alpha) = \bigcap_{r \in Row(T^C\alpha)} D_{T^C}(r)$. Using these facts it is not difficult to prove that $iU(T) = U(T^C)^-$ and $h(U(T^C)) = h(iU(T))$. From Theorem 4.34 it follows that $h(U(T^C)) \le M(T^C)\ln B(T^C) + 1$. From Lemmas 4.1 and 4.2 it follows that $B(T^C) = iB(T, T)$ and $M(T^C) = iM(T)$. Therefore $h(iU(T)) \le iM(T)\ln(iB(T, T)) + 1$. $\square$

Corollary 4.11 *For any nondegenerate binary decision table T with many-valued decisions,*

$$h(iU(T)) \le ih(T)\ln(iB(T, T)) + 1.$$

Proof By Theorem 4.14, $iM(T) \le ih(T)$. From here and from Theorem 4.36 it follows that $h(iU(T)) \le ih(T)\ln(iB(T, T)) + 1$. $\square$

Proposition 4.11 *The problem of minimization of inhibitory tree depth for binary decision tables with many-valued decisions is NP-hard.*

Proof Let T be a nondegenerate binary decision table with many-valued decisions. We denote $\Theta = T^C$. It is clear that Θ is nondegenerate and $\Theta^C = T$. From Proposition 4.1 it follows that a decision tree Γ over $\Theta^C = T$ is a decision tree for $\Theta^C = T$ if and only if the inhibitory tree Γ^- over $\Theta = T^C$ is an inhibitory tree for $\Theta = T^C$. By Corollary 4.1, $ih(\Theta) = h(\Theta^C)$. So we have a polynomial time reduction of the problem of minimization of decision tree depth for nondegenerate binary decision table T with many-valued decisions to the problem of minimization of inhibitory tree depth for nondegenerate binary decision table T^C with many-valued decisions. By Proposition 4.10, the problem of minimization of decision tree depth for nondegenerate binary decision tables with many-valued decisions is NP-hard. Therefore the problem of minimization of inhibitory tree depth for nondegenerate binary decision tables with many-valued decisions is NP-hard. $\square$

Theorem 4.37 *If $NP \nsubseteq DTIME(n^{O(\log \log n)})$ then, for any $\varepsilon > 0$, there is no a polynomial algorithm which, for a given nondegenerate binary decision table T with many-valued decisions, constructs an inhibitory tree for T which depth is at most*

$$(1 - \varepsilon)ih(T)\ln(iB(T, T)).$$

Proof Let us assume that, for some $\varepsilon > 0$, there exists a polynomial algorithm $\mathscr{A}$ which, for a given nondegenerate binary decision table T with many-valued decisions, constructs an inhibitory tree for T which depth is at most

$$(1 - \varepsilon)ih(T)\ln(iB(T,T)).$$

Let T be a nondegenerate binary decision table with many-valued decisions. We denote $\Theta = T^C$. It is clear that Θ is nondegenerate and $\Theta^C = T$. We apply the algorithm $\mathscr{A}$ to the table $\Theta = T^C$. As a result, we obtain an inhibitory tree Γ for Θ such that $h(\Gamma) \leq (1 - \varepsilon)ih(\Theta)\ln(iB(\Theta, \Theta))$. Let us consider a decision tree Γ^+ over $\Theta^C = T$. It is clear that $(\Gamma^+)^- = \Gamma$. From Proposition 4.1 it follows that Γ^+ is a decision tree for $\Theta^C = T$. We have $h(\Gamma^+) \leq (1 - \varepsilon)ih(\Theta)\ln(iB(\Theta, \Theta))$. By Corollary 4.1, $ih(\Theta) = h(\Theta^C)$. By Lemma 4.2, $iB(\Theta, \Theta) = B(\Theta^C)$. Therefore $h(\Gamma^+) \leq (1 - \varepsilon)h(T)\ln B(T)$.

As a result, there is a polynomial algorithm that, for a given nondegenerate binary decision table T with many-valued decisions, constructs a decision tree Γ for T such that $h(\Gamma) \leq (1 - \varepsilon)h(T)\ln B(T)$ which is impossible if $NP \nsubseteq DTIME(n^{O(\log\log n)})$ (see Theorem 4.35). Thus, if $NP \nsubseteq DTIME(n^{O(\log\log n)})$ then, for any $\varepsilon > 0$, there is no a polynomial algorithm that, for a given nondegenerate binary decision table T with many-valued decisions, constructs an inhibitory tree for T which depth is at most $(1 - \varepsilon)ih(T)\ln(iB(T,T))$. $\qquad\square$

4.7 Exact Algorithms for Optimization of Decision Trees, Rules, and Tests

In this section, we discuss exact algorithms for minimization of decision and inhibitory tree depth, rule length, and test cardinality.

4.7.1 Decision Interpretation

A dynamic programming algorithm W for minimization of decision tree depth was described in [2].

Theorem 4.38 ([2]) *For any nondegenerate binary decision table T with many-valued decisions, the algorithm W constructs a decision tree $W(T)$ for the table T such that $h(W(T)) = h(T)$, and makes exactly $2|SEP(T)| + 3$ steps. The time of the algorithm W work is bounded from below by $|SEP(T)|$, and bounded from above by a polynomial on $|SEP(T)|$ and the number of columns in the table T.*

A dynamic programming algorithm V for minimization of decision rule length was described in [2].

Theorem 4.39 ([2]) *For any nondegenerate binary decision table T with many-valued decisions and any row r of T, the algorithm V constructs a decision rule $V(T, r)$ which is true for T, realizable for r, and has minimum length $l(T, r)$. During the construction of optimal rules for rows of T the algorithm V makes exactly*

$2|SEP(T)| + 3$ *steps. The time of the algorithm V work is bounded from below by* $|SEP(T)|$, *and bounded from above by a polynomial on* $|SEP(T)|$ *and on the number of columns in the table* T.

The situation with decision test cardinality minimization is another.

Theorem 4.40 ([2]) *If* $P \neq NP$ *then there is no algorithm which, for a given binary decision table* T *with many-valued decisions, constructs a decision test for* T *with minimum cardinality, and for which the time of work is bounded from above by a polynomial depending on the number of columns in* T *and the number of separable subtables of* T.

4.7.2 Inhibitory Interpretation

In the next chapters, we will show that there exist dynamic programming algorithms for minimization of inhibitory tree depth and inhibitory rule length which time complexity is polynomial depending on the size of input table T and the number of separable subtables of T. The situation with inhibitory test cardinality minimization is another.

Theorem 4.41 *If* $P \neq NP$ *then there is no algorithm which, for a given nondegenerate binary decision table* T *with many-valued decisions, constructs an inhibitory test for* T *with minimum cardinality, and for which the time of work is bounded from above by a polynomial depending on the number of columns in* T *and the number of separable subtables of* T.

Proof Let us assume that there exists an algorithm $\mathscr{A}$ which, for a given nondegenerate binary decision table T with many-valued decisions, constructs an inhibitory test for T with minimum cardinality, and for which the time of work is bounded from above by a polynomial depending on the number of columns in T and the number of separable subtables of T.

Let T be a nondegenerate binary decision table with many-valued decisions. We denote $\Theta = T^C$. It is clear that Θ is nondegenerate and $\Theta^C = T$. We apply the algorithm $\mathscr{A}$ to the table $\Theta = T^C$. As a result, we obtain an inhibitory test Q for Θ which cardinality is equal to $iR(\Theta)$. From Proposition 4.1 it follows that Q is a decision test for $\Theta^C = T$. By Corollary 4.1, $iR(\Theta) = R(\Theta^C)$. It is clear that the number of columns in Θ is equal to the number of columns in Θ^C and $|SEP(\Theta)| = |SEP(\Theta^C)|$.

As a result, there exists an algorithm that, for a given nondegenerate binary decision table T with many-valued decisions, constructs a decision test for T with minimum cardinality, and for which the time of work is bounded from above by a polynomial depending on the number of columns in T and the number of separable subtables of T. But this is impossible if $P \neq NP$ (see Theorem 4.40). Thus, if

$P \neq NP$ then there is no algorithm which, for a given nondegenerate binary decision table T with many-valued decisions, constructs an inhibitory test for T with minimum cardinality, and for which the time of work is bounded from above by a polynomial depending on the number of columns in T and the number of separable subtables of T. $\qquad\square$

References

1. Johnson, D.S.: Approximation algorithms for combinatorial problems. J. Comput. Syst. Sci. **9**(3), 256–278 (1974)
2. Moshkov, M., Zielosko, B.: Combinatorial Machine Learning - A Rough Set Approach, Studies in Computational Intelligence, vol. 360. Springer, Heidelberg (2011)
3. Nigmatullin, R.: Method of steepest descent in problems on cover. In: Symposium Problems of Precision and Efficiency of Computing Algorithms (in Russian), vol. 5, pp. 116–126. Kiev, USSR (1969)

Part II
Extensions of Dynamic Programming for Decision and Inhibitory Trees

This part is devoted to the development of extensions of dynamic programming for decision and inhibitory trees. It consists of six chapters.

In Chap. 5, we consider some notions related to decision tables with many-valued decisions (the notions of table, directed acyclic graph for this table, uncertainty and completeness measures, and restricted information system) and discuss tools (statements and algorithms) for the work with Pareto optimal points. The definitions and results from this chapter are used both in Part I and Part II.

In Chap. 6, we define various types of decision and inhibitory trees. We discuss the notion of a cost function for trees, the notion of decision tree uncertainty, and the notion of inhibitory tree completeness. We also design an algorithm for counting the number of decision trees represented by a directed acyclic graph.

In Chap. 7, we consider multi-stage optimization of decision and inhibitory trees relative to a sequence of cost functions, and two applications of this technique: study of decision trees for sorting problem and study of totally optimal (simultaneously optimal relative to a number of cost functions) decision and inhibitory trees for modified decision tables from the UCI ML Repository.

In Chap. 8, we study bi-criteria optimization problem cost versus cost for decision and inhibitory trees. We design an algorithm which constructs the set of Pareto optimal points for bi-criteria optimization problem for decision trees, and show how the constructed set can be transformed into the graphs of functions that describe the relationships between the studied cost functions. We extend the obtained results to the case of inhibitory trees. As applications of bi-criteria optimization for two cost functions, we compare 12 greedy heuristics for construction of decision and inhibitory trees as single-criterion and bi-criteria optimization algorithms, and study two relationships for decision trees related to knowledge representation–number of nodes versus depth and number of nodes versus average depth.

In Chap. 9, we consider bi-criteria optimization problems cost versus uncertainty for decision trees and cost versus completeness for inhibitory trees, and discuss illustrative examples. The created tools allow us to understand complexity versus accuracy trade-off for decision and inhibitory trees and to choose appropriate trees.

In Chap. 10, we consider so-called multi-pruning approach based on dynamic programming algorithms for bi-criteria optimization of CART-like decision trees relative to the number of nodes and the number of misclassifications. This approach allows us to construct the set of all Pareto optimal points and to derive, for each such point, decision trees with parameters corresponding to that point. Experiments with decision tables from the UCI ML Repository show that, very often, we can find a suitable Pareto optimal point and derive a decision tree with small number of nodes at the expense of small increment in the number of misclassifications. Such decision trees can be used for knowledge representation. Multi-pruning approach includes a procedure which constructs decision trees that, as classifiers, often outperform decision trees constructed by CART. We also consider a modification of multi-pruning approach (restricted multi-pruning) that requires less memory and time but usually keeps the quality of constructed trees as classifiers or as a way for knowledge representation. We extend the considered approaches to the case of decision tables with many-valued decisions.

Chapter 5
Decision Tables and Tools for Study of Pareto-Optimal Points

In this chapter, we describe main notions related to decision tables with many-valued decisions, consider uncertainty and completeness measures for decision tables, define the structure of subtables of a given decision table represented as a directed acyclic graph (DAG), and discuss time complexity of algorithms on DAGs. We consider classes of decision tables over restricted information systems for which these algorithms have polynomial time complexity depending on the number of conditional attributes in the input decision table. At the end of this chapter, we discuss tools (statements and algorithms) which are used for the study of Pareto optimal points for bi-criteria optimization problems related to decision trees, decision rules, and decision rule systems. The notions and results from this chapter will be used both in Part II and Part III.

Note that many definitions and results considered in this chapter were obtained jointly with Hassan AbouEisha, Talha Amin, and Shahid Hussain and already published in the book [1].

5.1 Decision Tables

A *decision table with many-valued decisions* is a rectangular table T with $n \geq 1$ columns filled with numbers from the set $\omega = \{0, 1, 2, \ldots\}$ of nonnegative integers. Columns of the table are labeled with *conditional attribute* $f_1, \ldots, f_n$. Rows of the table are pairwise different, and each row r is labeled with a finite nonempty subset $D(r) = D_T(r)$ of ω which is interpreted as a *set of decisions*. Rows of the table are interpreted as tuples of values of conditional attributes. We denote by $Row(T)$ the set of rows of T and $D(T) = \bigcup_{r \in Row(T)} D(r)$.

© Springer Nature Switzerland AG 2020
F. Alsolami et al., *Decision and Inhibitory Trees and Rules for Decision Tables with Many-valued Decisions*, Intelligent Systems Reference Library 156,
https://doi.org/10.1007/978-3-030-12854-8_5

We denote by T^C *complementary to* T decision table obtained from the table T by changing, for each row $r \in Row(T)$, the set $D(r)$ with the set $D(T) \setminus D(r)$. When we consider complementary to T table T^C or when we study inhibitory rules or trees for T, we will assume that, for any row r of T, $D(r) \neq D(T)$. All results for decision rules and trees continue to be true if in the considered decision table T there are rows r such that $D(r) = D(T)$.

A decision table can be represented by a word over the alphabet $\{0, 1, ; , :, |\}$ in which numbers from ω are in binary representation (are represented by words over the alphabet $\{0, 1\}$), the symbol ";" is used to separate two numbers from ω, and the symbol "|" is used to separate two rows (we add numbers from $D(r)$ at the end of each row r and separate these numbers from r by the symbol ": "). The length of this word will be called the *size* of the decision table.

A decision table is called *empty* if it has no rows. We denote by $\mathscr{T}$ the set of all decision tables with many-valued decisions and by $\mathscr{T}^+$ – the set of nonempty decision tables with many-valued decisions. Let $T \in \mathscr{T}$. The table T is called *degenerate* if it is empty or has a *common decision* – a decision $d \in D(T)$ such that $d \in D(r)$ for any row r of T. We denote by $\dim(T)$ the number of columns (conditional attributes) in T. We denote by $N(T)$ the number of rows in the table T and, for any $d \in \omega$, we denote by $N_d(T)$ the number of rows r of T such that $d \in D(r)$. By $mcd(T)$ we denote the *most common decision* for T which is the minimum decision d_0 from $D(T)$ such that $N_{d_0}(T) = \max\{N_d(T) : d \in D(T)\}$. If T is empty then $mcd(T) = 0$. A nonempty decision table is called *diagnostic* if $D(r_1) \cap D(r_2) = \emptyset$ for any rows r_1, r_2 of T such that $r_1 \neq r_2$.

For any conditional attribute $f_i \in \{f_1, \ldots, f_n\}$, we denote by $E(T, f_i)$ the set of values of the attribute f_i in the table T. We denote by $E(T)$ the set of conditional attributes for which $|E(T, f_i)| \geq 2$. Let $range(T) = \max\{|E(T, f_i)| : i = 1, \ldots, n\}$.

Let T be a nonempty decision table. A *subtable* of T is a table obtained from T by removal of some rows. We denote by $Word(T)$ the set of all finite words over the alphabet $\{(f_i, a) : f_i \in \{f_1, \ldots, f_n\}, a \in \omega\}$ including the empty word λ. Let $\alpha \in Word(T)$. We define now a subtable $T\alpha$ of the table T. If $\alpha = \lambda$ then $T\alpha = T$. If $\alpha \neq \lambda$ and $\alpha = (f_{i_1}, a_1) \ldots (f_{i_m}, a_m)$ then $T\alpha = T(f_{i_1}, a_1) \ldots (f_{i_m}, a_m)$ is the subtable of the table T containing the rows from T which at the intersection with the columns $f_{i_1}, \ldots, f_{i_m}$ have numbers $a_1, \ldots, a_m$, respectively. Such nonempty subtables, including the table T, are called *separable* subtables of T. We denote by $SEP(T)$ the set of separable subtables of the table T. Note that $N(T) \leq |SEP(T)|$. It is clear that the size of each subtable of T is at most the size of T.

Let Θ be a subtable of T. The subtable Θ is called *incomplete relative to T* if $D(\Theta) \subset D(T)$. By $lcd(T, \Theta)$ we denote the *least common decision for Θ relative to T* which is the minimum decision d_0 from $D(T)$ such that $N_{d_0}(\Theta) = \min\{N_d(\Theta) : d \in D(T)\}$.

5.2 Uncertainty Measures

Let $\mathbb{R}$ be the set of real numbers and $\mathbb{R}_+$ be the set of all nonnegative real numbers. An *uncertainty measure* is a function $U : \mathscr{T} \to \mathbb{R}$ such that $U(T) \geq 0$ for any $T \in \mathscr{T}$, and $U(T) = 0$ if and only if T is a degenerate table. One can show that the following functions (we assume that, for any empty table T, the value of each of the considered functions is equal to 0) are uncertainty measures:

- *Misclassification error:* $me(T) = N(T) - N_{mcd(T)}(T)$.
- *Relative misclassification error:* $rme(T) = (N(T) - N_{mcd(T)}(T))/N(T)$.
- *Absence:* $abs(T) = \prod_{d \in D(T)}(1 - N_d(T)/N(T))$.

We assume that each of the following numerical operations (we call these operations *basic*) has time complexity $O(1)$: $\max(x, y), x + y, x \times y, x \div y, \log_2 x$. This assumption is reasonable for computations with floating-point numbers. Under this assumption, each of the considered three uncertainty measures has polynomial time complexity depending on the size of decision tables.

5.3 Completeness Measures

A *completeness measure* is a function W defined on the pairs nonempty decision table T and its subtable Θ such that $W(T, \Theta) \geq 0$ for any $T \subset \mathscr{T}^+$ and its subtable Θ, and $W(T, \Theta) = 0$ if and only if Θ is incomplete relative to T. One can show that the following functions (we assume that, for any empty subtable Θ, the value of each of the considered functions is equal to 0) are completeness measures:

- *Inhibitory misclassification error:* $ime(T, \Theta) = N_{lcd(T,\Theta)}(\Theta)$.
- *Inhibitory relative misclassification error:*

$$irme(T, \Theta) = N_{lcd(T,\Theta)}(\Theta)/N(\Theta).$$

- *Inhibitory absence:* $iabs(T, \Theta) = \prod_{d \in D(T)}(N_d(\Theta)/N(\Theta))$.

Each of the considered three completeness measures has polynomial time complexity depending on the size of decision tables.

It is not difficult to prove the following statement.

Lemma 5.1 *Let T be a nondegenerate decision table and $\alpha \in Word(T)$. Then*

1. $D(T) = D(T^C)$;
2. $N(T\alpha) = N(T^C\alpha)$;
3. $T\alpha$ *is incomplete relative to T if and only if $T^C\alpha$ is degenerate;*

4. For any $d \in D(T)$, $N_d(T\alpha) = N(T^C\alpha) - N_d(T^C\alpha)$;
5. $lcd(T, T\alpha) = mcd(T^C\alpha)$.

Let U be an uncertainty measure and W be a completeness measure. We will say that W and U are *dual* if, for any nondegenerate $T \in \mathcal{T}$ and any $\alpha \in Word(T)$, $W(T, T\alpha) = U(T^C\alpha)$.

Proposition 5.1 *The following pairs of completeness and uncertainty measures are dual: ime and me, irme and rme, iabs and abs.*

Proof Let T be a nondegenerate decision table and $\alpha \in Word(T)$. From Lemma 5.1 it follows that

$$ime(T, T\alpha) = N_{lcd(T,T\alpha)}(T\alpha) = N(T^C\alpha) - N_{mcd(T^C\alpha)}(T^C\alpha) = me(T^C\alpha) ,$$

$$irme(T, T\alpha) = N_{lcd(T,T\alpha)}(T\alpha)/N(T\alpha)$$
$$= (N(T^C\alpha) - N_{mcd(T^C\alpha)}(T^C\alpha))/N(T^C\alpha) = rme(T^C\alpha) ,$$

$$iabs(T, T\alpha) = \prod_{d \in D(T)} (N_d(T\alpha)/N(T\alpha)) = \prod_{d \in D(T^C)} (1 - N_d(T^C\alpha)/N(T^C\alpha))$$
$$= abs(T^C\alpha) .$$

$\square$

5.4 Directed Acyclic Graph $\Delta_{U,\alpha}(T)$

Let T be a nonempty decision table with n conditional attributes $f_1, \ldots, f_n$, U be an uncertainty measure, and $\alpha \in \mathbb{R}_+$. We now consider an algorithm $\mathscr{A}_1$ for the construction of a directed acyclic graph (DAG) $\Delta_{U,\alpha}(T)$ which will be used for the description and study of decision rules and decision trees. Nodes of this graph are some separable subtables of the table T. During each iteration we process one node. We start with the graph that consists of one node T which is not processed and finish when all nodes of the graph are processed.

Algorithm $\mathscr{A}_1$ (construction of DAG $\Delta_{U,\alpha}(T)$).
Input: A nonempty decision table T with n conditional attributes $f_1, \ldots, f_n$, an uncertainty measure U, and a number $\alpha \in \mathbb{R}_+$.
Output: Directed acyclic graph $\Delta_{U,\alpha}(T)$.

1. Construct the graph that consists of one node T which is not marked as processed.
2. If all nodes of the graph are processed then the work of the algorithm is finished. Return the resulting graph as $\Delta_{U,\alpha}(T)$. Otherwise, choose a node (table) Θ that has not been processed yet.
3. If $U(\Theta) \leq \alpha$ mark the node Θ as processed and proceed to step 2.
4. If $U(\Theta) > \alpha$ then, for each $f_i \in E(\Theta)$, draw a bundle of edges from the node Θ (this bundle of edges will be called f_i-bundle). Let $E(\Theta, f_i) = \{a_1, \ldots, a_k\}$. Then draw k edges from Θ and label these edges with the pairs $(f_i, a_1), \ldots, (f_i, a_k)$. These edges enter nodes $\Theta(f_i, a_1), \ldots, \Theta(f_i, a_k)$, respectively. If some of the nodes $\Theta(f_i, a_1), \ldots, \Theta(f_i, a_k)$ are not present in the graph then add these nodes to the graph. Mark the node Θ as processed and return to step 2.

We now analyze the time complexity of the algorithm $\mathscr{A}_1$. By $L(\Delta_{U,\alpha}(T))$ we denote the number of nodes in the graph $\Delta_{U,\alpha}(T)$.

Proposition 5.2 *Let the algorithm $\mathscr{A}_1$ use an uncertainty measure U which has polynomial time complexity depending on the size of the input decision table T. Then the time complexity of the algorithm $\mathscr{A}_1$ is bounded from above by a polynomial on the size of the input table T and the number $|SEP(T)|$ of different separable subtables of T.*

Proof Since the uncertainty measure U has polynomial time complexity depending on the size of decision tables, each step of the algorithm $\mathscr{A}_1$ has polynomial time complexity depending on the size of the table T and the number $L(\Delta_{U,\alpha}(T))$. The number of steps is $O(L(\Delta_{U,\alpha}(T)))$. Therefore the time complexity of the algorithm $\mathscr{A}_1$ is bounded from above by a polynomial on the size of the input table T and the number $L(\Delta_{U,\alpha}(T))$. The number $L(\Delta_{U,\alpha}(T))$ is bounded from above by the number $|SEP(T)|$ of different separable subtables of T. $\qquad\square$

In Sect. 5.5, we describe classes of decision tables such that the number of separable subtables in tables from the class is bounded from above by a polynomial on the number of conditional attributes in the table, and the size of tables is bounded from above by a polynomial on the number of conditional attributes. For each such class, the time complexity of the algorithm $\mathscr{A}_1$ is polynomial depending on the number of conditional attributes in decision tables.

Remark 5.1 Note that, for any decision table T, the graph $\Delta_{U,0}(T)$ does not depend on the uncertainty measure U. We denote this graph $\Delta(T)$. Note also that $L(\Delta(T)) = |SEP(T)|$ for any diagnostic decision table T.

A node of directed graph is called *terminal* if there are no edges starting in this node. A *bundle-preserving subgraph* of the graph $\Delta_{U,\alpha}(T)$ is a graph G obtained from $\Delta_{U,\alpha}(T)$ by removal of some bundles of edges such that each nonterminal node of $\Delta_{U,\alpha}(T)$ keeps at least one bundle of edges starting in this node. By definition, $\Delta_{U,\alpha}(T)$ is a bundle-preserving subgraph of $\Delta_{U,\alpha}(T)$. A node Θ of the graph G is terminal if and only if $U(\Theta) \leq \alpha$. We denote by $L(G)$ the number of nodes in the graph G.

5.5   Restricted Information Systems

In this section, we describe classes of decision tables for which algorithms that deal with the graphs $\Delta_{U,\alpha}(T)$ have polynomial time complexity depending on the number of conditional attributes in the input table T.

Let A be a nonempty set and F be a nonempty set of non-constant functions from A to $E_k = \{0, \ldots, k-1\}$, $k \geq 2$. Functions from F are called *attributes*, and the triple $\mathscr{U} = (A, E_k, F)$ is called a *k-valued information system*. In Parts II and III, we will consider only such information systems. They are close enough to the information systems proposed by Zdzislaw Pawlak [5].

Let $f_1, \ldots, f_n \in F$ and ν be a mapping that corresponds to each tuple $(\delta_1, \ldots, \delta_n)$ $\in E_k^n$ a nonempty subset $\nu(\delta_1, \ldots, \delta_n)$ of the set $\{0, \ldots, k^n - 1\}$ which cardinality is at most n^k. We denote by $T_\nu(f_1, \ldots, f_n)$ the decision table with n conditional attributes $f_1, \ldots, f_n$ which contains the row $(\delta_1, \ldots, \delta_n) \in E_k^n$ if and only if the system of equations

$$\{f_1(x) = \delta_1, \ldots, f_n(x) = \delta_n\}$$

is *consistent* (has a solution over the set A). This row is labeled with the set of decisions $\nu(\delta_1, \ldots, \delta_n)$. The table $T_\nu(f_1, \ldots, f_n)$ is called a *decision table with many-valued decisions over the information system* $\mathscr{U}$. We denote by $\mathscr{T}(\mathscr{U})$ the set of decision tables with many-valued decisions over $\mathscr{U}$.

Let us consider the function

$$SEP_{\mathscr{U}}(n) = \max\{|SEP(T)| : T \in \mathscr{T}(\mathscr{U}), \dim(T) \leq n\},$$

where $\dim(T)$ is the number of conditional attributes in T, which characterizes the maximum number of separable subtables depending on the number of conditional attributes in decision tables over $\mathscr{U}$.

Let r be a natural number. The information system $\mathscr{U}$ will be called *r-restricted* if, for each consistent system of equations of the kind

$$\{f_1(x) = \delta_1, \ldots, f_m(x) = \delta_m\}$$

where $m \in \omega \setminus \{0\}$, $f_1, \ldots, f_m \in F$ and $\delta_1, \ldots, \delta_m \in E_k$, there exists a subsystem of this system which has the same set of solutions over the set A and contains at most r equations. The information system $\mathscr{U}$ will be called *restricted* if it is r-restricted for some natural r.

Theorem 5.1 ([4]) *Let* $\mathscr{U} = (A, E_k, F)$ *be a k-valued information system. Then the following statements hold:*

1. *If* $\mathscr{U}$ *is an r-restricted information system for some natural r then* $SEP_{\mathscr{U}}(n) \leq$ $(nk)^r + 1$ *for any natural n.*
2. *If* $\mathscr{U}$ *is not a restricted information system then* $SEP_{\mathscr{U}}(n) \geq 2^n$ *for any natural n.*

We now evaluate the time complexity of the algorithm $\mathscr{A}_1$ for decision tables over a restricted information system $\mathscr{U}$ under the assumption that the uncertainty measure U used by $\mathscr{A}_1$ has polynomial time complexity.

Lemma 5.2 *Let $\mathscr{U}$ be a restricted information system. Then, for decision tables from $\mathscr{T}(\mathscr{U})$, both the size and the number of separable subtables are bounded from above by polynomials on the number of conditional attributes.*

Proof Let $\mathscr{U}$ be a k-valued information system which is r-restricted. For any decision table $T \in \mathscr{T}(\mathscr{U})$, each value of each conditional attribute is at most k, the value of each decision is at most $k^{\dim(T)}$, the cardinality of each set of decisions attached to rows of T is at most $\dim(T)^k$ and, by Theorem 5.1, $N(T) \leq |SEP(T)| \leq (\dim(T)k)^r + 1$. From here it follows that the size of decision tables from $\mathscr{T}(\mathscr{U})$ is bounded from above by a polynomial on the number of conditional attributes in decision tables. By Theorem 5.1, the number of separable subtables for decision tables from $\mathscr{T}(\mathscr{U})$ is bounded from above by a polynomial on the number of conditional attributes in decision tables. □

Proposition 5.3 *Let $\mathscr{U}$ be a restricted information system, and the uncertainty measure U used by the algorithm $\mathscr{A}_1$ have polynomial time complexity depending on the size of decision tables. Then the algorithm $\mathscr{A}_1$ has polynomial time complexity for decision tables from $\mathscr{T}(\mathscr{U})$ depending on the number of conditional attributes.*

Proof By Proposition 5.2, the time complexity of the algorithm $\mathscr{A}_1$ is bounded from above by a polynomial on the size of the input table T and the number $|SEP(T)|$ of different separable subtables of T. From Lemma 5.2 it follows that, for decision tables from $\mathscr{T}(\mathscr{U})$, both the size and the number of separable subtables are bounded from above by polynomials on the number of conditional attributes. □

Remark 5.2 Let $\mathscr{U}$ be an information system which is not restricted. Using Remark 5.1 and Theorem 5.1 one can show that there is no algorithm which constructs the graph $\Delta(T)$ for decision tables $T \in \mathscr{T}(\mathscr{U})$ and which time complexity is bounded from above by a polynomial on the number of conditional attributes in the considered decision tables.

Let us consider a family of restricted information systems. Let d and t be natural numbers, $f_1, \ldots, f_t$ be functions from $\mathbb{R}^d$ to $\mathbb{R}$, and s be a function from $\mathbb{R}$ to $\{0, 1\}$ such that $s(x) = 0$ if $x < 0$ and $s(x) = 1$ if $x \geq 0$. Then the 2-valued information system $(\mathbb{R}^d, E_2, F)$ where $F = \{s(f_i + c) : i = 1, \ldots, t, c \in \mathbb{R}\}$ is a $2t$-restricted information system.

If $f_1, \ldots, f_t$ are linear functions then we deal with attributes corresponding to t families of parallel hyperplanes in $\mathbb{R}^d$. This is usual for decision trees constructed by CART [2] for datasets with t numerical attributes only (in this case, $d = t$).

We consider now a class of so-called linear information systems for which all restricted systems are known. Let P be the set of all points in the two-dimensional Euclidean plane and l be a straight line (line in short) in the plane. This line divides the plane into two open half-planes H_1 and H_2, and the line l. Two attributes correspond to the line l. The first attribute takes value 0 on points from H_1, and value 1 on points from H_2 and l. The second one takes value 0 on points from H_2, and value 1 on points from H_1 and l. We denote by $\mathscr{L}$ the set of all attributes corresponding to lines in the plane. Information systems of the kind (P, E_2, F) where $F \subseteq \mathscr{L}$, will be called *linear* information systems. We describe all restricted linear information systems.

Let l be a line in the plane. Let us denote by $\mathscr{L}(l)$ the set of all attributes corresponding to lines which are parallel to l. Let p be a point in the plane. We denote by $\mathscr{L}(p)$ the set of all attributes corresponding to lines which pass through p. A set C of attributes from $\mathscr{L}$ is called a *clone* if $C \subseteq \mathscr{L}(l)$ for some line l or $C \subseteq \mathscr{L}(p)$ for some point p.

Theorem 5.2 ([3]) *A linear information system (P, E_2, F) is restricted if and only if F is the union of a finite number of clones.*

5.6   Time Complexity of Algorithms on $\Delta_{U,\alpha}(T)$

In this book, we consider a number of algorithms which deal with the graph $\Delta_{U,\alpha}(T)$. To evaluate time complexity of these algorithms, we will count the number of elementary operations made by the algorithms. These operations can either be basic numerical operations or computations of numerical parameters of decision tables. We assume, as we already mentioned, that each basic numerical operation ($\max(x, y)$, $x + y$, $x - y$, $x \times y$, $x \div y$, $\log_2 x$) has time complexity $O(1)$. This assumption is reasonable for computations with floating-point numbers. Furthermore, computing the considered parameters of decision tables usually has polynomial time complexity depending on the size of the decision table.

Proposition 5.4 *Let, for some algorithm $\mathscr{A}$ working with decision tables, the number of elementary operations (basic numerical operations and computations of numerical parameters of decision tables) be polynomial depending on the size of the input table T and on the number of separable subtables of T, and the computations of parameters of decision tables used by the algorithm $\mathscr{A}$ have polynomial time complexity depending on the size of decision tables. Then, for any restricted information system $\mathscr{U}$, the algorithm $\mathscr{A}$ has polynomial time complexity for decision tables from $\mathscr{T}(\mathscr{U})$ depending on the number of conditional attributes.*

Proof Let $\mathscr{U}$ be a restricted information system. From Lemma 5.2 it follows that the size and the number of separable subtables for decision tables from $\mathscr{T}(\mathscr{U})$ are bounded from above by polynomials on the number of conditional attributes in the tables. From here it follows that the algorithm $\mathscr{A}$ has polynomial time complexity for decision tables from $\mathscr{T}(\mathscr{U})$ depending on the number of conditional attributes. $\qquad\square$

5.7   Tools for Study of Pareto Optimal Points

In this section, we consider tools (statements and algorithms) which are used for the study of Pareto optimal points for bi-criteria optimization problems.

Let $\mathbb{R}^2$ be the set of pairs of real numbers (*points*). We consider a partial order $\leq$ on the set $\mathbb{R}^2$: $(c, d) \leq (a, b)$ if $c \leq a$ and $d \leq b$. Two points α and β are *comparable* if $\alpha \leq \beta$ or $\beta \leq \alpha$. A subset of $\mathbb{R}^2$ in which no two different points are comparable is called an *antichain*. We will write $\alpha < \beta$ if $\alpha \leq \beta$ and $\alpha \neq \beta$. If α and β are comparable then $\min(\alpha, \beta) = \alpha$ if $\alpha \leq \beta$ and $\min(\alpha, \beta) = \beta$ if $\alpha > \beta$.

Let A be a nonempty finite subset of $\mathbb{R}^2$. A point $\alpha \in A$ is called a *Pareto optimal point for* A if there is no a point $\beta \in A$ such that $\beta < \alpha$. We denote by $Par(A)$ the set of Pareto optimal points for A. It is clear that $Par(A)$ is an antichain.

Lemma 5.3 *Let A be a nonempty finite subset of the set $\mathbb{R}^2$. Then, for any point $\alpha \in A$, there is a point $\beta \in Par(A)$ such that $\beta \leq \alpha$.*

Proof Let $\beta = (a, b)$ be a point from A such that $(a, b) \leq \alpha$ and $a + b = \min\{c + d : (c, d) \in A, (c, d) \leq \alpha\}$. Then $(a, b) \in Par(A)$. ☐

Lemma 5.4 *Let A, B be nonempty finite subsets of the set $\mathbb{R}^2$, $A \subseteq B$, and, for any $\beta \in B$, there exists $\alpha \in A$ such that $\alpha \leq \beta$. Then $Par(A) = Par(B)$.*

Proof Let $\beta \in Par(B)$. Then there exists $\alpha \in A$ such that $\alpha \leq \beta$. By Lemma 5.3, there exists $\gamma \in Par(A)$ such that $\gamma \leq \alpha$. Therefore $\gamma \leq \beta$ and $\gamma = \beta$ since $\beta \in Par(B)$. Hence $Par(B) \subseteq Par(A)$.

Let $\alpha \in Par(A)$. By Lemma 5.3, there exists $\beta \in Par(B)$ such that $\beta \leq \alpha$. We know that there exists $\gamma \in A$ such that $\gamma \leq \beta$. Therefore $\gamma \leq \alpha$ and $\gamma = \alpha$ since $\alpha \in Par(A)$. As a result, we have $\beta = \alpha$ and $Par(A) \subseteq Par(B)$. Hence $Par(A) = Par(B)$. ☐

Lemma 5.5 *Let A be a nonempty finite subset of $\mathbb{R}^2$. Then*

$$|Par(A)| \leq \min\left(\left|A^{(1)}\right|, \left|A^{(2)}\right|\right)$$

where $A^{(1)} = \{a : (a, b) \in A\}$ and $A^{(2)} = \{b : (a, b) \in A\}$.

Proof Let $(a, b), (c, d) \in Par(A)$ and $(a, b) \neq (c, d)$. Then $a \neq c$ and $b \neq d$ (otherwise, (a, b) and (c, d) are comparable). Therefore $|Par(A)| \leq \min\left(\left|A^{(1)}\right|, \left|A^{(2)}\right|\right)$. ☐

Points from $Par(A)$ can be ordered in the following way: $(a_1, b_1), \ldots, (a_t, b_t)$ where $a_1 < \cdots < a_t$. Since points from $Par(A)$ are incomparable, $b_1 > \cdots > b_t$. We will refer to the sequence $(a_1, b_1), \ldots, (a_t, b_t)$ as the *normal representation of the set* $Par(A)$.

We now describe an algorithm which, for a given nonempty finite subset A of the set $\mathbb{R}^2$, constructs the normal representation of the set $Par(A)$. We assume that A is a multiset containing, possibly, repeating elements. The cardinality $|A|$ of A is the total number of elements in A.

Algorithm $\mathscr{A}_2$ (construction the normal representation of the set $Par(A)$).

Input: A nonempty finite subset A of the set $\mathbb{R}^2$ containing, possibly, repeating elements (multiset).

Output: Normal representation P of the set $Par(A)$ of Pareto optimal points for A.

1. Set P equal to the empty sequence.
2. Construct a sequence B of all points from A ordered according to the first coordinate in the ascending order.
3. If there is only one point in the sequence B, then add this point to the end of the sequence P, return P, and finish the work of the algorithm. Otherwise, choose the first $\alpha = (\alpha_1, \alpha_2)$ and the second $\beta = (\beta_1, \beta_2)$ points from B.
4. If α and β are comparable then remove α and β from B, add the point $\min(\alpha, \beta)$ to the beginning of B, and proceed to step 3.
5. If α and β are not comparable (in this case $\alpha_1 < \beta_1$ and $\alpha_2 > \beta_2$) then remove α from B, add the point α to the end of P, and proceed to step 3.

Proposition 5.5 *Let A be a nonempty finite subset of the set $\mathbb{R}^2$ containing, possibly, repeating elements (multiset). Then the algorithm $\mathscr{A}_2$ returns the normal representation of the set $Par(A)$ of Pareto optimal points for A and makes $O(|A| \log |A|)$ comparisons.*

Proof The step 2 of the algorithm requires $O(|A| \log |A|)$ comparisons. Each call to step 3 (with the exception of the last one) leads to two comparisons. The number of calls to step 3 is at most $|A|$. Therefore the algorithm $\mathscr{A}_2$ makes $O(|A| \log |A|)$ comparisons.

Let the output sequence P be equal to $(a_1, b_1), \ldots, (a_t, b_t)$ and let us set $Q = \{(a_1, b_1), \ldots, (a_t, b_t)\}$. It is clear that $a_1 < \cdots < a_t, b_1 > \cdots > b_t$ and, for any $\alpha \in A, \alpha \notin Q$, there exists $\beta \in Q$ such that $\beta < \alpha$. From here it follows that $Par(A) \subseteq Q$ and Q is an antichain. Let us assume that there exists $\gamma \in Q$ which does not belong to $Par(A)$. Then there exists $\alpha \in A$ such that $\alpha < \gamma$. Since Q is an antichain, $\alpha \notin Q$. We know that there exists $\beta \in Q$ such that $\beta \leq \alpha$. This results in two different points β and γ from Q being comparable, which is impossible. Therefore $Q = Par(A)$ and P is the normal representation of the set $Par(A)$. $\square$

Remark 5.3 Let A be a nonempty finite subset of $\mathbb{R}^2$, $(a_1, b_1), \ldots, (a_t, b_t)$ be the normal representation of the set $Par(A)$, and $rev(A) = \{(b, a) : (a, b) \in A\}$. Then $Par(rev(A)) = rev(Par(A))$ and $(b_t, a_t), \ldots, (b_1, a_1)$ is the normal representation of the set $Par(rev(A))$.

Lemma 5.6 *Let A be a nonempty finite subset of $\mathbb{R}^2$, $B \subseteq A$, and $Par(A) \subseteq B$. Then $Par(B) = Par(A)$.*

Proof It is clear that $Par(A) \subseteq Par(B)$. Let us assume that, for some β, $\beta \in Par(B)$ and $\beta \notin Par(A)$. Then there exists $\alpha \in A$ such that $\alpha < \beta$. By Lemma 5.3, there exists $\gamma \in Par(A) \subseteq B$ such that $\gamma \leq \alpha$. Therefore $\gamma < \beta$ and $\beta \notin Par(B)$. Hence $Par(B) = Par(A)$. $\qquad\qquad\qquad\qquad\qquad\qquad\qquad\qquad\qquad\qquad\qquad\quad\square$

Lemma 5.7 *Let* $A_1, \ldots, A_k$ *be nonempty finite subsets of* $\mathbb{R}^2$. *Then* $Par(A_1 \cup \cdots \cup A_k) \subseteq Par(A_1) \cup \cdots \cup Par(A_k)$.

Proof Let $\alpha \in (A_1 \cup \cdots \cup A_k) \setminus (Par(A_1) \cup \cdots \cup Par(A_k))$. Then there is $i \in \{1, \ldots, k\}$ such that $\alpha \in A_i$ but $\alpha \notin Par(A_i)$. Therefore there is $\beta \in A_i$ such that $\beta < \alpha$. Hence $\alpha \notin Par(A_1 \cup \cdots \cup A_k)$, and $Par(A_1 \cup \cdots \cup A_k) \subseteq Par(A_1) \cup \cdots \cup Par(A_k)$. $\qquad\qquad\qquad\square$

A function $f : \mathbb{R}^2 \to \mathbb{R}$ is called *increasing* if $f(x_1, y_1) \leq f(x_2, y_2)$ for any $x_1, x_2, y_1, y_2 \in \mathbb{R}$ such that $x_1 \leq x_2$ and $y_1 \leq y_2$. For example, $\text{sum}(x, y) = x + y$ and $\max(x, y)$ are increasing functions.

Let f, g be increasing functions from $\mathbb{R}^2$ to $\mathbb{R}$, and A, B be nonempty finite subsets of the set $\mathbb{R}^2$. We denote by $A \langle fg \rangle B$ the set $\{(f(a, c), g(b, d)) : (a, b) \in A, (c, d) \in B\}$.

Lemma 5.8 *Let* A, B *be nonempty finite subsets of* $\mathbb{R}^2$, *and* f, g *be increasing functions from* $\mathbb{R}^2$ *to* $\mathbb{R}$. *Then* $Par(A \langle fg \rangle B) \subseteq Par(A) \langle fg \rangle Par(B)$.

Proof Let $\beta \in Par(A \langle fg \rangle B)$ and $\beta = (f(a, c), g(b, d))$ where $(a, b) \in A$ and $(c, d) \in B$. Then, by Lemma 5.3, there exist $(a', b') \in Par(A)$ and $(c', d') \in Par(B)$ such that $(a', b') \leq (a, b)$ and $(c', d') \leq (c, d)$. It is clear that $\alpha = (f(a', c'), g(b', d')) \leq (f(a, c), g(b, d)) = \beta$, and $\alpha \in Par(A) \langle fg \rangle Par(B)$. Since $\beta \in Par(A \langle fg \rangle B)$, we have $\beta = \alpha$. Therefore $Par(A \langle fg \rangle B) \subseteq Par(A) \langle fg \rangle Par(B)$. $\qquad\qquad\qquad\qquad\qquad\qquad\square$

Let f, g be increasing functions from $\mathbb{R}^2$ to $\mathbb{R}$, $P_1, \ldots, P_t$ be nonempty finite subsets of $\mathbb{R}^2$, $Q_1 = P_1$, and, for $i = 2, \ldots, t$, $Q_i = Q_{i-1} \langle fg \rangle P_i$. We assume that, for $i = 1, \ldots, t$, the sets $Par(P_1), \ldots, Par(P_t)$ are already constructed. We now describe an algorithm that constructs the sets $Par(Q_1), \ldots, Par(Q_t)$ and returns $Par(Q_t)$. This algorithm will be used by algorithms for bi-criteria optimization of decision trees and decision rule systems.

Algorithm $\mathscr{A}_3$ (fusion of sets of POPs).

Input: Increasing functions f, g from $\mathbb{R}^2$ to $\mathbb{R}$, and sets $Par(P_1), \ldots, Par(P_t)$ for some nonempty finite subsets $P_1, \ldots, P_t$ of $\mathbb{R}^2$.

Output: The set $Par(Q_t)$ where $Q_1 = P_1$, and, for $i = 2, \ldots, t$, $Q_i = Q_{i-1} \langle fg \rangle P_i$.

1. Set $B_1 = Par(P_1)$ and $i = 2$.
2. Construct the multiset

$$A_i = B_{i-1} \langle fg \rangle Par(P_i) = \{(f(a, c), g(b, d)) : (a, b) \in B_{i-1}, (c, d) \in Par(P_i)\}$$

(we will not remove equal pairs from the constructed set).

3. Using the algorithm $\mathscr{A}_2$, construct the set $B_i = Par(A_i)$.
4. If $i = t$ then return B_i and finish the work of the algorithm. Otherwise, set $i = i + 1$ and proceed to step 2.

Proposition 5.6 *Let f, g be increasing functions from $\mathbb{R}^2$ to $\mathbb{R}$, $P_1, \ldots, P_t$ be nonempty finite subsets of $\mathbb{R}^2$, $Q_1 = P_1$, and, for $i = 2, \ldots, t$, $Q_i = Q_{i-1} \langle fg \rangle P_i$. Then the algorithm $\mathscr{A}_3$ returns the set $Par(Q_t)$.*

Proof We will prove by induction on i that, for $i = 1, \ldots, t$, the set B_i (see the description of the algorithm $\mathscr{A}_3$) is equal to the set $Par(Q_i)$. Since $B_1 = Par(P_1)$ and $Q_1 = P_1$, we have $B_1 = Par(Q_1)$. Let for some $i - 1, 2 \leq i \leq t$, the considered statement hold, i.e., $B_{i-1} = Par(Q_{i-1})$. Then $B_i = Par(B_{i-1} \langle fg \rangle Par(P_i)) = Par(Par(Q_{i-1}) \langle fg \rangle Par(P_i))$. We know that $Q_i = Q_{i-1} \langle fg \rangle P_i$. By Lemma 5.8, $Par(Q_i) \subseteq Par(Q_{i-1}) \langle fg \rangle Par(P_i)$. By Lemma 5.6,

$$Par(Q_i) = Par(Par(Q_{i-1}) \langle fg \rangle Par(P_i)) \,.$$

Therefore $B_i = Par(Q_i)$. So we have $B_t = Par(Q_t)$, and the algorithm $\mathscr{A}_3$ returns the set $Par(Q_i)$. $\qquad\square$

Proposition 5.7 *Let f, g be increasing functions from $\mathbb{R}^2$ to $\mathbb{R}$,*

$$f \in \{x + y, \max(x, y)\} \,,$$

$P_1, \ldots, P_t$ be nonempty finite subsets of $\mathbb{R}^2$, $Q_1 = P_1$, and, for $i = 2, \ldots, t$, $Q_i = Q_{i-1} \langle fg \rangle P_i$. Let $P_i^{(1)} = \{a : (a, b) \in P_i\}$ for $i = 1, \ldots, t$, $m \in \omega$, and

$$P_i^{(1)} \subseteq \{0, 1, \ldots, m\}$$

for $i = 1, \ldots, t$, or $P_i^{(1)} \subseteq \{0, -1, \ldots, -m\}$ for $i = 1, \ldots, t$. Then, during the construction of the set $Par(Q_t)$, the algorithm $\mathscr{A}_3$ makes

$$O(t^2 m^2 \log(tm))$$

elementary operations (computations of f, g and comparisons) if $f = x + y$, and

$$O(tm^2 \log m)$$

elementary operations (computations of f, g and comparisons) if $f = \max(x, y)$. If $f = x + y$, then $|Par(Q_t)| \leq tm + 1$, and if $f = \max(x, y)$, then $|Par(Q_t)| \leq m + 1$.

Proof For $i = 1, \ldots, t$, we denote $p_i = |Par(P_i)|$ and $q_i = |Par(Q_i)|$. Let $i \in \{2, \ldots, t\}$. To construct the multiset $A_i = Par(Q_{i-1}) \langle fg \rangle Par(P_i)$, we need to compute the values of f and g a number of times equal to $q_{i-1} p_i$. The cardinality of A_i is equal to $q_{i-1} p_i$. We apply to A_i the algorithm $\mathscr{A}_2$ which makes

$O(q_{i-1} p_i \log(q_{i-1} p_i))$ comparisons. As a result, we find the set $Par(A_i) = Par(Q_i)$. To construct the sets $Par(Q_1), \ldots, Par(Q_t)$, the algorithm $\mathscr{A}_3$ makes $\sum_{i=2}^{t} q_{i-1} p_i$ computations of f, $\sum_{i=2}^{t} q_{i-1} p_i$ computations of g, and $O\left(\sum_{i=2}^{t} q_{i-1} p_i \log(q_{i-1} p_i)\right)$ comparisons.

We know that $P_i^{(1)} \subseteq \{0, 1, \ldots, m\}$ for $i = 1, \ldots, t$, or $P_i^{(1)} \subseteq \{0, -1, \ldots, -m\}$ for $i = 1, \ldots, t$. Then, by Lemma 5.5, $p_i \le m + 1$ for $i = 1, \ldots, t$.

Let $f = x + y$. Then, for $i = 1, \ldots, t$, $Q_i^{(1)} = \{a : (a, b) \in Q_i\} \subseteq \{0, 1, \ldots, im\}$ or, for $i = 1, \ldots, t$, $Q_i^{(1)} \subseteq \{0, -1, \ldots, -im\}$ and, by Lemma 5.5, $q_i \le im + 1$. In this case, to construct the sets $Par(Q_1), \ldots, Par(Q_t)$ the algorithm $\mathscr{A}_3$ makes $O(t^2 m^2)$ computations of f, $O(t^2 m^2)$ computations of g, and $O(t^2 m^2 \log(tm))$ comparisons, i.e.,

$$O(t^2 m^2 \log(tm))$$

elementary operations (computations of f, g, and comparisons). Since $q_t \le tm + 1$, $|Par(Q_t)| \le tm + 1$.

Let $f = \max(x, y)$. Then, for $i = 1, \ldots, t$, $Q_i^{(1)} = \{a : (a, b) \in Q_i\} \subseteq \{0, 1, \ldots, m\}$ or, for $i = 1, \ldots, t$, $Q_i^{(1)} \subseteq \{0, -1, \ldots, -m\}$ and, by Lemma 5.5, $q_i \le m + 1$. In this case, to construct the sets $Par(Q_1), \ldots, Par(Q_t)$ the algorithm $\mathscr{A}_3$ makes $O(tm^2)$ computations of f, $O(tm^2)$ computations of g, and $O(tm^2 \log m)$ comparisons, i.e.,

$$O(tm^2 \log m)$$

elementary operations (computations of f, g, and comparisons). Since $q_t \le m + 1$, $|Par(Q_t)| \le m + 1$. $\qquad \square$

Similar analysis can be done for the sets $P_i^{(2)} = \{b : (a, b) \in P_i\}$, $Q_i^{(2)} = \{b : (a, b) \in Q_i\}$, and the function g.

A function $p : \mathbb{R} \to \mathbb{R}$ is called *strictly increasing* if $p(x) < p(y)$ for any $x, y \in \mathbb{R}$ such that $x < y$. Let p and q be strictly increasing functions from $\mathbb{R}$ to $\mathbb{R}$. For $(a, b) \in \mathbb{R}^2$, we denote by $(a, b)^{pq}$ the pair $(p(a), q(b))$. For a nonempty finite subset A of $\mathbb{R}^2$, we denote $A^{pq} = \{(a, b)^{pq} : (a, b) \in A\}$.

Lemma 5.9 *Let A be a nonempty finite subset of $\mathbb{R}^2$, and p, q be strictly increasing functions from $\mathbb{R}$ to $\mathbb{R}$. Then $Par(A^{pq}) = Par(A)^{pq}$.*

Proof Let $(a, b), (c, d) \in A$. It is clear that $(c, d) = (a, b)$ if and only if $(c, d)^{pq} = (a, b)^{pq}$, and $(c, d) < (a, b)$ if and only if $(c, d)^{pq} < (a, b)^{pq}$. From here it follows that $Par(A^{pq}) = Par(A)^{pq}$. $\qquad \square$

Let A be a nonempty finite subset of $\mathbb{R}^2$. We correspond to A a partial function $\mathscr{F}_A : \mathbb{R} \to \mathbb{R}$ defined in the following way: $\mathscr{F}_A(x) = \min\{b : (a, b) \in A, a \le x\}$.

Lemma 5.10 *Let A be a nonempty finite subset of $\mathbb{R}^2$, and $(a_1, b_1), \ldots, (a_t, b_t)$ be the normal representation of the set $Par(A)$. Then, for any $x \in \mathbb{R}$, $\mathscr{F}_A(x) = \mathscr{F}(x)$ where*

$$\mathscr{F}(x) = \begin{cases} undefined, & x < a_1 \\ b_1, & a_1 \le x < a_2 \\ \ldots & \ldots \\ b_{t-1}, & a_{t-1} \le x < a_t \\ b_t, & a_t \le x \end{cases}.$$

Proof One can show that $a_1 = \min\{a : (a, b) \in A\}$. Therefore the value $\mathscr{F}_A(x)$ is undefined if $x < a_1$. Let $x \ge a_1$. Then both values $\mathscr{F}(x)$ and $\mathscr{F}_A(x)$ are defined. It is easy to check that $\mathscr{F}(x) = \mathscr{F}_{Par(A)}(x)$. Since $Par(A) \subseteq A$, we have $\mathscr{F}_A(x) \le \mathscr{F}(x)$. By Lemma 5.3, for any point $(a, b) \in A$, there is a point $(a_i, b_i) \in Par(A)$ such that $(a_i, b_i) \le (a, b)$. Therefore $\mathscr{F}(x) \le \mathscr{F}_A(x)$ and $\mathscr{F}_A(x) = \mathscr{F}(x)$. $\square$

Remark 5.4 We can consider not only function $\mathscr{F}_A$ but also function $\mathscr{F}_{rev(A)} : \mathbb{R} \to \mathbb{R}$ defined in the following way:

$$\mathscr{F}_{rev(A)}(x) = \min\{a : (b, a) \in rev(A), b \le x\} = \min\{a : (a, b) \in A, b \le x\}.$$

From Remark 5.3 and Lemma 5.10 it follows that

$$\mathscr{F}_{rev(A)}(x) = \begin{cases} undefined, & x < b_t \\ a_t, & b_t \le x < b_{t-1} \\ \ldots & \ldots \\ a_2, & b_2 \le x < b_1 \\ a_1, & b_1 \le x \end{cases}.$$

References

1. AbouEisha, H., Amin, T., Chikalov, I., Hussain, S., Moshkov, M.: Extensions of Dynamic Programming for Combinatorial Optimization and Data Mining. Intelligent Systems Reference Library, vol. 146. Springer, Berlin (2019)
2. Breiman, L., Friedman, J.H., Olshen, R.A., Stone, C.J.: Classification and Regression Trees. Wadsworth and Brooks, Monterey (1984)
3. Moshkov, M.: On the class of restricted linear information systems. Discret. Math. **307**(22), 2837–2844 (2007)
4. Moshkov, M., Chikalov, I.: On algorithm for constructing of decision trees with minimal depth. Fundam. Inform. **41**(3), 295–299 (2000)
5. Pawlak, Z.: Information systems theoretical foundations. Inf. Syst. **6**(3), 205–218 (1981)

Chapter 6
Decision and Inhibitory Trees

Decision trees are widely used as predictors [1], as tools for data mining and knowledge representation (see, for example, [3]), and as algorithms for problem solving (see, for example, [2]). Inhibitory tree is a less known object. However, examples considered in Chap. 2 show that the use of inhibitory trees can give us additional possibilities in comparison with decisions trees.

In this chapter, we consider various types of decision and inhibitory trees. We discuss the notion of a cost function for trees, the notion of decision tree uncertainty, and the notion of inhibitory tree completeness. We consider also an algorithm for counting the number of decision trees represented by a directed acyclic graph.

6.1 Different Kinds of Decision Trees

In this section, we discuss main notions related to decision trees.

6.1.1 Decision Trees for T

Let T be a decision table with n conditional attributes $f_1, \ldots, f_n$.

A *decision tree over* T is a finite directed tree with root in which nonterminal nodes are labeled with attributes from the set $\{f_1, \ldots, f_n\}$, terminal nodes are labeled with numbers from ω, and, for each nonterminal node, edges starting from this node are labeled with pairwise different numbers from ω.

Let Γ be a decision tree over T and v be a node of Γ. We denote by $\Gamma(v)$ the subtree of Γ for which v is the root. We define now a subtable $T(v) = T_\Gamma(v)$ of

© Springer Nature Switzerland AG 2020
F. Alsolami et al., *Decision and Inhibitory Trees and Rules for Decision Tables with Many-valued Decisions*, Intelligent Systems Reference Library 156, https://doi.org/10.1007/978-3-030-12854-8_6

the table T. If v is the root of Γ then $T(v) = T$. Let v be not the root of Γ and $v_1, e_1, \ldots, v_m, e_m, v_{m+1} = v$ be the directed path from the root of Γ to v in which nodes $v_1, \ldots, v_m$ are labeled with attributes $f_{i_1}, \ldots, f_{i_m}$ and edges $e_1, \ldots, e_m$ are labeled with numbers $a_1, \ldots, a_m$, respectively. Then $T(v) = T(f_{i_1}, a_1) \ldots (f_{i_m}, a_m)$.

A decision tree Γ over T is called a *decision tree for T* if, for any node v of Γ,

- If $T(v)$ is a degenerate table then v is a terminal node labeled with $mcd(T(v))$.
- If $T(v)$ is not degenerate then either v is a terminal node labeled with $mcd(T(v))$, or v is a nonterminal node which is labeled with an attribute $f_i \in E(T(v))$ and, if $E(T(v), f_i) = \{a_1, \ldots, a_t\}$, then t edges start from the node v that are labeled with $a_1, \ldots, a_t$, respectively.

We denote by $DT(T)$ the set of decision trees for T.

For $b \in \omega$, we denote by $tree(b)$ the decision tree that contains only one (terminal) node labeled with b.

Let $f_i \in \{f_1, \ldots, f_n\}$, $a_1, \ldots, a_t$ be pairwise different numbers from ω, and $\Gamma_1, \ldots, \Gamma_t$ be decision trees over T. We denote by $tree(f_i, a_1, \ldots, a_t, \Gamma_1, \ldots, \Gamma_t)$ the following decision tree over T: the root of the tree is labeled with f_i, and t edges start from the root which are labeled with $a_1, \ldots, a_t$ and enter the roots of decision trees $\Gamma_1, \ldots, \Gamma_t$, respectively.

Let $f_i \in E(T)$ and $E(T, f_i) = \{a_1, \ldots, a_t\}$. We denote

$$DT(T, f_i) = \{tree(f_i, a_1, \ldots, a_t, \Gamma_1, \ldots, \Gamma_t) : \Gamma_j \in DT(T(f_i, a_j)), j = 1, \ldots, t\}.$$

Proposition 6.1 *Let T be a decision table. Then $DT(T) = \{tree(mcd(T))\}$ if T is degenerate, and $DT(T) = \{tree(mcd(T))\} \cup \bigcup_{f_i \in E(T)} DT(T, f_i)$ if T is nondegenerate.*

Proof Let T be a degenerate decision table. From the definition of a decision tree for T it follows that $tree(mcd(T))$ is the only decision tree for T.

Let T be a nondegenerate decision table. It is clear that $tree(mcd(T))$ is the only decision tree for T with one node.

Let $\Gamma \in DT(T)$ and Γ have more than one node. Then, by definition of the set $DT(T)$, $\Gamma = tree(f_i, a_1, \ldots, a_t, \Gamma_1, \ldots, \Gamma_t)$ where $f_i \in E(T)$ and $\{a_1, \ldots, a_t\} = E(T, f_i)$. Using the fact that $\Gamma \in DT(T)$ it is not difficult to show that $\Gamma_j \in DT(T(f_i, a_j))$ for $j = 1, \ldots, t$. From here it follows that $\Gamma \in DT(T, f_i)$ for $f_i \in E(T)$. Therefore $DT(T) \subseteq \{tree(mcd(T))\} \cup \bigcup_{f_i \in E(T)} DT(T, f_i)$.

Let, for some $f_i \in E(T)$, $\Gamma \in DT(T, f_i)$. Then

$$\Gamma = tree(f_i, a_1, \ldots, a_t, \Gamma_1, \ldots, \Gamma_t)$$

where $\{a_1, \ldots, a_t\} = E(T, f_i)$ and $\Gamma_j \in DT(T(f_i, a_j))$ for $j = 1, \ldots, t$. Using these facts it is not difficult to show that $\Gamma \in DT(T)$. Therefore

$$\{tree(mcd(T))\} \cup \bigcup_{f_i \in E(T)} DT(T, f_i) \subseteq DT(T).$$

For each node Θ of the directed acyclic graph $\Delta(T)$, we define the set □

$$Tree^*(\Delta(T), \Theta)$$

of decision trees in the following way. If Θ is a terminal node of $\Delta(T)$ (in this case Θ is degenerate) then $Tree^*(\Delta(T), \Theta) = \{tree(mcd(\Theta))\}$. Let Θ be a nonterminal node of $\Delta(T)$ (in this case Θ is nondegenerate), $f_i \in E(T)$, and $E(\Theta, f_i) = \{a_1, \ldots, a_t\}$. We denote by $Tree^*(\Delta(T), \Theta, f_i)$ the set of decision trees $\{tree(f_i, a_1, \ldots, a_t, \Gamma_1, \ldots, \Gamma_t) : \Gamma_j \in Tree^*(\Delta(T), \Theta(f_i, a_j)), j = 1, \ldots, t\}$. Then

$$Tree^*(\Delta(T), \Theta) = \{tree(mcd(\Theta))\} \cup \bigcup_{f_i \in E(\Theta)} Tree^*(\Delta(T), \Theta, f_i) .$$

Proposition 6.2 *For any decision table T and any node Θ of the graph $\Delta(T)$, the equality $Tree^*(\Delta(T), \Theta) = DT(\Theta)$ holds.*

Proof We prove this statement by induction on nodes of $\Delta(T)$. Let Θ be a terminal node of $\Delta(T)$. Then $Tree^*(\Delta(T), \Theta) = \{tree(mcd(\Theta))\} = DT(\Theta)$. Let now Θ be a nonterminal node of $\Delta(T)$, and let us assume that $Tree^*(\Delta(T), \Theta(f_i, a_j)) = DT(\Theta(f_i, a_j))$ for any $f_i \in E(\Theta)$ and $a_j \in E(\Theta, f_i)$. Then, for any $f_i \in E(\Theta)$, we have $Tree^*(\Delta(T), \Theta, f_i) = DT(\Theta, f_i)$. Using Proposition 6.1, we obtain

$$Tree^*(\Delta(T), \Theta) = DT(\Theta) .$$

□

6.1.2 (U, α)-Decision Trees for T

Let U be an uncertainty measure and $\alpha \in \mathbb{R}_+$. A decision tree Γ over T is called a (U, α)-*decision tree for* T if, for any node v of Γ,

- If $U(T(v)) \leq \alpha$ then v is a terminal node which is labeled with $mcd(T(v))$.
- If $U(T(v)) > \alpha$ then v is a nonterminal node labeled with an attribute $f_i \in E(T(v))$, and if $E(T(v), f_i) = \{a_1, \ldots, a_t\}$ then t edges start from the node v which are labeled with $a_1, \ldots, a_t$, respectively.

We denote by $DT_{U,\alpha}(T)$ the set of (U, α)-decision trees for T. It is easy to show that $DT_{U,\alpha}(T) \subseteq DT(T)$. Let $f_i \in E(T)$ and $E(T, f_i) = \{a_1, \ldots, a_t\}$. We denote

$$DT_{U,\alpha}(T, f_i) = \{tree(f_i, a_1, \ldots, a_t, \Gamma_1, \ldots, \Gamma_t) : \Gamma_j \in DT_{U,\alpha}(T(f_i, a_j)), j = 1, \ldots, t\} .$$

Proposition 6.3 *Let U be an uncertainty measure, $\alpha \in \mathbb{R}_+$, and T be a decision table. Then $DT_{U,\alpha}(T) = \{tree(mcd(T))\}$ if $U(T) \leq \alpha$, and $DT_{U,\alpha}(T) = \bigcup_{f_i \in E(T)} DT_{U,\alpha}(T, f_i)$ if $U(T) > \alpha$.*

Proof Let $U(T) \leq \alpha$. Then $tree(mcd(T))$ is the only (U, α)-decision tree for T. Let $U(T) > \alpha$ and $\Gamma \in DT_{U,\alpha}(T)$. Then, by definition, $\Gamma = tree(f_i, a_1, \ldots, a_t,$ $\Gamma_1, \ldots, \Gamma_t)$ where $f_i \in E(T)$ and $\{a_1, \ldots, a_t\} = E(T, f_i)$. Using the fact that $\Gamma \in DT_{U,\alpha}(T)$ it is not difficult to show that $\Gamma_j \in DT_{U,\alpha}(T(f_i, a_j))$ for $j = 1, \ldots, t$. From here it follows that $\Gamma \in DT_{U,\alpha}(T, f_i)$ for $f_i \in E(T)$. Therefore $DT_{U,\alpha}(T) \subseteq \bigcup_{f_i \in E(T)} DT_{U,\alpha}(T, f_i)$.

Let, for some $f_i \in E(T)$, $\Gamma \in DT_{U,\alpha}(T, f_i)$. Then $\Gamma = tree(f_i, a_1, \ldots, a_t,$ $\Gamma_1, \ldots, \Gamma_t)$ where $\{a_1, \ldots, a_t\} = E(T, f_i)$, and $\Gamma_j \in DT_{U,\alpha}(T(f_i, a_j))$ for $j = 1, \ldots, t$. Using these facts it is not difficult to show that $\Gamma \in DT_{U,\alpha}(T)$. Therefore $\bigcup_{f_i \in E(T)} DT_{U,\alpha}(T, f_i) \subseteq DT_{U,\alpha}(T)$. $\qquad\square$

Let G be a bundle-preserving subgraph of the graph $\Delta_{U,\alpha}(T)$. For each non-terminal node Θ of the graph G, we denote by $E_G(\Theta)$ the set of attributes f_i from $E(\Theta)$ such that f_i-bundle of edges starts from Θ in G. For each terminal node Θ, $E_G(\Theta) = \emptyset$. For each node Θ of the graph G, we define the set $Tree(G, \Theta)$ of decision trees in the following way. If Θ is a terminal node of G (in this case $U(\Theta) \leq \alpha$), then $Tree(G, \Theta) = \{tree(mcd(\Theta))\}$. Let Θ be a nonterminal node of G (in this case $U(\Theta) > \alpha$), $f_i \in E_G(\Theta)$, and $E(\Theta, f_i) = \{a_1, \ldots, a_t\}$. We denote $Tree(G, \Theta, f_i) = \{tree(f_i, a_1, \ldots, a_t, \Gamma_1, \ldots, \Gamma_t) : \Gamma_j \in Tree(G, \Theta(f_i, a_j)), j = 1, \ldots, t\}$. Then

$$Tree(G, \Theta) = \bigcup_{f_i \in E_G(T)} Tree(G, \Theta, f_i) .$$

Proposition 6.4 *Let U be an uncertainty measure, $\alpha \in \mathbb{R}_+$, and T be a decision table. Then, for any node Θ of the graph $\Delta_{U,\alpha}(T)$, the following equality holds:* $Tree(\Delta_{U,\alpha}(T), \Theta) = DT_{U,\alpha}(\Theta)$.

Proof We prove this statement by induction on nodes of $\Delta_{U,\alpha}(T)$. Let Θ be a terminal node of $\Delta_{U,\alpha}(T)$. Then $Tree(\Delta_{U,\alpha}(T), \Theta) = \{tree(mcd(\Theta))\} = DT_{U,\alpha}(\Theta)$. Let now Θ be a nonterminal node of $\Delta_{U,\alpha}(T)$, and let us assume that, for any $f_i \in E(\Theta)$ and $\alpha_j \in E(\Theta, f_i)$, $Tree(\Delta_{U,\alpha}(T), \Theta(f_i, a_j)) = DT_{U,\alpha}(\Theta(f_i, a_j))$. Then, for any $f_i \in E(\Theta)$, we have $Tree(\Delta_{U,\alpha}(T), \Theta, f_i) = DT_{U,\alpha}(\Theta, f_i)$. Using Proposition 6.3, we obtain $Tree(\Delta_{U,\alpha}(T), \Theta) = DT_{U,\alpha}(\Theta)$. $\qquad\square$

6.1.3 Cardinality of the Set $Tree(G, \Theta)$

Let U be an uncertainty measure, $\alpha \in \mathbb{R}_+$, T be a decision table with n attributes $f_1, \ldots, f_n$, and G be a bundle-preserving subgraph of the graph $\Delta_{U,\alpha}(T)$. We describe now an algorithm which counts, for each node Θ of the graph G, the cardinality $C(\Theta)$ of the set $Tree(G, \Theta)$, and returns the number $C(T) = |Tree(G, T)|$.

Algorithm $\mathcal{A}_4$ (counting of decision trees).

Input: A bundle-preserving subgraph G of the graph $\Delta_{U,\alpha}(T)$ for some decision table T, uncertainty measure U, and number $\alpha \in \mathbb{R}_+$.

Output: The number $|Tree(G, T)|$.

1. If all nodes of the graph G are processed then return the number $C(T)$ and finish the work of the algorithm. Otherwise, choose a node Θ of the graph G which is not processed yet and which is either a terminal node of G or a nonterminal node of G such that, for each $f_i \in E_G(T)$ and $a_j \in E(\Theta, f_i)$, the node $\Theta(f_i, a_j)$ is processed.
2. If Θ is a terminal node then set $C(\Theta) = 1$, mark the node Θ as processed, and proceed to step 1.
3. If Θ is a nonterminal node then set

$$C(\Theta) = \sum_{f_i \in E_G(\Theta)} \prod_{a_j \in E(\Theta, f_i)} C(\Theta(f_i, a_j)),$$

mark the node Θ as processed, and proceed to step 1.

Proposition 6.5 *Let U be an uncertainty measure, $\alpha \in \mathbb{R}_+$, T be a decision table with n attributes $f_1, \ldots, f_n$, and G be a bundle-preserving subgraph of the graph $\Delta_{U,\alpha}(T)$. Then the algorithm $\mathscr{A}_4$ returns the number $|Tree(G, T)|$ and makes at most*

$$nL(G)range(T)$$

operations of addition and multiplication.

Proof We prove by induction on the nodes of G that $C(\Theta) = |Tree(G, \Theta)|$ for each node Θ of G. Let Θ be a terminal node of G. Then $Tree(G, \Theta) = \{tree(mcd(\Theta))\}$ and $|Tree(G, \Theta)| = 1$. Therefore the considered statement holds for Θ. Let now Θ be a nonterminal node of G such that the considered statement holds for its children. By definition, $Tree(G, \Theta) = \bigcup_{f_i \in E_G(\Theta)} Tree(G, \Theta, f_i)$, and, for $f_i \in E_G(\Theta)$,

$$Tree(G, \Theta, f_i) = \{tree(f_i, a_1, \ldots, a_t, \Gamma_1, \ldots, \Gamma_t)$$
$$: \Gamma_j \in Tree(G, \Theta(f_i, a_j)), j = 1, \ldots, t\}$$

where $\{a_1, \ldots, a_t\} = E(\Theta, f_i)$. One can show that, for any $f_i \in E_G(\Theta)$,

$$|Tree(G, \Theta, f_i)| = \prod_{a_j \in E(\Theta, f_i)} |Tree(G, \Theta(f_i, a_j))|,$$

and $|Tree(G, \Theta)| = \sum_{f_i \in E_G(\Theta)} |Tree(G, \Theta, f_i)|$. By the induction hypothesis,

$$C(\Theta(f_i, a_j)) = |Tree(G, \Theta(f_i, a_j))|$$

for any $f_i \in E_G(\Theta)$ and $a_j \in E(\Theta, f_i)$. Therefore $C(\Theta) = |Tree(G, \Theta)|$. Hence, the considered statement holds. From here it follows that $C(T) = |Tree(G, T)|$, and the Algorithm $\mathscr{A}_4$ returns the cardinality of the set $Tree(G, T)$.

It is easy to see that the considered algorithm makes at most $nL(G)range(T)$ operations of addition and multiplication where $L(G)$ is the number of nodes in the graph G and $range(T) = \max\{|E(T, f_i)| : i = 1, \ldots, n\}$.   □

Proposition 6.6 *Let $\mathscr{U}$ be a restricted information system. Then the algorithm $\mathscr{A}_4$ has polynomial time complexity for decision tables from $\mathscr{T}(\mathscr{U})$ depending on the number of conditional attributes in these tables.*

Proof All operations made by the algorithm $\mathscr{A}_4$ are basic numerical operations. From Proposition 6.5 it follows that the number of these operations is bounded from above by a polynomial depending on the size of input table T and on the number of separable subtables of T.

According to Proposition 5.4, the algorithm $\mathscr{A}_4$ has polynomial time complexity for decision tables from $\mathscr{T}(\mathscr{U})$ depending on the number of conditional attributes in these tables.   □

6.1.4 $U^{\max}$-Decision Trees for T

Let U be an uncertainty measure, T be a decision table, and Γ be a decision tree for T. We denote by $V_t(\Gamma)$ the set of terminal nodes of Γ, and by $V_n(\Gamma)$ we denote the set of nonterminal nodes of Γ. We denote by $U^{\max}(T, \Gamma)$ the number $\max\{U(T(v)) : v \in V_t(\Gamma)\}$ which will be interpreted as a kind of uncertainty of Γ.

Let $V_n(\Gamma) \neq \emptyset$. For a nonterminal node v of Γ, we denote by Γ^v a decision tree for T which is obtained from Γ by removal all nodes and edges of the subtree $\Gamma(v)$ with the exception of v. Instead of an attribute we attach to v the number $mcd(T(v))$. The operation of transformation of Γ into Γ^v will be called *pruning of the subtree* $\Gamma(v)$. Let $v_1, \ldots, v_m$ be nonterminal nodes of Γ such that, for any $i, j \in \{1, \ldots, m\}$, $i \neq j$, the node v_i does not belong to the subtree $\Gamma(v_j)$. We denote by $\Gamma^{v_1 \cdots v_m}$ the tree obtained from Γ by sequential pruning of subtrees $\Gamma(v_1), \ldots, \Gamma(v_m)$.

A decision tree Γ for T is called a $U^{\max}$-*decision tree for T* if either $V_n(\Gamma) = \emptyset$ or $U^{\max}(T, \Gamma^v) > U^{\max}(T, \Gamma)$ for any node $v \in V_n(\Gamma)$. We denote by $DT_U^{\max}(T)$ the set of $U^{\max}$-decision trees for T. These trees can be considered as irredundant decision trees for T relative to the uncertainty $U^{\max}$ of decision trees. According to the definition, $DT_U^{\max}(T) \subseteq DT(T)$.

Proposition 6.7 *Let U be an uncertainty measure and T be a decision table. Then $DT_U^{\max}(T) = \bigcup_{\alpha \in \mathbb{R}_+} DT_{U,\alpha}(T)$.*

Proof From Propositions 6.1 and 6.3 it follows that $tree(mcd(T))$ is the only decision tree with one node in $DT_U^{\max}(T)$ and the only decision tree with one node in $\bigcup_{\alpha \in \mathbb{R}_+} DT_{U,\alpha}(T)$. We now consider decision trees Γ from $DT(T)$ that contain more than one node.

Let $\Gamma \in DT_U^{\max}(T)$ and $\alpha = U^{\max}(T, \Gamma)$. Let $v \in V_n(\Gamma)$. Since $U^{\max}(T, \Gamma^v) > \alpha$, we have $U(T(v)) > \alpha$. So, for each terminal node v of Γ, $U(T(v)) \leq \alpha$ and, for each nonterminal node v of Γ, $U(T(v)) > \alpha$. Taking into account that $\Gamma \in DT(T)$, we obtain $\Gamma \in DT_{U,\alpha}(T)$.

Let $\Gamma \in DT_{U,\alpha}(T)$ for some $\alpha \in \mathbb{R}_+$. Then $U^{\max}(T, \Gamma) \leq \alpha$ and, for each node $v \in V_n(\Gamma)$, the inequality $U(T(v)) > \alpha$ holds. Therefore, for each node $v \in V_n(\Gamma)$, we have $U^{\max}(T, \Gamma^v) > U^{\max}(T, \Gamma)$, and $\Gamma \in DT_U^{\max}(T)$.   □

Proposition 6.8 *Let U be an uncertainty measure, T be a decision table, and $\Gamma \in DT(T) \setminus DT_U^{\max}(T)$. Then there exist nodes $v_1, \ldots, v_m \in V_n(\Gamma)$ such that, for any $i, j \in \{1, \ldots, m\}$, $i \neq j$, the node v_i does not belong to the subtree $\Gamma(v_j)$, $\Gamma^{v_1 \ldots v_m} \in DT_U^{\max}(T)$ and $U^{\max}(T, \Gamma^{v_1 \ldots v_m}) \leq U^{\max}(T, \Gamma)$.*

Proof Let $\alpha = U^{\max}(T, \Gamma)$ and $v_1, \ldots, v_m$ be all nonterminal nodes v in Γ such that $U(T(v)) \leq \alpha$ and there is no a nonterminal node v' such that $v \neq v'$, $U(T(v')) \leq \alpha$ and v belongs to $\Gamma(v')$. One can show that, for any $i, j \in \{1, \ldots, m\}$, $i \neq j$, the node v_i does not belong to the subtree $\Gamma(v_j)$, $\Gamma^{v_1 \ldots v_m} \in DT_U^{\max}(T)$ and $U^{\max}(T, \Gamma^{v_1 \ldots v_m}) \leq \alpha$.   □

6.1.5 U^{sum}-Decision Trees

Let U be an uncertainty measure, T be a decision table, and Γ be a decision tree for T. We denote by $U^{\mathrm{sum}}(T, \Gamma)$ the number $\sum_{v \in V_t(\Gamma)} U(T(v))$ which will be interpreted as a kind of uncertainty of Γ.

A decision tree Γ for T is called a U^{sum}-*decision tree for T* if either $V_n(\Gamma) = \emptyset$ or $U^{\mathrm{sum}}(T, \Gamma^v) > U^{\mathrm{sum}}(T, \Gamma)$ for any node $v \in V_n(\Gamma)$. We denote by $DT_U^{\mathrm{sum}}(T)$ the set of U^{sum}-decision trees for T. These trees can be considered as irredundant decision trees for T relative to the uncertainty U^{sum} of decision trees. According to the definition, $DT_U^{\mathrm{sum}}(T) \subseteq DT(T)$.

Proposition 6.9 *Let U be an uncertainty measure, T be a decision table, and $\Gamma \in DT(T) \setminus DT_U^{\mathrm{sum}}(T)$. Then there exist nodes $v_1, \ldots, v_m \in V_n(\Gamma)$ such that, for any $i, j \in \{1, \ldots, m\}$, $i \neq j$, the node v_i does not belong to the subtree $\Gamma(v_j)$, $\Gamma^{v_1 \ldots v_m} \in DT_U^{\mathrm{sum}}(T)$ and $U^{\mathrm{sum}}(T, \Gamma^{v_1 \ldots v_m}) \leq U^{\mathrm{sum}}(T, \Gamma)$.*

Proof Let $v_1, \ldots, v_m$ be all nonterminal nodes v in Γ such that

$$U(T(v)) \leq \sum_{u \in V_t(\Gamma(v))} U(T(u))$$

and there is no a nonterminal node v' such that $U(T(v')) \leq \sum_{u \in V_t(\Gamma(v'))} U(T(u))$, $v \neq v'$, and v belongs to $\Gamma(v')$. It is clear that, for any $i, j \in \{1, \ldots, m\}$, $i \neq j$, the node v_i does not belong to the subtree $\Gamma(v_j)$ and, for any nonterminal node v of the tree $\Gamma' = \Gamma^{v_1 \ldots v_m}$, $\sum_{u \in V_t(\Gamma'(v))} U(T(u)) \leq \sum_{u \in V_t(\Gamma(v))} U(T(u))$. Using this fact, one can show that $\Gamma^{v_1 \ldots v_m} \in DT_U^{\mathrm{sum}}(T)$ and $U^{\mathrm{sum}}(T, \Gamma^{v_1 \ldots v_m}) \leq U^{\mathrm{sum}}(T, \Gamma)$.   □

6.1.6 Cost Functions for Decision Trees

Let n be a natural number. We consider a partial order $\leq$ on the set $\mathbb{R}^n$: $(x_1, \ldots, x_n) \leq$ $(y_1, \ldots, y_n)$ if $x_1 \leq y_1, \ldots, x_n \leq y_n$. A function $g : \mathbb{R}^n_+ \to \mathbb{R}_+$ is called *increasing* if $g(x) \leq g(y)$ for any $x, y \in \mathbb{R}^n_+$ such that $x \leq y$. A function $g : \mathbb{R}^n_+ \to \mathbb{R}_+$ is called *strictly increasing* if $g(x) < g(y)$ for any $x, y \in \mathbb{R}^n_+$ such that $x \leq y$ and $x \neq y$. If g is strictly increasing then, evidently, g is increasing. For example $\max(x_1, x_2)$ is increasing and $x_1 + x_2$ is strictly increasing.

Let f be a function from $\mathbb{R}^2_+$ to $\mathbb{R}_+$. We can extend f to a function with arbitrary number of variables in the following way: $f(x_1) = x_1$ and, if $n > 2$ then $f(x_1, \ldots, x_n) = f(f(x_1, \ldots, x_{n-1}), x_n)$.

Proposition 6.10 *Let f be an increasing function from $\mathbb{R}^2_+$ to $\mathbb{R}_+$. Then, for any natural n, the function $f(x_1, \ldots, x_n)$ is increasing.*

Proof We prove the considered statement by induction on n. If $n = 1$ then, evidently, the function $f(x_1) = x_1$ is increasing. We know that the function $f(x_1, x_2)$ is increasing. Let us assume that, for some $n \geq 2$, the function $f(x_1, \ldots, x_n)$ is increasing. We now show that the function $f(x_1, \ldots, x_{n+1})$ is increasing. Let $x = (x_1, \ldots, x_{n+1})$, $y = (y_1, \ldots, y_{n+1}) \in \mathbb{R}^{n+1}_+$ and $x \leq y$. By induction hypothesis,

$$f(x_1, \ldots, x_n) \leq f(y_1, \ldots, y_n) \ .$$

Since $x_{n+1} \leq y_{n+1}$, we obtain $f(f(x_1, \ldots, x_n), x_{n+1}) \leq f(f(y_1, \ldots, y_n), y_{n+1})$. Therefore the function $f(x_1, \ldots, x_{n+1})$ is increasing. □

Proposition 6.11 *Let f be a strictly increasing function from $\mathbb{R}^2_+$ to $\mathbb{R}_+$. Then, for any natural n, the function $f(x_1, \ldots, x_n)$ is strictly increasing.*

Proof We prove the considered statement by induction on n. If $n = 1$ then the function $f(x_1) = x_1$ is strictly increasing. We know that the function $f(x_1, x_2)$ is strictly increasing. Let us assume that, for some $n \geq 2$, the function $f(x_1, \ldots, x_n)$ is strictly increasing. We now show that the function $f(x_1, \ldots, x_{n+1})$ is strictly increasing. Let $x = (x_1, \ldots, x_{n+1})$, $y = (y_1, \ldots, y_{n+1}) \in \mathbb{R}^{n+1}_+$, $x \leq y$ and $x \neq y$. It is clear that $x' = (x_1, \ldots, x_n) \leq y' = (y_1, \ldots, y_n)$. If $x' \neq y'$ then, by induction hypothesis, $f(x_1, \ldots, x_n) < f(y_1, \ldots, y_n)$ and, since $x_{n+1} \leq y_{n+1}$, $f(f(x_1, \ldots, x_n), x_{n+1}) <$ $f(f(y_1, \ldots, y_n), y_{n+1})$. Let now $x' = y'$. Then $x_{n+1} < y_{n+1}$ and

$$f(f(x_1, \ldots, x_n), x_{n+1}) < f(f(y_1, \ldots, y_n), y_{n+1}) \ .$$

Therefore the function $f(x_1, \ldots, x_{n+1})$ is strictly increasing. □

A *cost function for decision trees* is a function $\psi(T, \Gamma)$ which is defined on pairs decision table T and a decision tree Γ for T, and has values from $\mathbb{R}_+$. The function ψ is given by three functions $\psi^0 : \mathscr{T} \to \mathbb{R}_+$, $F : \mathbb{R}^2_+ \to \mathbb{R}_+$, and $w : \mathscr{T} \to \mathbb{R}_+$.

The value of $\psi(T, \Gamma)$ is defined by induction:

- If $\Gamma = tree(mcd(T))$ then $\psi(T, \Gamma) = \psi^0(T)$.
- If $\Gamma = tree(f_i, a_1, \ldots, a_t, \Gamma_1, \ldots, \Gamma_t)$ then

$$\psi(T, \Gamma) = F(\psi(T(f_i, a_1), \Gamma_1), \ldots, \psi(T(f_i, a_t), \Gamma_t)) + w(T) .$$

The cost function ψ is called *increasing* if F is an increasing function. The cost function ψ is called *strictly increasing* if F is a strictly increasing function. The cost function ψ is called *integral* if, for any $T \in \mathscr{T}$ and any $x, y \in \omega = \{0, 1, 2, \ldots\}$, $\psi^0(T) \in \omega$, $F(x, y) \in \omega$, and $w(T) \in \omega$.

We now consider examples of cost functions for decision trees:

- *Depth* $h(T, \Gamma) = h(\Gamma)$ of a decision tree Γ for a decision table T is the maximum length of a path in Γ from the root to a terminal node. For this cost function, $\psi^0(T) = 0$, $F(x, y) = \max(x, y)$, and $w(T) = 1$. This is an increasing integral cost function.
- *Total path length* $tpl(T, \Gamma)$ of a decision tree Γ for a decision table T is equal to $\sum_{r \in Row(T)} l_\Gamma(r)$ where $Row(T)$ is the set of rows of T, and $l_\Gamma(r)$ is the length of a path in Γ from the root to a terminal node v such that the row r belongs to $T_\Gamma(v)$. For this cost function, $\psi^0(T) = 0$, $F(x, y) = x + y$, and $w(T) = N(T)$. This is a strictly increasing integral cost function. Let T be a nonempty decision table. The value $tpl(T, \Gamma)/N(T)$ is the *average depth* $h_{avg}(T, \Gamma)$ of a decision tree Γ for a decision table T.
- *Number of nodes* $L(T, \Gamma) = L(\Gamma)$ of a decision tree Γ for a decision table T. For this cost function, $\psi^0(T) = 1$, $F(x, y) = x + y$, and $w(T) = 1$. This is a strictly increasing integral cost function.
- *Number of nonterminal nodes* $L_n(T, \Gamma) = L_n(\Gamma)$ of a decision tree Γ for a decision table T. For this cost function, $\psi^0(T) = 0$, $F(x, y) = x + y$, and $w(T) = 1$. This is a strictly increasing integral cost function.
- *Number of terminal nodes* $L_t(T, \Gamma) = L_t(\Gamma)$ of a decision tree Γ for a decision table T. For this cost function, $\psi^0(T) = 1$, $F(x, y) = x + y$, and $w(T) = 0$. This is a strictly increasing integral cost function.

For each of the considered cost functions, corresponding functions $\psi^0(T)$ and $w(T)$ have polynomial time complexity depending on the size of decision tables.

Note that the functions $U^{sum}(T, \Gamma)$ and $U^{max}(T, \Gamma)$ where U is an uncertainty measure can be represented in the form of cost functions for decision trees:

- For $U^{max}(T, \Gamma)$, $\psi^0(T) = U(T)$, $F(x, y) = \max(x, y)$, and $w(T) = 0$.
- For $U^{sum}(T, \Gamma)$, $\psi^0(T) = U(T)$, $F(x, y) = x + y$, and $w(T) = 0$.

We will say that a cost function ψ is *bounded* if, for any decision table T and any decision tree Γ for T, $\psi(T, tree(mcd(T))) \leq \psi(T, \Gamma)$.

Lemma 6.1 *The cost functions h, tpl, L, L_t, and L_n are bounded cost functions for decision trees.*

Proof If T is a degenerate decision table then there is only one decision tree for T which is equal to $tree(mcd(T))$. Let T be a nondegenerate decision table, $\Gamma_0 = tree(mcd(T))$, and Γ be a decision tree for T such that $\Gamma \neq \Gamma_0$. Since Γ is a decision tree for T different from Γ_0, the root of Γ is a nonterminal node, and there are at least two edges starting from the root. From here it follows that $h(\Gamma) \geq 1$, $tpl(T, \Gamma) \geq 2$, $L(\Gamma) \geq 3$, $L_n(\Gamma) \geq 1$, and $L_t(\Gamma) \geq 2$. Since $h(\Gamma_0) = 0$, $tpl(T, \Gamma_0) = 0$, $L(\Gamma_0) = 1$, $L_n(\Gamma_0) = 0$, and $L_t(\Gamma_0) = 1$, we have h, tpl, L, L_t, and L_n are bounded cost functions. $\square$

Let ψ be an integral cost function and T be a decision table. We denote

$$ub(\psi, T) = \max\{\psi(\Theta, \Gamma) : \Theta \in SEP(T), \Gamma \in DT(\Theta)\} .$$

It is clear that, for any separable subtable Θ of T and for any decision tree Γ for Θ, $\psi(\Theta, \Gamma) \in \{0, \dots, ub(\psi, T)\}$.

Lemma 6.2 *Let T be a decision table with n conditional attributes. Then $ub(h, T) \leq n$, $ub(tpl, T) \leq nN(T)$, $ub(L, T) \leq 2N(T)$, $ub(L_n, T) \leq N(T)$, and $ub(L_t, T) \leq N(T)$.*

Proof Let Θ be a separable subtable of the table T and Γ be a decision tree for Θ. From the definition of a decision tree for a decision table it follows that, for any node v of Γ, the subtable $\Theta_\Gamma(v)$ is nonempty, for any nonterminal node v of Γ, at least two edges start in v, and, in any path from the root to a terminal node of Γ, nonterminal nodes are labeled with pairwise different attributes. Therefore $h(\Theta, \Gamma) \leq n$, $L_t(\Theta, \Gamma) \leq N(\Theta) \leq N(T)$, $tpl(\Theta, \Gamma) \leq nN(\Theta) \leq nN(T)$, $L_n(\Theta, \Gamma) \leq N(\Theta) \leq N(T)$, and $L(\Theta, \Gamma) \leq 2N(\Theta) \leq 2N(T)$. $\square$

Proposition 6.12 *Let U be an uncertainty measure, ψ be bounded and increasing cost function for decision trees, T be a decision table, and $\Gamma \in DT(T) \setminus DT_U^{\max}(T)$. Then there is a decision tree $\Gamma' \in DT_U^{\max}(T)$ such that $U^{\max}(T, \Gamma') \leq U^{\max}(T, \Gamma)$ and $\psi(T, \Gamma') \leq \psi(T, \Gamma)$.*

Proof By Proposition 6.8, there are nodes $v_1, \dots, v_m \in V_n(\Gamma)$ such that the decision tree $\Gamma' = \Gamma^{v_1 \dots v_m}$ belongs to $DT_U^{\max}(T)$ and $U^{\max}(T, \Gamma') \leq U^{\max}(T, \Gamma)$. Since ψ is bounded and increasing, $\psi(T, \Gamma') \leq \psi(T, \Gamma)$. $\square$

Proposition 6.13 *Let U be an uncertainty measure, ψ be bounded and increasing cost function for decision trees, T be a decision table, and $\Gamma \in DT(T) \setminus DT_U^{\text{sum}}(T)$. Then there is a decision tree $\Gamma' \in DT_U^{\text{sum}}(T)$ such that $U^{\text{sum}}(T, \Gamma') \leq U^{\text{sum}}(T, \Gamma)$ and $\psi(T, \Gamma') \leq \psi(T, \Gamma)$.*

Proof By Proposition 6.9, there are nodes $v_1, \dots, v_m \in V_n(\Gamma)$ such that the decision tree $\Gamma' = \Gamma^{v_1 \dots v_m}$ belongs to $DT_U^{\text{sum}}(T)$ and $U^{\text{sum}}(T, \Gamma') \leq U^{\text{sum}}(T, \Gamma)$. Since ψ is bounded and increasing, $\psi(T, \Gamma') \leq \psi(T, \Gamma)$. $\square$

6.2 Different Kinds of Inhibitory Trees

In this section, we discuss main notions related to inhibitory trees.

Let T be a nondegenerate decision table with n conditional attributes $f_1, \ldots, f_n$. *An inhibitory tree over* T is a finite directed tree with root in which nonterminal nodes are labeled with attributes from the set $\{f_1, \ldots, f_n\}$, terminal nodes are labeled with expressions of the kind $\neq t$, $t \in \omega$, and, for each nonterminal node, edges starting from this node are labeled with pairwise different numbers from ω.

Let Γ be an inhibitory tree over T and v be a node of Γ. We denote by $\Gamma(v)$ the subtree of Γ for which v is the root. We define now a subtable $T(v) = T_\Gamma(v)$ of the table T. If v is the root of Γ then $T(v) = T$. Let v be not the root of Γ and $v_1, e_1, \ldots, v_m, e_m, v_{m+1} = v$ be the directed path from the root of Γ to v in which nodes $v_1, \ldots, v_m$ are labeled with attributes $f_{i_1}, \ldots, f_{i_m}$ and edges $e_1, \ldots, e_m$ are labeled with numbers $a_1, \ldots, a_m$, respectively. Then $T(v) = T(f_{i_1}, a_1) \ldots (f_{i_m}, a_m)$.

An inhibitory tree Γ over T is called an *inhibitory tree for* T if, for any node v of Γ,

- If $T(v)$ is an incomplete table relative to T then v is a terminal node labeled with $\neq lcd(T, T(v))$.
- If $T(v)$ is not an incomplete table relative to T then either v is a terminal node labeled with $\neq lcd(T, T(v))$, or v is a nonterminal node which is labeled with an attribute $f_i \in E(T(v))$ and, if $E(T(v), f_i) = \{a_1, \ldots, a_t\}$, then t edges start from the node v that are labeled with $a_1, \ldots, a_t$, respectively.

We denote by $IT(T)$ the set of inhibitory trees for T.

A *cost function for inhibitory trees* is a function $\psi(T, \Gamma)$ which is defined on pairs decision table T and an inhibitory tree Γ for T, and has values from $\mathbb{R}_+$.

We now consider examples of cost functions for inhibitory trees:

- *Depth* $h(T, \Gamma) = h(\Gamma)$ of an inhibitory tree Γ for a decision table T is the maximum length of a path in Γ from the root to a terminal node.
- *Total path length* $tpl(T, \Gamma)$ of an inhibitory tree Γ for a decision table T is equal to $\sum_{r \in Row(T)} l_\Gamma(r)$ where $Row(T)$ is the set of rows of T, and $l_\Gamma(r)$ is the length of a path in Γ from the root to a terminal node v such that the row r belongs to $T_\Gamma(v)$. Let T be a nonempty decision table. The value $tpl(T, \Gamma)/N(T)$ is the *average depth* $h_{avg}(T, \Gamma)$ of an inhibitory tree Γ for a decision table T.
- *Number of nodes* $L(T, \Gamma) = L(\Gamma)$ of an inhibitory tree Γ for a decision table T.
- *Number of nonterminal nodes* $L_n(T, \Gamma) = L_n(\Gamma)$ of an inhibitory tree Γ for a decision table T.
- *Number of terminal nodes* $L_t(T, \Gamma) = L_t(\Gamma)$ of an inhibitory tree Γ for a decision table T.

Let W be a completeness measure and $\alpha \in \mathbb{R}_+$. An inhibitory tree Γ over T is called a (W, α)-*inhibitory tree for* T if, for any node v of Γ,

- If $W(T, T_\Gamma(v)) \leq \alpha$ then v is a terminal node which is labeled with $\neq lcd(T, T(v))$.

- If $W(T, T_\Gamma(v)) > \alpha$ then v is a nonterminal node labeled with an attribute $f_i \in E(T_\Gamma(v))$, and if $E(T_\Gamma(v), f_i) = \{a_1, \ldots, a_t\}$, then t edges start from the node v which are labeled with $a_1, \ldots, a_t$, respectively.

We denote by $IT_{W,\alpha}(T)$ the set of (W, α)-inhibitory trees for T. It is easy to show that $IT_{W,\alpha}(T) \subseteq IT(T)$.

Let W be a completeness measure, T be a decision table, and Γ be an inhibitory tree for T. We denote by $V_t(\Gamma)$ the set of terminal nodes of Γ, and by $V_n(\Gamma)$ we denote the set of nonterminal nodes of Γ. We denote by $W^{\max}(T, \Gamma)$ the number $\max\{W(T, T_\Gamma(v)) : v \in V_t(\Gamma)\}$ which will be interpreted as a kind of completeness of Γ. We denote by $W^{\mathrm{sum}}(T, \Gamma)$ the number $\sum_{v \in V_t(\Gamma)} W(T, T_\Gamma(v))$ which will be also interpreted as a kind of completeness of Γ.

Let Γ be an inhibitory tree for T and $V_n(\Gamma) \neq \emptyset$. For a nonterminal node v of Γ, we denote by Γ^v an inhibitory tree for T which is obtained from Γ by removal all nodes and edges of the subtree $\Gamma(v)$ with the exception of v. Instead of an attribute we attach to v the expression $\neq lcd(T, T_\Gamma(v))$. The operation of transformation of Γ into Γ^v will be called *pruning of the subtree* $\Gamma(v)$.

An inhibitory tree Γ for T is called a $W^{\max}$-*inhibitory tree for* T if either $V_n(\Gamma) = \emptyset$ or $W^{\max}(T, \Gamma^v) > W^{\max}(T, \Gamma)$ for any node $v \in V_n(\Gamma)$. We denote by $IT_W^{\max}(T)$ the set of $W^{\max}$-inhibitory trees for T. These trees can be considered as irredundant inhibitory trees for T relative to the completeness $W^{\max}$ of inhibitory trees. According to the definition, $IT_W^{\max}(T) \subseteq IT(T)$.

An inhibitory tree Γ for T is called a W^{sum}-*inhibitory tree for* T if either $V_n(\Gamma) = \emptyset$ or $W^{\mathrm{sum}}(T, \Gamma^v) > W^{\mathrm{sum}}(T, \Gamma)$ for any node $v \in V_n(\Gamma)$. We denote by $IT_W^{\mathrm{sum}}(T)$ the set of W^{sum}-inhibitory trees for T. These trees can be considered as irredundant inhibitory trees for T relative to the completeness W^{sum} of inhibitory trees. According to the definition, $IT_W^{\mathrm{sum}}(T) \subseteq IT(T)$.

Let Γ_1 be an inhibitory tree over T and Γ_2 be a decision tree over T^C. We denote by Γ_1^+ a decision tree over T^C obtained from Γ_1 by changing expressions attached to terminal nodes: if a terminal node in Γ_1 is labeled with $\neq t$ then the corresponding node in Γ_1^+ is labeled with t. We denote by Γ_2^- an inhibitory tree over T obtained from Γ_2 by changing expressions attached to terminal nodes: if a terminal node in Γ_2 is labeled with t then the corresponding node in Γ_2^- is labeled with $\neq t$. It is clear that $(\Gamma_1^+)^- = \Gamma_1$ and $(\Gamma_2^-)^+ = \Gamma_2$. Let A be a set of inhibitory trees over T. We denote $A^+ = \{\Gamma^+ : \Gamma \in A\}$. Let B be a set of decision trees over T^C. We denote $B^- = \{\Gamma^- : \Gamma \in B\}$.

Proposition 6.14 *Let T be a nondegenerate decision table with attributes $f_1, \ldots, f_n$, Γ be a decision tree over T^C, W be a completeness measure, U be an uncertainty measure, W and U be dual, and $\alpha \in \mathbb{R}_+$. Then*

1. *$W^H(T, \Gamma^-) = U^H(T^C, \Gamma)$ for any $H \in \{\mathrm{sum}, \max\}$;*
2. *$\Gamma \in DT(T^C)$ if and only if $\Gamma^- \in IT(T)$;*
3. *For any $H \in \{\mathrm{sum}, \max\}$, $\Gamma \in DT_U^H(T^C)$ if and only if $\Gamma^- \in IT_W^H(T)$;*
4. *$\Gamma \in DT_{U,\alpha}(T^C)$ if and only if $\Gamma^- \in IT_{W,\alpha}(T)$;*
5. *If $\Gamma \in DT(T^C)$ then $\psi(T, \Gamma^-) = \psi(T^C, \Gamma)$ for any $\psi \in \{h, tpl, L, L_n, L_t\}$.*

Proof Let v be a node of Γ (we keep the notation v for corresponding to v node in Γ^-). Then there exists a word $\alpha \in Word(T)$ such that $T_{\Gamma^-}(v) = T\alpha$ and $T_{\Gamma}^C(v) = T^C\alpha$. Since W and U are dual, $W(T, T\alpha) = U(T^C\alpha)$. From here it follows that the statement 1 of the proposition holds. By Lemma 5.1, $mcd(T^C\alpha) = lcd(T, T\alpha)$. It is clear that $E(T\alpha) = E(T^C\alpha)$ and $E(T\alpha, f_i) = E(T^C\alpha, f_i)$ for any $f_i \in \{f_1, \ldots, f_n\}$. Since W and U are dual, $W(T, T\alpha) = 0$ if and only if $U(T^C\alpha) = 0$, i.e., $T\alpha$ is incomplete if and only if $T^C\alpha$ is degenerate. Using these facts it is not difficult to show that statements 2, 3, and 4 hold. If $\Gamma \in DT(T)$ and $\psi \in \{h, tpl, L, L_n, L_t\}$, then it is easy to show that $\psi(T, \Gamma^-) = \psi(T^C, \Gamma)$, i.e., the statement 5 holds. $\square$

Corollary 6.1 *Let T be a nondegenerate decision table, W be a completeness measure, U be an uncertainty measure, W and U be dual, and $\alpha \in \mathbb{R}_+$. Then*

1. *$IT(T) = DT(T^C)^-$;*
2. *For any $H \in \{sum, max\}$, $IT_W^H(T) = DT_U^H(T^C)^-$;*
3. *$IT_{W,\alpha}(T) = DT_{U,\alpha}(T^C)^-$.*

Proof Let $\Gamma \in DT(T^C)$. Then, by the statement 2 of Proposition 6.14, $\Gamma^- \in IT(T)$. Therefore $DT(T^C)^- \subseteq IT(T)$. Let $\Gamma \in IT(T)$. Then Γ^+ is a decision tree over T^C and $\Gamma = (\Gamma^+)^-$. By the statement 2 of Proposition 6.14, $\Gamma^+ \in DT(T^C)$. Therefore $\Gamma \in DT(T^C)^-$ and $DT(T^C)^- \supseteq IT(T)$. Hence the statement 1 of the corollary holds. The statements 2 and 3 can be proven in a similar way. $\square$

Let G be a bundle-preserving subgraph of the graph $\Delta_{U,\alpha}(T^C)$. The algorithm $\mathscr{A}_4$ described in Sect. 6.1.3 allows us to find the cardinality of the set $Tree(G, T^C)$ containing some (U, α)-decision trees for T^C and, in the same time, the cardinality of the set $Tree(G, T^C)^-$ containing some (W, α)-inhibitory trees for T. It can be, for example, the set of all (W, α)-inhibitory trees for T with minimum number of nodes. We will discuss this and other interesting cases in Sect. 7.2 devoted to multi-stage optimization of inhibitory trees.

References

1. Breiman, L., Friedman, J.H., Olshen, R.A., Stone, C.J.: Classification and Regression Trees. Wadsworth and Brooks, Monterey (1984)
2. Moshkov, M.: Time complexity of decision trees. In: Peters, J.F., Skowron, A. (eds.) Trans. Rough Sets III. Lecture Notes in Computer Science, vol. 3400, pp. 244–459. Springer, Berlin (2005)
3. Rokach, L., Maimon, O.: Data Mining with Decision Trees: Theory and Applications. World Scientific Publishing, River Edge (2008)

Chapter 7
Multi-stage Optimization of Decision and Inhibitory Trees

In this chapter, we consider multi-stage optimization of decision (Sect. 7.1) and inhibitory (Sect. 7.2) trees relative to a sequence of cost functions, and two applications of this technique: study of decision trees for sorting problem (Sect. 7.3) and study of totally optimal (simultaneously optimal relative to a number of cost functions) decision and inhibitory trees for modified decision tables from the UCI ML Repository (Sect. 7.4).

We study an algorithm for optimization of decision trees for decision tables with many-valued decisions relative to both increasing and strictly increasing cost functions. After that, we extend the obtained results to the case of inhibitory trees.

We consider two applications of this technique. The first application is the study of decision trees for sorting problem. Usually, the problem under consideration is to sort n pairwise different elements $x_1, \ldots, x_n$ from a linearly ordered set [1, 9, 12]. In this first case, there is only one permutation $(p_1, \ldots, p_n)$ of the set $\{1, \ldots, n\}$ such that $x_{p_1} < \cdots < x_{p_n}$, and each comparison $x_i : x_j$ of two elements has only two results: $x_i < x_j$ and $x_i > x_j$. We consider also the second case when it is possible to have equal elements in the sequence $x_1, \ldots, x_n$. In this case, each comparison $x_i : x_j$ of two elements can have three possible results $x_i < x_j$, $x_i = x_j$, and $x_i > x_j$, it can be more than one permutation $(p_1, ..., p_n)$ such that $x_{p_1} \leq \cdots \leq x_{p_n}$, and we should find one of these permutations. For $n = 2, \ldots, 7$, we compare the minimum depth, average depth, and number of nodes of decision trees for the considered two cases. The minimum depth is the same for both cases, the minimum average depth for $n > 2$ is less for the second case, and the minimum number of nodes is less for the first case.

The second application is the study of totally optimal (simultaneously optimal relative to a number of cost functions) decision and inhibitory trees for modified decision tables from the UCI ML Repository. We consider various combinations of depth, average depth, and number of nodes and show that totally optimal trees exist in many cases.

© Springer Nature Switzerland AG 2020

F. Alsolami et al., *Decision and Inhibitory Trees and Rules for Decision Tables with Many-valued Decisions*, Intelligent Systems Reference Library 156, https://doi.org/10.1007/978-3-030-12854-8_7

Multi-stage optimization approach was created for decision tables with single-valued decisions in [3, 11] and developed later in the book [2]. The first results in this direction for decision tables with many-valued decisions were obtained in [6]. Totally optimal decision trees for Boolean functions were studied in [7].

Note that the paper [4] contains the most part of theoretical and experimental results considered in this chapter.

7.1 Multi-stage Optimization of Decision Trees

In this section, we discuss how to optimize (U, α)-decision trees represented by a bundle-preserving subgraph of the graph $\Delta_{U,\alpha}(T)$ relative to a cost function for decision trees. We explain also possibilities of multi-stage optimization of decision trees for different cost functions, and consider the notion of a totally optimal decision tree relative to a number of cost functions. This is a decision tree which is optimal simultaneously for each of the considered cost functions.

Let ψ be an increasing cost function for decision trees given by the triple of functions ψ^0, F and w, U be an uncertainty measure, $\alpha \in \mathbb{R}_+$, T be a decision table with n conditional attributes $f_1, \ldots, f_n$, and G be a bundle-preserving subgraph of the graph $\Delta_{U,\alpha}(T)$.

In Sect. 6.1.2, for each nonterminal node Θ of the graph G, we denoted by $E_G(\Theta)$ the set of attributes f_i from $E(\Theta)$ such that f_i-bundle of edges starts from Θ in G. For each node Θ of the graph G, we defined a set $Tree(G, \Theta)$ of (U, α)-decision trees for Θ in the following way. If Θ is a terminal node of G, then $Tree(G, \Theta) = \{tree(mcd(\Theta))\}$. Let Θ be a nonterminal node of G, $f_i \in E_G(\Theta)$, and $E(\Theta, f_i) = \{a_1, \ldots, a_t\}$. Then $Tree(G, \Theta, f_i) = \{tree(f_i, a_1, \ldots, a_t, \Gamma_1, \ldots, \Gamma_t) : \Gamma_j \in Tree(G, \Theta(f_i, a_j)), j = 1, \ldots, t\}$ and $Tree(G, \Theta) = \bigcup_{f_i \in E_G(T)} Tree(G, \Theta, f_i)$.

Let Θ be a node of G, $\Gamma \in Tree(G, \Theta)$, and v be a node of Γ. In Sect. 6.1.1, we defined a subtree $\Gamma(v)$ of Γ for which v is the root, and a subtable $\Theta_\Gamma(v)$ of Θ. If v is the root of Γ then $\Theta_\Gamma(v) = \Theta$. Let v be not the root of Γ and $v_1, e_1, \ldots, v_m, e_m, v_{m+1} = v$ be the directed path from the root of Γ to v in which nodes $v_1, \ldots, v_m$ are labeled with attributes $f_{i_1}, \ldots, f_{i_m}$ and edges $e_1, \ldots, e_m$ are labeled with numbers $a_1, \ldots, a_m$, respectively. Then $\Theta_\Gamma(v) = \Theta(f_{i_1}, a_1) \ldots (f_{i_m}, a_m)$. One can show that the decision tree $\Gamma(v)$ belongs to the set $Tree(G, \Theta_\Gamma(v))$.

A decision tree Γ from $Tree(G, \Theta)$ is called an *optimal decision tree for Θ relative to ψ and G* if $\psi(\Theta, \Gamma) = \min\{\psi(\Theta, \Gamma') : \Gamma' \in Tree(G, \Theta)\}$.

A decision tree Γ from $Tree(G, \Theta)$ is called a *strictly optimal decision tree for Θ relative to ψ and G* if, for any node v of Γ, the decision tree $\Gamma(v)$ is an optimal decision tree for $\Theta_\Gamma(v)$ relative to ψ and G.

We denote by $Tree_\psi^{opt}(G, \Theta)$ the set of optimal decision trees for Θ relative to ψ and G. We denote by $Tree_\psi^{s-opt}(G, \Theta)$ the set of strictly optimal decision trees for Θ relative to ψ and G. Let $\Gamma \in Tree_\psi^{opt}(G, \Theta)$ and $\Gamma =$

$tree(f_i, a_1, \ldots, a_t, \Gamma_1, \ldots, \Gamma_t)$. Then $\Gamma \in Tree_{\psi}^{s-opt}(G, \Theta)$ if and only if $\Gamma_j \in Tree_{\psi}^{s-opt}(G, \Theta(f_i, a_j))$ for $j = 1, \ldots, t$.

Proposition 7.1 *Let ψ be a strictly increasing cost function for decision trees, U be an uncertainty measure, $\alpha \in \mathbb{R}_+$, T be a decision table, and G be a bundle-preserving subgraph of the graph $\Delta_{U,\alpha}(T)$. Then, for any node Θ of the graph G, $Tree_{\psi}^{opt}(G, \Theta) = Tree_{\psi}^{s-opt}(G, \Theta)$.*

Proof It is clear that $Tree_{\psi}^{s-opt}(G, \Theta) \subseteq Tree_{\psi}^{opt}(G, \Theta)$. Let $\Gamma \in Tree_{\psi}^{opt}(G, \Theta)$ and let us assume that $\Gamma \notin Tree_{\psi}^{s-opt}(G, \Theta)$. Then there is a node v of Γ such that $\Gamma(v) \notin Tree_{\psi}^{opt}(G, \Theta_{\Gamma}(v))$. Let $\Gamma_0 \in Tree_{\psi}^{opt}(G, \Theta_{\Gamma}(v))$ and Γ' be the decision tree obtained from Γ by replacing $\Gamma(v)$ with Γ_0. One can show that $\Gamma' \in Tree(G, \Theta)$. Since ψ is strictly increasing and $\psi(\Theta_{\Gamma}(v), \Gamma_0) < \psi(\Theta_{\Gamma}(v), \Gamma(v))$, we have $\psi(\Theta, \Gamma') < \psi(\Theta, \Gamma)$. Therefore $\Gamma \notin Tree_{\psi}^{opt}(G, \Theta)$ which is impossible. Thus $Tree_{\psi}^{opt}(G, \Theta) \subseteq Tree_{\psi}^{s-opt}(G, \Theta)$. $\qquad\square$

We describe now an algorithm $\mathscr{A}_5$ (*a procedure of decision tree optimization relative to the cost function ψ*). The algorithm $\mathscr{A}_5$ attaches to each node Θ of G the number $c(\Theta) = \min\{\psi(\Theta, \Gamma) : \Gamma \in Tree(G, \Theta)\}$ and, probably, remove some f_i-bundles of edges starting from nonterminal nodes of G. As a result, we obtain a bundle-preserving subgraph G^{ψ} of the graph G. It is clear that G^{ψ} is also a bundle-preserving subgraph of the graph $\Delta_{U,\alpha}(T)$.

Algorithm $\mathscr{A}_5$ (procedure of decision tree optimization).
Input: A bundle-preserving subgraph G of the graph $\Delta_{U,\alpha}(T)$ for some decision table T, uncertainty measure U, and number $\alpha \in \mathbb{R}_+$, and an increasing cost function ψ for decision trees given by the triple of functions ψ^0, F, and w.
Output: The bundle-preserving subgraph G^{ψ} of the graph G.

1. If all nodes of the graph G are processed then return the obtained graph as G^{ψ} and finish the work of the algorithm. Otherwise, choose a node Θ of the graph G which is not processed yet and which is either a terminal node of G or a nonterminal node of G for which all children are processed.
2. If Θ is a terminal node then set $c(\Theta) = \psi^0(\Theta)$, mark node Θ as processed and proceed to step 1.
3. If Θ is a nonterminal node then, for each $f_i \in E_G(\Theta)$, compute the value $c(\Theta, f_i) = F(c(\Theta(f_i, a_1)), \ldots, c(\Theta(f_i, a_t))) + w(\Theta)$ where

$$\{a_1, \ldots, a_t\} = E(\Theta, f_i)$$

and set $c(\Theta) = \min\{c(\Theta, f_i) : f_i \in E_G(\Theta)\}$. Remove all f_i-bundles of edges starting from Θ for which $c(\Theta) < c(\Theta, f_i)$. Mark the node Θ as processed and proceed to step 1.

Proposition 7.2 *Let G be a bundle-preserving subgraph of the graph $\Delta_{U,\alpha}(T)$ for some decision table T with n conditional attributes $f_1, \ldots, f_n$, uncertainty measure*

U, and number $\alpha \in \mathbb{R}_+$, and ψ be an increasing cost function for decision trees given by the triple of functions ψ^0, F, and w. Then, to construct the graph G^{ψ}, the algorithm $\mathcal{A}_5$ makes

$$O(nL(G)range(T))$$

elementary operations (computations of F, w, ψ^0, comparisons, and additions).

Proof In each terminal node of the graph G, the algorithm $\mathcal{A}_5$ computes the value of ψ^0. In each nonterminal node of G, the algorithm $\mathcal{A}_5$ computes the value of F (as function with two variables) at most $range(T)n$ times, where $range(T) = \max\{|E(T, f_i)| : i = 1, \ldots, n\}$, and the value of w at most n times, makes at most n additions and at most $2n$ comparisons. Therefore the algorithm $\mathcal{A}_5$ makes

$$O(nL(G)range(T))$$

elementary operations.   □

Proposition 7.3 *Let $\psi \in \{h, tpl, L, L_n, L_t\}$ and $\mathcal{U}$ be a restricted information system. Then the algorithm $\mathcal{A}_5$ has polynomial time complexity for decision tables from $\mathcal{T}(\mathcal{U})$ depending on the number of conditional attributes in these tables.*

Proof Since $\psi \in \{h, tpl, L, L_n, L_t\}$, ψ^0 is a constant, F is either $\max(x, y)$ or $x + y$, and w is either a constant or $N(T)$. Therefore the elementary operations used by the algorithm $\mathcal{A}_5$ are either basic numerical operations or computations of numerical parameters of decision tables which have polynomial time complexity depending on the size of decision tables. From Proposition 7.2 it follows that the number of elementary operations is bounded from above by a polynomial depending on the size of input table T and on the number of separable subtables of T.

According to Proposition 5.4, the algorithm $\mathcal{A}_5$ has polynomial time complexity for decision tables from $\mathcal{T}(\mathcal{U})$ depending on the number of conditional attributes in these tables.   □

For any node Θ of the graph G and for any $f_i \in E_G(\Theta)$, we denote $\psi_G(\Theta) = \min\{\psi(\Theta, \Gamma) : \Gamma \in Tree(G, \Theta)\}$ and

$$\psi_G(\Theta, f_i) = \min\{\psi(\Theta, \Gamma) : \Gamma \in Tree(G, \Theta, f_i)\}.$$

Lemma 7.1 *Let G be a bundle-preserving subgraph of the graph $\Delta_{U,\alpha}(T)$ for some decision table T with n conditional attributes, uncertainty measure U, and number $\alpha \in \mathbb{R}_+$, and ψ be an increasing cost function for decision trees given by the triple of functions ψ^0, F, and w. Then, for any node Θ of the graph G and for any attribute $f_i \in E_G(\Theta)$, the algorithm $\mathcal{A}_5$ computes values $c(\Theta) = \psi_G(\Theta)$ and $c(\Theta, f_i) = \psi_G(\Theta, f_i)$.*

Proof We prove the considered statement by induction on the nodes of the graph G. Let Θ be a terminal node of G. Then $Tree(G, \Theta) = \{tree(mcd(\Theta))\}$ and

$\psi_G(\Theta) = \psi^0(\Theta)$. Therefore $c(\Theta) = \psi_G(\Theta)$. Since $E_G(\Theta) = \emptyset$, the considered statement holds for Θ.

Let now Θ be a nonterminal node of G such that the considered statement holds for each node $\Theta(f_i, a_j)$ with $f_i \in E_G(\Theta)$ and $a_j \in E(\Theta, f_i)$. By definition, $Tree(G, \Theta) = \bigcup_{f_i \in E_G(\Theta)} Tree(G, \Theta, f_i)$ and, for each $f_i \in E_G(\Theta)$, $Tree(G, \Theta, f_i) = \{tree(f_i, a_1, \dots, a_t, \Gamma_1, \dots, \Gamma_t): \Gamma_j \in Tree(G, \Theta(f_i, a_j)), j = 1, \dots, t\}$ where

$$\{a_1, \dots, a_t\} = E(\Theta, f_i) \, .$$

Since ψ is an increasing cost function,

$$\psi_G(\Theta, f_i) = F(\psi_G(\Theta(f_i, a_1)), \dots, \psi_G(\Theta(f_i, a_t))) + w(\Theta)$$

where $\{a_1, \dots, a_t\} = E(\Theta, f_i)$. It is clear that $\psi_G(\Theta) = \min\{\psi_G(\Theta, f_i): f_i \in E_G(\Theta)\}$. By the induction hypothesis, $\psi_G(\Theta(f_i, a_j)) = c(\Theta(f_i, a_j))$ for each $f_i \in E_G(\Theta)$ and $a_j \in E(\Theta, f_i)$. Therefore $c(\Theta, f_i) = \psi_G(\Theta, f_i)$ for each $f_i \in E_G(\Theta)$, and $c(\Theta) = \psi_G(\Theta)$. $\qquad\qquad\square$

Theorem 7.1 *Let ψ be an increasing cost function for decision trees, U be an uncertainty measure, $\alpha \in \mathbb{R}_+$, T be a decision table, and G be a bundle-preserving subgraph of the graph $\Delta_{U,\alpha}(T)$. Then, for any node Θ of the graph G^ψ, the following equality holds: $Tree(G^\psi, \Theta) = Tree_\psi^{s-opt}(G, \Theta)$.*

Proof We prove the considered statement by induction on nodes of G^ψ. We use Lemma 7.1 which shows that, for any node Θ of the graph G and for any $f_i \in E_G(\Theta)$, $c(\Theta) = \psi_G(\Theta)$ and $c(\Theta, f_i) = \psi_G(\Theta, f_i)$.

Let Θ be a terminal node of G^ψ. Then $Tree(G^\psi, \Theta) = \{tree(mcd(\Theta))\}$. It is clear that $Tree(G^\psi, \Theta) = Tree_\psi^{s-opt}(G, \Theta)$. Therefore the considered statement holds for Θ.

Let Θ be a nonterminal node of G^ψ such that the considered statement holds for each node $\Theta(f_i, a_j)$ with $f_i \in E_G(\Theta)$ and $a_j \in E(\Theta, f_i)$. By definition,

$$Tree(G^\psi, \Theta) = \bigcup_{f_i \in E_{G^\psi}(\Theta)} Tree(G^\psi, \Theta, f_i)$$

and, for each $f_i \in E_{G^\psi}(\Theta)$, $Tree(G^\psi, \Theta, f_i) = \{tree(f_i, a_1, \dots, a_t, \Gamma_1, \dots, \Gamma_t): \Gamma_j \in Tree(G^\psi, \Theta(f_i, a_j)), j = 1, \dots, t\}$, where $\{a_1, \dots, a_t\} = E(\Theta, f_i)$.

We know that $E_{G^\psi}(\Theta) = \{f_i: f_i \in E_G(\Theta), \psi_G(\Theta, f_i) = \psi_G(\Theta)\}$. Let $f_i \in E_{G^\psi}(\Theta)$ and $\Gamma \in Tree(G^\psi, \Theta, f_i)$. Then $\Gamma = tree(f_i, a_1, \dots, a_t, \Gamma_1, \dots, \Gamma_t)$, where

$$\{a_1, \dots, a_t\} = E(\Theta, f_i)$$

and $\Gamma_j \in Tree(G^\psi, \Theta(f_i, a_j))$ for $j = 1, \dots, t$. According to the induction hypothesis, $Tree(G^\psi, \Theta(f_i, a_j)) = Tree_\psi^{s-opt}(G, \Theta(f_i, a_j))$ and $\Gamma_j \in Tree_\psi^{s-opt}(G^\psi, \Theta(f_i, a_j))$ for $j = 1, \dots, t$. In particular, $\psi(\Theta(f_i, a_j), \Gamma_j) = \psi_G(\Theta(f_i, a_j))$ for $j = 1, \dots, t$. Since $\psi_G(\Theta, f_i) = \psi_G(\Theta)$, we have

$F(\psi_G(\Theta(f_i, a_1)), \ldots, \psi_G(\Theta(f_i, a_t))) + w(\Theta) = \psi_G(\Theta)$ and $\psi(\Theta, \Gamma) = \psi_G(\Theta)$. Therefore $\Gamma \in Tree_\psi^{opt}(G, \Theta)$, $\Gamma \in Tree_\psi^{s-opt}(G, \Theta)$ and $Tree(G^\psi, \Theta) \subseteq Tree_\psi^{s-opt}(G, \Theta)$.

Let $\Gamma \in Tree_\psi^{s-opt}(G, \Theta)$. Since Θ is a nonterminal node, Γ can be represented in the form $\Gamma = tree(f_i, a_1, \ldots, a_t, \Gamma_1, \ldots, \Gamma_t)$, where $f_i \in E_G(\Theta)$, $\{a_1, \ldots, a_t\} = E(\Theta, f_i)$, and $\Gamma_j \in Tree_\psi^{s-opt}(G, \Theta(f_i, a_j))$ for $j = 1, \ldots, t$. Since

$$\Gamma \in Tree_\psi^{s-opt}(G, \Theta) ,$$

$\psi_G(\Theta, f_i) = \psi_G(\Theta)$ and $f_i \in E_{G^\psi}(T)$. According to the induction hypothesis,

$$Tree(G^\psi, \Theta(f_i, a_j)) = Tree_\psi^{s-opt}(G, \Theta(f_i, a_j))$$

for $j = 1, \ldots, t$. Therefore $\Gamma \in Tree(G^\psi, \Theta, f_i) \subseteq Tree(G^\psi, \Theta)$. As a result, we have $Tree_\psi^{s-opt}(G, \Theta) \subseteq Tree(G^\psi, \Theta)$. □

Corollary 7.1 *Let ψ be a strictly increasing cost function, U be an uncertainty measure, $\alpha \in \mathbb{R}_+$, T be a decision table, and G be a bundle-preserving subgraph of the graph $\Delta_{U,\alpha}(T)$. Then, for any node Θ of the graph G^ψ, $Tree(G^\psi, \Theta) = Tree_\psi^{opt}(G, \Theta)$.*

This corollary follows immediately from Proposition 7.1 and Theorem 7.1.

We can make multi-stage optimization of (U, α)-decision trees for T relative to a sequence of strictly increasing cost functions $\psi_1, \psi_2, \ldots$. We begin from the graph $G = \Delta_{U,\alpha}(T)$ and apply to it the procedure of optimization relative to the cost function ψ_1 (the algorithm $\mathcal{A}_5$). As a result, we obtain a bundle-preserving subgraph G^{ψ_1} of the graph G.

By Proposition 6.4, the set $Tree(G, T)$ is equal to the set $DT_{U,\alpha}(T)$ of all (U, α)-decision trees for T. Using Corollary 7.1, we obtain that the set $Tree(G^{\psi_1}, T)$ coincides with the set $Tree_{\psi_1}^{opt}(G, T)$ of all decision trees from $Tree(G, T)$ which have minimum cost relative to ψ_1 among all trees from the set $Tree(G, T)$. Next we apply to G^{ψ_1} the procedure of optimization relative to the cost function ψ_2. As a result, we obtain a bundle-preserving subgraph G^{ψ_1, ψ_2} of the graph G^{ψ_1} (and of the graph $G = \Delta_\alpha(T)$). By Corollary 7.1, the set $Tree(G^{\psi_1, \psi_2}, T)$ coincides with the set $Tree_{\psi_2}^{opt}(G^{\psi_1}, T)$ of all decision trees from $Tree(G^{\psi_1}, T)$ which have minimum cost relative to ψ_2 among all trees from $Tree(G^{\psi_1}, T)$, etc.

If one of the cost functions ψ_i is increasing and not strictly increasing then the set $Tree(G^{\psi_1, \ldots, \psi_i}, T)$ coincides with the set $Tree_{\psi_i}^{s-opt}(G^{\psi_1, \ldots, \psi_{i-1}}, T)$ which is a subset of the set of all decision trees from $Tree(G^{\psi_1, \ldots, \psi_{i-1}}, T)$ that have minimum cost relative to ψ_i among all trees from $Tree(G^{\psi_1, \ldots, \psi_{i-1}}, T)$.

For a cost function ψ, we denote $\psi^{U,\alpha}(T) = \min\{\psi(T, \Gamma) : \Gamma \in DT_{U,\alpha}(T)\}$, i.e., $\psi^{U,\alpha}(T)$ is the minimum cost of a (U, α)-decision tree for T relative to the cost function ψ. Let $\psi_1, \ldots, \psi_m$ be cost functions and $m \geq 2$. A (U, α)-decision tree Γ for T is called a *totally optimal (U, α)-decision tree for T relative to the cost* *functions* $\psi_1, \ldots, \psi_m$ if $\psi_1(T, \Gamma) = \psi_1^{U,\alpha}(T), \ldots, \psi_m(T, \Gamma) = \psi_m^{U,\alpha}(T)$, i.e., Γ is optimal relative to $\psi_1, \ldots, \psi_m$ simultaneously.

Assume that $\psi_1, \ldots, \psi_{m-1}$ are strictly increasing cost functions and ψ_m is increasing or strictly increasing. We now describe how to recognize the existence of a (U, α)-decision tree for T which is a totally optimal (U, α)-decision tree for T relative to the cost functions $\psi_1, \ldots, \psi_m$.

First, we construct the graph $G = \Delta_{U,\alpha}(T)$ using the algorithm $\mathscr{A}_1$. For $i = 1, \ldots, m$, we apply to G the procedure of optimization relative to ψ_i (the algorithm $\mathscr{A}_5$). As a result, we obtain for $i = 1, \ldots, m$ the graph G^{ψ_i} and the number $\psi_i^{U,\alpha}(T)$ attached to the node T of G^{ψ_i}. Next, we apply to G sequentially the procedures of optimization relative to the cost functions $\psi_1, \ldots, \psi_m$. As a result, we obtain graphs $G^{\psi_1}, G^{\psi_1,\psi_2}, \ldots, G^{\psi_1,\ldots,\psi_m}$ and numbers $\varphi_1, \varphi_2, \ldots, \varphi_m$ attached to the node T of these graphs. It is clear that $\varphi_1 = \psi_1^{U,\alpha}(T)$. For $i = 2, \ldots, m$, $\varphi_i = \min\{\psi_i(T, \Gamma) : \Gamma \in Tree(G^{\psi_1,\ldots,\psi_{i-1}}, T)\}$. One can show that a totally optimal (U, α)-decision tree for T relative to the cost functions $\psi_1, \ldots, \psi_m$ exists if and only if $\varphi_i = \psi_i^{U,\alpha}(T)$ for $i = 1, \ldots, m$.

7.2 Multi-stage Optimization of Inhibitory Trees

In this section, we consider possibilities of optimization of inhibitory trees including multi-stage optimization relative to a sequence of cost functions. We discuss also the notion of totally optimal inhibitory tree.

Let T be a nondegenerate decision table with n conditional attributes $f_1, \ldots, f_n$, T^C be the decision table complementary to T, U be an uncertainty measure, W be a completeness measure, U be dual to W, and $\alpha \in \mathbb{R}_+$.

Let G be a bundle-preserving subgraph of the graph $\Delta_{U,\alpha}(T^C)$. We correspond to the node T of G a set $Tree(G, T^C)$ of (U, α)-decision trees for T^C. If $G = \Delta_{U,\alpha}(T^C)$ then, by Proposition 6.4, the set $Tree(G, T^C)$ is equal to the set $DT_{U,\alpha}(T^C)$ of all (U, α)-decision trees for T. In general case, $Tree(G, T^C) \subseteq DT_{U,\alpha}(T^C)$.

Let us consider the set $Tree(G, T^C)^-$. By Corollary 6.1, $IT_{W,\alpha}(T) = DT_{U,\alpha}(T^C)^-$. Therefore, if $G = \Delta_{U,\alpha}(T^C)$ then $Tree(G, T^C)^- = IT_{W,\alpha}(T)$. In general case,

$$Tree(G, T^C)^- \subseteq DT_{U,\alpha}(T^C)^- = IT_{W,\alpha}(T) .$$

In Sect. 7.1, the algorithm $\mathscr{A}_5$ is considered which, for the graph G and increasing cost function ψ for decision trees, constructs a bundle-preserving subgraph G^ψ of the graph G.

Let $\psi \in \{tpl, L, L_n, L_t\}$. Then ψ is strictly increasing and, by Corollary 7.1, the set $Tree(G^\psi, T^C)$ is equal to the set of all trees from $Tree(G, T^C)$ which have minimum cost relative to ψ among all decision trees from the set $Tree(G, T^C)$. From Proposition 6.14 it follows that, for any $\Gamma \in Tree(G, T^C)$, $\psi(T^C, \Gamma) = \psi(T, \Gamma^-)$. Therefore, the set $Tree(G^\psi, T^C)^-$ is equal to the set of all trees from $Tree(G, T^C)^-$

which have minimum cost relative to ψ among all inhibitory trees from the set $Tree(G, T^C)^-$.

Let $\psi = h$. Then ψ is increasing and, by Theorem 7.1, the set $Tree(G^\psi, T^C)$ is a subset of the set of all trees from $Tree(G, T^C)$ which have minimum cost relative to ψ among all decision trees from the set $Tree(G, T^C)$. From here it follows that the set $Tree(G^\psi, T^C)^-$ is a subset of the set of all trees from $Tree(G, T^C)^-$ which have minimum cost relative to ψ among all inhibitory trees from the set $Tree(G, T^C)^-$. More accurate analysis can be done if we consider the notion of a strictly optimal inhibitory tree.

We can make multi-stage optimization of inhibitory trees relative to a sequence of cost functions $\psi_1, \psi_2, \ldots$ from $\{h, tpl, L, L_n, L_t\}$. We begin from the graph $G = \Delta_{U,\alpha}(T^C)$ and apply to it the procedure of optimization relative to the cost function ψ_1 (the algorithm $\mathscr{A}_5$). As a result, we obtain a bundle-preserving subgraph G^{ψ_1} of the graph G. The set $Tree(G, T^C)^-$ is equal to the set $IT_{W,\alpha}(T)$ of all (W, α)-inhibitory trees for T. If $\psi_1 \neq h$ then the set $Tree(G^{\psi_1}, T^C)^-$ coincides with the set of all inhibitory trees from $Tree(G, T^C)^-$ which have minimum cost relative to ψ_1 among all trees from the set $Tree(G, T^C)^-$. Next we apply to G^{ψ_1} the procedure of optimization relative to the cost function ψ_2. As a result, we obtain a bundle-preserving subgraph G^{ψ_1, ψ_2} of the graph G^{ψ_1}. If $\psi_2 \neq h$ then the set $Tree(G^{\psi_1, \psi_2}, T^C)^-$ coincides with the set of inhibitory trees from $Tree(G^{\psi_1}, T^C)^-$ which have minimum cost relative to ψ_2 among all trees from $Tree(G^{\psi_1}, T^C)^-$, etc.

If one of the cost functions ψ_i is equal to h then the set $Tree(G^{\psi_1, \ldots, \psi_i}, T^C)^-$ is a subset of the set of all inhibitory trees from $Tree(G^{\psi_1, \ldots, \psi_{i-1}}, T^C)^-$ that have minimum cost relative to h among all trees from $Tree(G^{\psi_1, \ldots, \psi_{i-1}}, T^C)^-$.

We can study also totally optimal inhibitory trees relative to various combinations of cost functions. For a cost function ψ, we denote $\psi^{W,\alpha}(T) = \min\{\psi(T, \Gamma) : \Gamma \in IT_{W,\alpha}(T)\}$, i.e., $\psi^{W,\alpha}(T)$ is the minimum cost of a (W, α)-inhibitory tree for T relative to the cost function ψ. Let $\psi_1, \ldots, \psi_m$ be cost functions and $m \geq 2$. A (W, α)-inhibitory tree Γ for T is called a *totally optimal (W, α)-inhibitory tree for T relative to the cost functions* $\psi_1, \ldots, \psi_m$ if $\psi_1(T, \Gamma) = \psi_1^{W,\alpha}(T), \ldots, \psi_m(T, \Gamma) = \psi_m^{W,\alpha}(T)$, i.e., Γ is optimal relative to $\psi_1, \ldots, \psi_m$ simultaneously.

Assume that $\psi_1, \ldots, \psi_{m-1} \in \{tpl, L, L_n, L_t\}$ and $\psi_m \in \{h, tpl, L, L_n, L_t\}$. We now describe how to recognize the existence of a (W, α)-inhibitory tree for T which is a totally optimal (W, α)-inhibitory tree for T relative to the cost functions $\psi_1, \ldots, \psi_m$.

First, we construct the graph $G = \Delta_{U,\alpha}(T^C)$ using the algorithm $\mathscr{A}_1$. For $i = 1, \ldots, m$, we apply to G the procedure of optimization relative to ψ_i (the algorithm $\mathscr{A}_5$). As a result, we obtain, for $i = 1, \ldots, m$, the graph G^{ψ_i} and the number $\psi_i^{U,\alpha}(T^C)$ attached to the node T^C of G^{ψ_i}. Next, we apply to G sequentially the procedures of optimization relative to the cost functions $\psi_1, \ldots, \psi_m$. As a result, we obtain graphs $G^{\psi_1}, G^{\psi_1, \psi_2}, \ldots, G^{\psi_1, \ldots, \psi_m}$ and numbers $\varphi_1, \varphi_2, \ldots, \varphi_m$ attached to the node T^C of these graphs. We know (see Sect. 7.1) that a totally optimal (U, α)-

Table 7.1 Minimum depth, minimum average depth, and minimum number of nodes in decision trees for decision table $T_{sort}^2(n)$, $n = 2, \ldots, 8$

n	2	3	4	5	6	7	8
$h(T_{sort}^2(n))$	1	3	5	7	10	13	16
$h_{avg}(T_{sort}^2(n))$	1	2.6667	4.6667	6.9333	9.5778	12.384	15.381
$L(T_{sort}^2(n))$	3	11	47	239	1439	10079	80639

decision tree for T^C relative to the cost functions $\psi_1, \ldots, \psi_m$ exists if and only if $\varphi_i = \psi_i^{U,\alpha}(T^C)$ for $i = 1, \ldots, m$. Using Proposition 6.14 one can show that a totally optimal (W, α)-inhibitory tree for T relative to the cost functions $\psi_1, \ldots, \psi_m$ exists if and only if a totally optimal (U, α)-decision tree for T^C relative to the cost functions $\psi_1, \ldots, \psi_m$ exists.

7.3 Decision Trees for Sorting

In theoretical investigations (see [1, 9, 12]), the problem of sorting usually means to sort a sequence of n pairwise different elements $x_1, \ldots, x_n$ from a linearly ordered set. In this case, there is only one permutation $(p_1, \ldots, p_n)$ of the set $\{1, \ldots, n\}$ such that $x_{p_1} < \cdots < x_{p_n}$, and each comparison $x_i : x_j$ of two elements has only two possible results: $x_i < x_j$ and $x_i > x_j$. Our aim is to find, for a given sequence $x_1, \ldots, x_n$, the permutation $(p_1, \ldots, p_n)$ such that $x_{p_1} < \cdots < x_{p_n}$.

For a given n, we can construct the decision table with single-valued decisions $T_{sort}^2(n)$, in which columns correspond to attributes $x_i : x_j$, $1 \le i < j \le n$, and rows are all possible tuples of values of these attributes for sequences of pairwise different elements $x_1, \ldots, x_n$. Each row is labeled with the corresponding permutation. The index 2 in the notation $T_{sort}^2(n)$ means that we consider two-valued attributes.

We denote by $h(T_{sort}^2(n))$ the minimum depth of a decision tree for the decision table $T_{sort}^2(n)$, by $h_{avg}(T_{sort}^2(n))$—the minimum average depth of a decision tree for the decision table $T_{sort}^2(n)$, and by $L(T_{sort}^2(n))$—the minimum number of nodes in a decision tree for the decision table $T_{sort}^2(n)$. Using results obtained in [1, 5, 8, 9, 12], we can fill in Table 7.1. Moreover, for $n = 2, \ldots, 8$, there exists a decision tree Γ_n^2 for the table $T_{sort}^2(n)$ such that $h(\Gamma_n^2) = h(T_{sort}^2(n))$, $h_{avg}(T_{sort}^2(n), \Gamma_n^2) = h_{avg}(T_{sort}^2(n))$, and $L(\Gamma_n^2) = L(T_{sort}^2(n))$. It means that the decision tree Γ_n^2 is totally optimal relative to the depth, average depth, and number of nodes.

We now consider the case when it is possible to have equal elements in the sequence $x_1, \ldots, x_n$. In this case, it can be more than one permutation $(p_1, \ldots, p_n)$ such that $x_{p_1} \le \cdots \le x_{p_n}$, and each comparison $x_i : x_j$ of two elements can have three possible results: $x_i < x_j$, $x_i = x_j$, and $x_i > x_j$. Our aim is to find, for a given sequence $x_1, \ldots, x_n$, a permutation $(p_1, \ldots, p_n)$ such that $x_{p_1} \le \cdots \le x_{p_n}$.

For a given n, we can construct the decision table with many-valued decisions $T_{sort}^3(n)$, in which columns correspond to attributes $x_i : x_j$, $1 \le i < j \le n$, rows are

Table 7.2 Minimum depth, minimum average depth, and minimum number of nodes in decision trees for decision table $T_{sort}^3(n)$, $n = 2, \ldots, 7$

n	2	3	4	5	6	7
$h(T_{sort}^3(n))$	1	3	5	7	10	13
$h_{avg}(T_{sort}^3(n))$	1	2.4615	4.1733	6.1479	8.3850	10.821
$L(T_{sort}^3(n))$	4	19	112	811	7024	70939

all possible tuples of values of these attributes for sequences of elements $x_1, \ldots, x_n$ which can contain equal elements. Each row is labeled with the set of corresponding permutations. The index 3 in the notation $h(T_{sort}^3(n))$ means that we consider three-valued attributes.

We denote by $h(T_{sort}^3(n))$ the minimum depth of a decision tree for the decision table $T_{sort}^3(n)$, by $h_{avg}(T_{sort}^3(n))$—the minimum average depth of a decision tree for the decision table $T_{sort}^3(n)$, and by $L(T_{sort}^3(n))$—the minimum number of nodes in a decision tree for the decision table $T_{sort}^3(n)$. The considered parameters for $n = 2, \ldots, 7$ can be found in Table 7.2. Moreover, for $n = 2, \ldots, 7$, there exists a decision tree Γ_n^3 for the table $T_{sort}^3(n)$ such that $h(\Gamma_n^3) = h(T_{sort}^2(n))$, $h_{avg}(T_{sort}^3(n), \Gamma_n^3) = h_{avg}(T_{sort}^3(n))$, and $L(\Gamma_n^3) = L(T_{sort}^3(n))$. It means that the decision tree Γ_n^3 is totally optimal relative to the depth, average depth, and number of nodes.

To obtain these results, we used multi-stage optimization of decision trees for the table $T_{sort}^3(n)$ relative to the depth, average depth, and number of nodes for $n = 1, \ldots, 7$. Note that the directed acyclic graph $\Delta(T_{sort}^3(7))$ contains 9,535,241 nodes and 265,549,158 edges.

Comparing Table 7.1 (the first case) and Table 7.2 (the second case) we obtain that, for the second case, the minimum depth is the same as for the first case, the minimum average depth is less for $n > 2$, and the minimum number of nodes is greater than for the first case.

7.4　Experimental Study of Totally Optimal Trees

We did experiments to study the existence of totally optimal decision and inhibitory trees relative to the depth, average depth, and number of nodes. Instead of the investigation of inhibitory trees for a decision table with many-valued decisions T, we studied decision trees for the table T^C complementary to T.

7.4.1　Decision Tables Used in Experiments

We took data sets (decision tables) from the UCI ML Repository [10] and remove one or more conditional attributes from them. As a result, for some tables, there are multiple rows that have equal values of conditional attributes but different decisions which are then merged into a single row labeled with the set of decisions from the group of equal rows. Before the experiment work, some preprocessing procedures are performed. An attribute is removed if it has unique value for each row. The missing value for an attribute is filled up with the most common value for that attribute.

In Table 7.3, the first column "Decision table T" refers to the name of the new decision table T (that we get after removing attributes from the data set from the UCI ML Repository), the second column "Original data set—removed attributes" refers

Table 7.3 Decision tables with many-valued decisions used in experiments

Decision table T	Original data set—removed attributes	Rows	Attr	$|D(T)|$	Spectrum #1, #2, #3, ...
CARS-1	CARS 1	432	5	4	258, 161, 13
FLAGS-4	FLAGS 1, 2, 3, 19	176	22	6	168, 8
FLAGS-5	FLAGS 1, 2, 3, 5, 15	177	21	6	166, 10, 1
FLAGS-3	FLAGS 1, 2, 3	184	23	6	178, 6
FLAGS-1	FLAGS 1	190	25	6	188, 2
LYMPH-5	LYMPHOGRAPHY 1, 13, 14, 15, 18	122	13	4	113, 9
LYMPH-4	LYMPHOGRAPHY 13, 14, 15, 18	136	14	4	132, 4
NURSERY-4	NURSERY 1, 5, 6, 7	240	4	5	97, 96, 47
NURSERY-1	NURSERY 1	4320	7	5	2858, 1460, 2
POKER 5A	POKER-HAND 1, 2, 4, 6, 8	3324	5	10	128, 1877, 1115, 198, 5, 1
POKER-5B	POKER-HAND 2, 3, 4, 6, 8	3323	5	10	130, 1850, 1137, 199, 6, 1
POKER-5C	POKER-HAND 2, 4, 6, 8, 10	1024	5	10	0, 246, 444, 286, 44, 4
ZOO-5	ZOO-DATA 2, 6, 8, 9, 13	43	11	7	40, 1, 2
ZOO-4	ZOO-DATA 2, 9, 13, 14	44	12	7	40, 4
ZOO-2	ZOO-DATA 6, 13	46	14	7	44, 2

Table 7.4 Existence of totally optimal decision and inhibitory trees for two cost functions

Decision table	Has totally optimal decision trees?			Has totally optimal inhibitory trees?		
	(h_{avg}, h)	(L, h)	(L, h_{avg})	(h_{avg}, h)	(L, h)	(L, h_{avg})
CARS-1	No	Yes	No	Yes	Yes	Yes
FLAGS-4	Yes	No	No	Yes	Yes	No
FLAGS-5	No	No	No	Yes	Yes	No
FLAGS-3	No	No	No	Yes	Yes	No
FLAGS-1	Yes	No	No	Yes	Yes	Yes
LYMPH-5	No	No	No	Yes	Yes	Yes
LYMPH-4	No	No	No	Yes	Yes	Yes
NURSERY-4	Yes	Yes	Yes	Yes	Yes	Yes
NURSERY-1	Yes	Yes	No	Yes	Yes	Yes
POKER-5A	Yes	Yes	No	Yes	No	No
POKER-5B	Yes	Yes	No	Yes	No	No
POKER-5C	Yes	Yes	Yes	Yes	Yes	Yes
ZOO-5	No	No	Yes	Yes	Yes	Yes
ZOO-4	No	Yes	No	Yes	Yes	Yes
ZOO-2	No	No	Yes	Yes	Yes	Yes

to the name of the original data set along with the indexes of attributes removed from the original data set, the column "Rows" refers to the number of rows, the column "Attr" refers to the number of conditional attributes, the column "$|D(T)|$" refers to the total number of decisions in T, and the column "Spectrum" refers to a sequence #1, #2, #3, . . ., where #i means the number of rows in T that are labeled with sets of decisions containing i decisions. For each of the considered decision tables T, $D_T(r) \neq D(T)$ for any row r of T.

7.4.2 Totally Optimal Trees Relative to Two Cost Functions

In Table 7.4, we show the results for the existence of totally optimal decision and inhibitory trees relative to two cost functions. The results are grouped according to the three pairs of cost functions: (average depth, depth), (number of nodes, depth), and (number of nodes, average depth).

For the case of decision trees, for each pair of cost functions, more than half of the considered decision tables do not have totally optimal decision trees. On the other hand, for the case of inhibitory trees, we have totally optimal inhibitory trees for all the decision tables for the pair (average depth, depth), for almost all the decision tables except two cases for the pair (number of nodes, depth), and for ten cases for the pair (number of nodes, average depth).

Table 7.5 Existence of totally optimal decision trees for three cost functions

Decision table	Non-sequential			Sequential			Has totally optimal decision trees?
	L	h_{avg}	h	L	h_{avg}	h	
CARS-1	28	1.96	4	28	2.06	4	No
FLAGS-5	152	3.6	5	152	5.93	10	No
FLAGS-4	149	3.73	6	149	6.7	11	No
FLAGS-3	151	3.64	5	151	6.6	11	No
FLAGS-1	112	2.77	5	112	3.9	7	No
LYMPH-5	56	3.65	5	56	4.72	9	No
LYMPH-4	67	3.73	5	67	4.79	7	No
NURSERY-4	9	1.33	2	9	1.33	2	Yes
NURSERY-1	117	2.127	7	117	2.134	7	No
POKER-5A	230	2.62	4	230	2.63	4	No
POKER-5B	226	2.529	5	226	2.531	5	No
POKER-5C	69	2.33	5	69	2.33	5	Yes
ZOO-5	17	2.77	4	17	2.77	6	No
ZOO-4	23	3.34	5	23	3.41	7	No
ZOO-2	13	2.72	4	13	2.72	5	No

7.4.3 Totally Optimal Trees Relative to Three Cost Functions

In Tables 7.5 and 7.6, we show the results of the experiments with multi-stage optimization for the three cost functions L, h_{avg}, and h. We listed first the values of the cost functions for non-sequential, i.e., individual cost optimization, then the values of cost functions for sequential, i.e., multi-stage optimization, and then whether the decision tables have totally optimal trees or not. A totally optimal tree exists if and only if the values after non-sequential optimization are equal to the values after sequential optimization.

We can see that, for the case of decision trees, there are only two decision tables that have totally optimal trees. On the other hand, for the case of inhibitory trees, there are ten decision tables that have totally optimal trees. Moreover, the optimal inhibitory trees have usually smaller depth, average depth, and number of nodes compared to optimal decision trees.

Table 7.6 Existence of totally optimal inhibitory trees for three cost functions

Decision table	Non-sequential			Sequential			Has totally optimal inhibitory trees?
	L	h_{avg}	h	L	h_{avg}	h	
CARS-1	14	1.47	4	14	1.47	4	Yes
FLAGS-5	17	1.9	3	17	2.43	3	No
FLAGS-4	17	1.81	3	17	2.43	3	No
FLAGS-3	17	1.83	3	17	2.45	3	No
FLAGS-1	9	1	1	9	1	1	Yes
LYMPH-5	3	1	1	3	1	1	Yes
LYMPH-4	3	1	1	3	1	1	Yes
NURSERY-4	4	1	1	4	1	1	Yes
NURSERY-1	4	1	1	4	1	1	Yes
POKER-5A	13	1	1	13	1.5	2	No
POKER-5B	13	1	1	13	1.5	2	No
POKER-5C	13	1.5	2	13	1.5	2	Yes
ZOO-5	3	1	1	3	1	1	Yes
ZOO-4	3	1	1	3	1	1	Yes
ZOO-2	3	1	1	3	1	1	Yes

References

1. AbouEisha, H., Chikalov, I., Moshkov, M.: Decision trees with minimum average depth for sorting eight elements. Discret. Appl. Math. **204**, 203–207 (2016)
2. AbouEisha, H., Amin, T., Chikalov, I., Hussain, S., Moshkov, M.: Extensions of Dynamic Programming for Combinatorial Optimization and Data Mining. Intelligent Systems Reference Library, vol. 146. Springer, Berlin (2019)
3. Alkhalid, A., Amin, T., Chikalov, I., Hussain, S., Moshkov, M., Zielosko, B.: Optimization and analysis of decision trees and rules: dynamic programming approach. Int. J. Gen. Syst. **42**(6), 614–634 (2013)
4. Azad, M., Moshkov, M.: Multi-stage optimization of decision and inhibitory trees for decision tables with many-valued decisions. Eur. J. Oper. Res. **263**(3), 910–921 (2017)
5. Césari, Y.: Questionnaire, codage et tris. Ph.D. thesis, Institut Blaise Pascal, Centre National de la Recherche (1968)
6. Chikalov, I., Moshkov, M., Zelentsova, M.: On optimization of decision trees. In: Peters, J.F., Skowron, A. (eds.) Trans. Rough Sets IV. Lecture Notes in Computer Science, vol. 3700, pp. 18–36. Springer, Berlin (2005)
7. Chikalov, I., Hussain, S., Moshkov, M.: Totally optimal decision trees for Boolean functions. Discret. Appl. Math. **215**, 1–13 (2016)
8. Knuth, D.E.: The Art of Computer Programming: Sorting and Searching, vol. 3, 2nd edn. Pearson Education, Boston (1998)
9. Kollár, L.: Optimal sorting of seven element sets. In: Gruska, J., Rovan, B., Wiedermann, J. (eds.) Mathematical Foundations of Computer Science 1986, Bratislava, Czechoslovakia, 25–

29 Aug 1986. Lecture Notes in Computer Science, vol. 233, pp. 449–457. Springer, Berlin (1986)

10. Lichman, M.: UCI Machine Learning Repository. University of California, Irvine, School of Information and Computer Sciences (2013). http://archive.ics.uci.edu/ml

11. Moshkov, M., Chikalov, I.: Consecutive optimization of decision trees concerning various complexity measures. Fundam. Inform. **61**(2), 87–96 (2004)

12. Peczarski, M.: New results in minimum-comparison sorting. Algorithmica **40**(2), 133–145 (2004)

Chapter 8
Bi-criteria Optimization Problem for Decision and Inhibitory Trees: Cost Versus Cost

In this chapter, we study bi-criteria optimization problem cost versus cost for decision (Sect. 8.1) and inhibitory (Sect. 8.2) trees. We design an algorithm which constructs the set of Pareto optimal points for bi-criteria optimization problem for decision trees, and show how the constructed set can be transformed into the graphs of functions that describe the relationships between the studied cost functions. We extend the obtained results to the case of inhibitory trees.

We consider two applications of the created methods. The first application is the comparison of 12 greedy heuristics for construction of decision and inhibitory trees (Sects. 8.3 and 8.4) as single-criterion and bi-criteria optimization algorithms. For single-criterion optimization, we not only rank the heuristics based on the cost of constructed trees, but we also find relative difference between the cost of trees constructed by heuristics and the cost of optimal trees. For bi-criteria optimization, we rank heuristics based on the minimum distance from the set of Pareto optimal points to the heuristic point coordinates of which are values of the two cost functions for the tree constructed by the heuristic.

The second application is related to knowledge representation (Sect. 8.5). When decision trees are used for knowledge representation, the usual goal is to minimize the number of nodes in the tree. However, the depth and the average depth are also important: we need to understand conjunctions of conditions corresponding to paths in the tree from the root to terminal nodes. We construct the sets of Pareto optimal points and study tradeoffs number of nodes versus depth and number of nodes versus average depth for some decision tables with many-valued decisions. We show that, at the cost of a minor increase in the number of nodes, we can decrease essentially the depth or average depth of decision trees.

Algorithms for analysis of relationships between various pairs of cost functions for decision trees and decision tables with single-valued decisions were previously discussed and presented in papers such as [5, 10, 14] and in the book [1], and implemented in the Dagger system [2]. Some initial results regarding comparison of

© Springer Nature Switzerland AG 2020
F. Alsolami et al., *Decision and Inhibitory Trees and Rules for Decision Tables with Many-valued Decisions*, Intelligent Systems Reference Library 156,
https://doi.org/10.1007/978-3-030-12854-8_8

greedy heuristics for construction of decision trees for decision tables with single-valued decisions can be found in [3, 4, 13]. Papers [6–9] contain initial results related to comparison of greedy heuristics for construction of decision trees for decision tables with many-valued decisions.

8.1   Bi-criteria Optimization Problem for Decision Trees: Cost Versus Cost

In this section, we consider an algorithm which constructs the sets of Pareto optimal points for bi-criteria optimization problems for decision trees relative to two cost functions. We also show how the constructed set of Pareto optimal points can be transformed into the graphs of functions which describe the relationships between the considered cost functions.

8.1.1   Pareto Optimal Points: Cost Versus Cost

We begin with the consideration of an algorithm for construction of the set of Pareto optimal points.

Let ψ and φ be increasing cost functions for decision trees, U be an uncertainty measure, $\alpha \in \mathbb{R}_+$, T be a decision table with n conditional attributes $f_1, \ldots, f_n$, and G be a bundle-preserving subgraph of the graph $\Delta_{U,\alpha}(T)$ (it is possible that $G = \Delta_{U,\alpha}(T)$). Interesting cases are when $G = \Delta_{U,\alpha}(T)$ or G is a result of application of the procedure of optimization of decision trees (the algorithm $\mathscr{A}_5$) relative to cost functions different from ψ and φ to the graph $\Delta_{U,\alpha}(T)$.

For each node Θ of the graph G, we denote $t_{\psi,\varphi}(G, \Theta) = \{(\psi(\Theta, \Gamma), \varphi(\Theta, \Gamma)) : \Gamma \in Tree(G, \Theta)\}$. Note that, by Proposition 6.4, if $G = \Delta_{U,\alpha}(T)$ then the set $Tree(G, \Theta)$ is equal to the set of (U, α)-decision trees for Θ. We denote by $Par(t_{\psi,\varphi}(G, \Theta))$ the set of Pareto optimal points for $t_{\psi,\varphi}(G, \Theta)$.

We now describe an algorithm $\mathscr{A}_6$ which constructs the set $Par(t_{\psi,\varphi}(G, T))$. In fact, this algorithm constructs, for each node Θ of the graph G, the set $B(\Theta) = Par(t_{\psi,\varphi}(G, \Theta))$.

Algorithm $\mathscr{A}_6$ (construction of POPs for decision trees, cost versus cost).

Input: Increasing cost functions for decision trees ψ and φ given by triples of functions ψ^0, F, w and φ^0, H, u, respectively, a decision table T with n conditional attributes $f_1, \ldots, f_n$, and a bundle-preserving subgraph G of the graph $\Delta_{U,\alpha}(T)$ where U is an uncertainty measure and $\alpha \in \mathbb{R}_+$.

Output: The set $Par(t_{\psi,\varphi}(G, T))$ of Pareto optimal points for the set of pairs $t_{\psi,\varphi}(G, T) = \{(\psi(T, \Gamma), \varphi(T, \Gamma)) : \Gamma \in Tree(G, T)\}$.

1. If all nodes in G are processed, then return the set $B(T)$. Otherwise, choose a node Θ in the graph G which is not processed yet and which is either a terminal node of G or a nonterminal node of G such that, for any $f_i \in E_G(\Theta)$ and any $a_j \in E(\Theta, f_i)$, the node $\Theta(f_i, a_j)$ is already processed, i.e., the set $B(\Theta(f_i, a_j))$ is already constructed.
2. If Θ is a terminal node, then set $B(\Theta) = \{(\psi^0(\Theta), \varphi^0(\Theta))\}$. Mark the node Θ as processed and proceed to step 1.
3. If Θ is a nonterminal node then, for each $f_i \in E_G(\Theta)$, apply the algorithm $\mathscr{A}_3$ to the functions F, H and the sets $B(\Theta(f_i, a_1)), \ldots, B(\Theta(f_i, a_t))$, where

$$\{a_1, \ldots, a_t\} = E(\Theta, f_i) .$$

Set $C(\Theta, f_i)$ the output of the algorithm $\mathscr{A}_3$ and

$$\begin{aligned} B(\Theta, f_i) &= C(\Theta, f_i) \langle ++ \rangle \{(w(\Theta), u(\Theta))\} \\ &= \{(a + w(\Theta), b + u(\Theta)) : (a, b) \in C(\Theta, f_i)\} . \end{aligned}$$

4. Construct the multiset $A(\Theta) = \bigcup_{f_i \in E_G(\Theta)} B(\Theta, f_i)$ by simple transcription of elements from the sets $B(\Theta, f_i)$, $f_i \in E_G(\Theta)$. Apply to the obtained multiset $A(\Theta)$ the algorithm $\mathscr{A}_2$ which constructs the set $Par(A(\Theta))$. Set $B(\Theta) = Par(A(\Theta))$. Mark the node Θ as processed and proceed to step 1.

Proposition 8.1 *Let ψ and φ be increasing cost functions for decision trees given by triples of functions ψ^0, F, w and φ^0, H, u, respectively, U be an uncertainty measure, $\alpha \in \mathbb{R}_+$, T be a decision table with n conditional attributes $f_1, \ldots, f_n$, and G be a bundle-preserving subgraph of the graph $\Delta_{U,\alpha}(T)$. Then, for each node Θ of the graph G, the algorithm $\mathscr{A}_6$ constructs the set $B(\Theta) = Par(t_{\psi,\varphi}(G, \Theta))$.*

Proof We prove the considered statement by induction on nodes of G. Let Θ be a terminal node of G. Then $Tree(G, \Theta) = \{tree(mcd(\Theta))\}$,

$$t_{\psi,\varphi}(G, \Theta) = Par(t_{\psi,\varphi}(G, \Theta)) = \{(\psi^0(\Theta), \varphi^0(\Theta))\} ,$$

and $B(\Theta) = Par(t_{\psi,\varphi}(G, \Theta))$.

Let Θ be a nonterminal node of G such that, for any $f_i \in E_G(\Theta)$ and any $a_j \in E(\Theta, f_i)$, the considered statement holds for the node $\Theta(f_i, a_j)$, i.e., $B(\Theta(f_i, a_j)) = Par(t_{\psi,\varphi}(G, \Theta(f_i, a_j)))$.

Let $f_i \in E_G(\Theta)$ and $E(\Theta, f_i) = \{a_1, \ldots, a_t\}$. We denote

$$\begin{aligned} P(f_i) = \{(F(b_1, \ldots, b_t) + w(\Theta), H(c_1, \ldots, c_t) + u(\Theta)) \\ : (b_j, c_j) \in t_{\psi,\varphi}(G, \Theta(f_i, a_j)), j = 1, \ldots, t\} , \end{aligned}$$

and, for $j = 1, \ldots, t$, we denote $P_j = t_{\psi,\varphi}(G, \Theta(f_i, a_j))$.

If we apply the algorithm $\mathscr{A}_3$ to the functions F, H and the sets

$$Par(P_1), \ldots, Par(P_t),$$

we obtain the set $Par(Q_t)$ where $Q_1 = P_1$, and, for $j = 2, \ldots, t$, $Q_j = Q_{j-1}\langle FH \rangle$ P_j. It is not difficult to show that $P(f_i) = Q_t \langle ++ \rangle \{(w(\Theta), u(\Theta))\} = \{(a + w(\Theta),$ $b + u(\Theta)) : (a, b) \in Q_t\}$ and $Par(P(f_i)) = Par(Q_t) \langle ++ \rangle \{(w(\Theta), u(\Theta))\}$.

According to the induction hypothesis, $B(\Theta(f_i, a_j)) = Par(P_j)$ for $j = 1, \ldots, t$. Therefore $C(\Theta, f_i) = Par(Q_t)$ and $B(\Theta, f_i) = Par(P(f_i))$.

One can show that $t_{\psi, \varphi}(G, \Theta) = \bigcup_{f_i \in E_G(\Theta)} P(f_i)$. By Lemma 5.7,

$$Par(t_{\psi, \varphi}(G, \Theta)) = Par\left(\bigcup_{f_i \in E_G(\Theta)} P(f_i)\right)$$

$$\subseteq \bigcup_{f_i \in E_G(\Theta)} Par(P(f_i)).$$

Using Lemma 5.6 we obtain $Par(t_{\psi, \varphi}(G, \Theta)) = Par\left(\bigcup_{f_i \in E_G(\Theta)} Par(P(f_i))\right)$. Since $B(\Theta, f_i) = Par(P(f_i))$ for any $f_i \in E_G(\Theta)$, $Par(t_{\psi, \varphi}(G, \Theta)) = Par(A(\Theta)) = B(\Theta)$. □

We now analyze the number of elementary operations made by the algorithm $\mathscr{A}_6$ during the construction of the set $Par(t_{\psi, \varphi}(G, T))$ for integral cost functions, for the cases when $F(x, y) = \max(x, y)$ (h is an example of such cost function) and when $F(x, y) = x + y$ (tpl, L, L_n, and L_t are examples of such cost functions).

Let us recall that $range(T) = \max\{|E(T, f_i)| : i = 1, \ldots, n\}$,

$$ub(\psi, T) = \max\{\psi(\Theta, \Gamma) : \Theta \in SEP(T), \Gamma \in DT(\Theta)\},$$

and $\psi(\Theta, \Gamma) \in \{0, 1, \ldots, ub(\psi, T)\}$ for any separable subtable Θ of T and for any decision tree Γ for Θ. Upper bounds on the value $ub(\psi, T)$ for $\psi \in \{h, tpl, L, L_n, L_t\}$ are given in Lemma 6.2: $ub(h, T) \le n$, $ub(tpl, T) \le nN(T)$, $ub(L, T) \le 2N(T)$, $ub(L_n, T) \le N(T)$, and $ub(L_t, T) \le N(T)$.

Proposition 8.2 *Let ψ and φ be increasing integral cost functions for decision trees given by triples of functions ψ^0, F, w and φ^0, H, u, respectively, $F \in \{\max(x, y),$ $x + y\}$, U be an uncertainty measure, $\alpha \in \mathbb{R}_+$, T be a decision table with n conditional attributes $f_1, \ldots, f_n$, and G be a bundle-preserving subgraph of the graph $\Delta_{U, \alpha}(T)$. Then, to construct the set $Par(t_{\psi, \varphi}(G, T))$, the algorithm $\mathscr{A}_6$ makes*

$$O(L(G)range(T)ub(\psi, T)^2 n \log(ub(\psi, T)n))$$

elementary operations (computations of F, H, w, u, ψ^0, φ^0, additions, and comparisons) if $F = \max(x, y)$, and

$$O(L(G)range(T)^2 ub(\psi, T)^2 n \log(range(T)ub(\psi, T)n))$$

elementary operations (computations of F, H, w, u, ψ^0, φ^0, additions, and comparisons) if $F = x + y$.

Proof It is clear that, for any node Θ of the graph G, $t_{\psi, \varphi}(G, \Theta)^{(1)} = \{a : (a, b) \in t_{\psi, \varphi}(G, \Theta)\} \subseteq \{0, \dots, ub(\psi, T)\}$. From Proposition 8.1 it follows that $B(\Theta) = Par(t_{\psi, \varphi}(G, \Theta))$ for any node Θ of the graph G. Therefore, for any node Θ of the graph G, $B(\Theta)^{(1)} = \{a : (a, b) \in B(\Theta)\} \subseteq \{0, \dots, ub(\psi, T)\}$.

Let $F(x, y) = \max(x, y)$. To process a terminal node Θ, the algorithm $\mathscr{A}_6$ makes two elementary operations (computations of ψ^0 and φ^0).

We now consider the processing of a nonterminal node Θ (see description of the algorithm $\mathscr{A}_6$). We know that $B(\Theta(f_i, a_j))^{(1)} \subseteq \{0, \dots, ub(\psi, T)\}$ for $j = 1, \dots, t$, and $t \leq range(T)$. From Proposition 5.7 it follows that $|C(\Theta, f_i)| \leq ub(\psi, T) + 1$, and to construct the set $C(\Theta, f_i)$, the algorithm $\mathscr{A}_3$ makes

$$O(range(T)ub(\psi, T)^2 \log ub(\psi, T))$$

elementary operations (computations of F, H and comparisons). To construct the set $B(\Theta, f_i)$ from the set $C(\Theta, f_i)$, the algorithm makes computations of w and u, and at most $2ub(\psi, T) + 2$ additions. From here it follows that, to construct the set $B(\Theta, f_i)$, the algorithm makes

$$O(range(T)ub(\psi, T)^2 \log ub(\psi, T))$$

elementary operations. It is clear that $|E_G(\Theta)| \leq n$. Therefore, to construct the set $B(\Theta, f_i)$ for each $f_i \in E_G(\Theta)$, the algorithm makes

$$O(range(T)ub(\psi, T)^2 n \log ub(\psi, T))$$

elementary operations.

Since $|C(\Theta, f_i)| \leq ub(\psi, T) + 1$, we have $|B(\Theta, f_i)| \leq ub(\psi, T) + 1$. Since

$$|E_G(\Theta)| \leq n ,$$

we have $|A(\Theta)| \leq n(ub(\psi, T) + 1)$ where $A(\Theta) = \bigcup_{f_i \in E(\Theta)} B(\Theta, f_i)$. From Proposition 5.5 it follows that, to construct the set $B(\Theta) = Par(A(\Theta))$, the algorithm $\mathscr{A}_2$ makes $O(ub(\psi, T)n \log(ub(\psi, T)n))$ comparisons. So, to process a nonterminal node Θ (to construct $B(\Theta) = Par(t_{\psi, \varphi}(G, \Theta))$) if

$$B(\Theta(f_i, a_j)) = Par(t_{\psi, \varphi}(G, \Theta(f_i, a_j)))$$

is known for all $f_i \in E_G(\Theta)$ and $a_j \in E(\Theta, f_i)$), the algorithm $\mathscr{A}_6$ makes

$$O(range(T)ub(\psi, T)^2 n \log(ub(\psi, T)n))$$

elementary operations.

To construct the set $Par(t_{\psi, \varphi}(G, T))$ for given decision table T with n conditional attributes and the graph G, it is enough to make

$$O(L(G)range(T)ub(\psi, T)^2 n \log(ub(\psi, T)n))$$

elementary operations (computations of F, H, w, u, ψ^0, φ^0, additions, and comparisons).

Let $F(x, y) = x + y$. To process a terminal node Θ, the algorithm $\mathscr{A}_6$ makes two elementary operations (computations of ψ^0 and φ^0).

We now consider the processing of a nonterminal node Θ (see description of the algorithm $\mathscr{A}_6$). We know that $B(\Theta(f_i, a_j))^{(1)} \subseteq \{0, \ldots, ub(\psi, T)\}$ for $j = 1, \ldots, t$, and $t \leq range(T)$. From Proposition 5.7 it follows that $|C(\Theta, f_i)| \leq t \times ub(\psi, T) + 1 \leq range(T)ub(\psi, T) + 1$, and to construct the set $C(\Theta, f_i)$, the algorithm $\mathscr{A}_3$ makes

$$O(range(T)^2 ub(\psi, T)^2 \log(range(T)ub(\psi, T)))$$

elementary operations (computations of F, H and comparisons). To construct the set $B(\Theta, f_i)$ from the set $C(\Theta, f_i)$, the algorithm makes computations of w and u, and at most $2range(T)ub(\psi, T) + 2$ additions. From here it follows that, to construct the set $B(\Theta, f_i)$ the algorithm makes

$$O(range(T)^2 ub(\psi, T)^2 \log(range(T)ub(\psi, T)))$$

elementary operations. It is clear that $|E_G(\Theta)| \leq n$. Therefore, to construct the set $B(\Theta, f_i)$ for each $f_i \in E_G(\Theta)$, the algorithm makes

$$O(range(T)^2 ub(\psi, T)^2 n \log(range(T)ub(\psi, T)))$$

elementary operations.

Since $|C(\Theta, f_i)| \leq range(T)ub(\psi, T) + 1$, we have

$$|B(\Theta, f_i)| \leq range(T)ub(\psi, T) + 1 .$$

Since $|E_G(\Theta)| \leq n$, we have $|A(\Theta)| \leq n(range(T)ub(\psi, T) + 1)$ where $A(\Theta) = \bigcup_{f_i \in E_G(\Theta)} B(\Theta, f_i)$. From Proposition 5.5 it follows that, to construct the set $B(\Theta) = Par(A(\Theta))$, the algorithm $\mathscr{A}_2$ makes

$$O(range(T)ub(\psi, T)n \log(range(T)ub(\psi, T)n))$$

comparisons. So, to process a nonterminal node Θ (to construct

$$B(\Theta) = Par(t_{\psi,\varphi}(G, \Theta))$$

if $B(\Theta(f_i, a_j)) = Par(t_{\psi,\varphi}(G, \Theta(f_i, a_j)))$ is known for all $f_i \in E_G(\Theta)$ and $a_j \in E(\Theta, f_i)$), the algorithm $\mathscr{A}_6$ makes

$$O(range(T)^2 ub(\psi, T)^2 n \log(range(T)ub(\psi, T)n))$$

elementary operations.

To construct the set $Par(t_{\psi,\varphi}(G, T))$ for given decision table T with n conditional attributes and the graph G, it is enough to make

$$O(L(G)range(T)^2 ub(\psi, T)^2 n \log(range(T)ub(\psi, T)n))$$

elementary operations (computations of F, H, w, u, ψ^0, φ^0, additions, and comparisons). □

Note that similar results can be obtained if $H \in \{\max(x, y), x + y\}$.

Proposition 8.3 *Let ψ and φ be cost functions for decision trees given by triples of functions ψ^0, F, w and φ^0, H, u, respectively, $\psi, \varphi \in \{h, tpl, L, L_n, L_t\}$, and $\mathscr{U}$ be a restricted information system. Then the algorithm $\mathscr{A}_6$ has polynomial time complexity for decision tables from $\mathscr{T}(\mathscr{U})$ depending on the number of conditional attributes in these tables.*

Proof Since $\psi, \varphi \in \{h, tpl, L, L_n, L_t\}$, ψ^0 and φ^0 arc constants, each of the functions F, H is either $\max(x, y)$ or $x + y$, and each of the functions w, u is either a constant or $N(T)$. From Proposition 8.2 and Lemma 6.2 it follows that, for the algorithm $\mathscr{A}_6$, the number of elementary operations (computations of F, H, w, u, ψ^0, φ^0, additions, and comparisons) is bounded from above by a polynomial depending on the size of input table T and on the number of separable subtables of T. All operations with numbers are basic ones. The computations of numerical parameters of decision tables used by the algorithm $\mathscr{A}_6$ (constants and $N(T)$) have polynomial time complexity depending on the size of decision tables.

According to Proposition 5.4, the algorithm $\mathscr{A}_6$ has polynomial time complexity for decision tables from $\mathscr{T}(\mathscr{U})$ depending on the number of conditional attributes in these tables. □

8.1.2 Relationships Between Two Cost Functions

We now show how to construct graphs describing relationships between two cost functions.

Let ψ and φ be increasing cost functions for decision trees, U be an uncertainty measure, $\alpha \in \mathbb{R}_+$, T be a decision table, and G be a bundle-preserving subgraph of the graph $\Delta_{U,\alpha}(T)$ (it is possible that $G = \Delta_{U,\alpha}(T)$).

To study relationships between cost functions ψ and φ on the set of decision trees $Tree(G, T)$ we consider partial functions $\mathcal{T}_{G,T}^{\psi,\varphi} : \mathbb{R} \to \mathbb{R}$ and $\mathcal{T}_{G,T}^{\varphi,\psi} : \mathbb{R} \to \mathbb{R}$ defined as follows:

$$\mathcal{T}_{G,T}^{\psi,\varphi}(x) = \min\{\varphi(T, \Gamma) : \Gamma \in Tree(G, T), \psi(T, \Gamma) \le x\},$$
$$\mathcal{T}_{G,T}^{\varphi,\psi}(x) = \min\{\psi(T, \Gamma) : \Gamma \in Tree(G, T), \varphi(T, \Gamma) \le x\}.$$

Let $(a_1, b_1), \ldots, (a_k, b_k)$ be the normal representation of the set $Par(t_{\psi,\varphi}(G, T))$ where $a_1 < \cdots < a_k$ and $b_1 > \cdots > b_k$. By Lemma 5.10 and Remark 5.4, for any $x \in \mathbb{R}$,

$$\mathcal{T}_{G,T}^{\psi,\varphi}(x) = \begin{cases} undefined, & x < a_1 \\ b_1, & a_1 \le x < a_2 \\ \cdots & \cdots \\ b_{k-1}, & a_{k-1} \le x < a_k \\ b_k, & a_k \le x \end{cases},$$

$$\mathcal{T}_{G,T}^{\varphi,\psi}(x) = \begin{cases} undefined, & x < b_k \\ a_k, & b_k \le x < b_{k-1} \\ \cdots & \cdots \\ a_2, & b_2 \le x < b_1 \\ a_1, & b_1 \le x \end{cases}.$$

8.2 Bi-criteria Optimization Problem for Inhibitory Trees: Cost Versus Cost

In this section, we consider bi-criteria optimization problem cost versus cost for inhibitory trees.

Let T be a nondegenerate decision table with n conditional attributes $f_1, \ldots, f_n$, T^C be the decision table complementary to T, U be an uncertainty measure, W be a completeness measure, U and W be dual, $\alpha \in \mathbb{R}_+$, and $\psi, \varphi \in \{h, tpl, L, L_n, L_t\}$.

Let G be a bundle-preserving subgraph of the graph $\Delta_{U,\alpha}(T^C)$ and

$$t_{\psi,\varphi}(G, T^C) = \{(\psi(T^C, \Gamma), \varphi(T^C, \Gamma)) : \Gamma \in Tree(G, T^C)\}.$$

In Sect. 8.1, the algorithm $\mathcal{A}_6$ is described which constructs the set

$$Par(t_{\psi,\varphi}(G, T^C))$$

of Pareto optimal points for the set of pairs $t_{\psi,\varphi}(G, T^C)$.

In Sect. 7.2, we show that $Tree(G, T^C)^- \subseteq IT_{W,\alpha}(T)$. In particular, if $G = \Delta_{U,\alpha}(T^C)$ then $Tree(G, T^C)^- = IT_{W,\alpha}(T)$. If $G = \Delta_{U,\alpha}(T^C)^\eta$ where

$$\eta \in \{tpl, L, L_t, L_n\}$$

then $Tree(G, T^C)^-$ is equal to the set of all trees from $IT_{W,\alpha}(T)$ which have minimum cost relative to η among all inhibitory trees from the set $IT_{W,\alpha}(T)$.

Denote $it_{\psi,\varphi}(G, T) = \{(\psi(T, \Gamma^-), \varphi(T, \Gamma^-)) : \Gamma^- \in Tree(G, T^C)^-\}$. From Proposition 6.14 it follows that $(\psi(T, \Gamma^-), \varphi(T, \Gamma^-)) = (\psi(T^C, \Gamma), \varphi(T^C, \Gamma))$ for any $\Gamma \in Tree(G, T^C)$. Therefore $t_{\psi,\varphi}(G, T^C) = it_{\psi,\varphi}(G, T)$ and

$$Par(t_{\psi,\varphi}(G, T^C)) = Par(it_{\psi,\varphi}(G, T)) .$$

To study relationships between cost functions ψ and φ on the set of inhibitory trees $Tree(G, T^C)^-$, we consider partial functions $\mathscr{I}\mathscr{T}_{G,T}^{\psi,\varphi} : \mathbb{R} \to \mathbb{R}$ and $\mathscr{I}\mathscr{T}_{G,T}^{\varphi,\psi} : \mathbb{R} \to \mathbb{R}$ defined as follows:

$$\mathscr{I}\mathscr{T}_{G,T}^{\psi,\varphi}(x) = \min\{\varphi(T, \Gamma^-) : \Gamma^- \in Tree(G, T^C)^-, \psi(T, \Gamma^-) \le x\},$$
$$\mathscr{I}\mathscr{T}_{G,T}^{\varphi,\psi}(x) = \min\{\psi(T, \Gamma^-) : \Gamma^- \in Tree(G, T^C)^-, \varphi(T, \Gamma^-) \le x\}.$$

Let $(a_1, b_1), \ldots, (a_k, b_k)$ be the normal representation of the set

$$Par(it_{\psi,\varphi}(G, T)) = Par(t_{\psi,\varphi}(G, T^C))$$

where $a_1 < \cdots < a_k$ and $b_1 > \cdots > b_k$. By Lemma 5.10 and Remark 5.4, for any $x \in \mathbb{R}$,

$$\mathscr{I}\mathscr{T}_{G,T}^{\psi,\varphi}(x) = \begin{cases} undefined, & x < a_1 \\ b_1, & a_1 \le x < a_2 \\ \ldots & \ldots \\ b_{k-1}, & a_{k-1} \le x < a_k \\ b_k, & a_k \le x \end{cases},$$

$$\mathscr{I}\mathscr{T}_{G,T}^{\varphi,\psi}(x) = \begin{cases} undefined, & x < b_k \\ a_k, & b_k \le x < b_{k-1} \\ \ldots & \ldots \\ a_2, & b_2 \le x < b_1 \\ a_1, & b_1 \le x \end{cases}.$$

8.3   Greedy Heuristics for Construction of Decision and Inhibitory Trees

In this section, we consider four uncertainty measures and three types of impurity functions. Each pair (uncertainty measure, type of impurity function) defines an impurity function which is used to choose attributes attached to nodes of decision trees.

The obtained 12 greedy heuristics for construction of decision trees are extended also to the construction of inhibitory trees.

8.3.1 Uncertainty Measures

Uncertainty measure U is a function from the set of nonempty decision tables with many-valued decisions to the set of real numbers such that $U(T) \geq 0$ for any decision table T, and $U(T) = 0$ if and only if T is degenerate.

Let T be a decision table with many-valued decisions, $D(T) = \{d_1, \ldots, d_m\}$ and $p_i = \frac{N_{d_i}(T)}{N(T)}$ for $i = 1, \ldots, m$. Let $d_1, \ldots, d_m$ be ordered such that $p_1 \geq \ldots \geq p_m$. For $i = 1, \ldots, m$, we denote by $N'_{d_i}(T)$ the number of rows in T such that the set of decisions attached to row contains d_i, and if $i > 1$ then this set does not contain $d_1, \ldots, d_{i-1}$, and $p'_i = \frac{N'_{d_i}(T)}{N(T)}$. We consider the following four uncertainty measures (we assume $0 \log_2 0 = 0$):

- Misclassification error: $me(T) = N(T) - N_{mcd}(T)$.
- Sorted entropy: $entSort(T) = -\sum_{i=1}^m p'_i \log_2 p'_i$ (see [12]).
- Multi-label entropy: $entML(T) = 0$, if and only if T is degenerate, otherwise, it is equal to $-\sum_{i=1}^m (p_i \log_2 p_i + q_i \log_2 q_i)$, where, $q_i = 1 - p_i$ (see [11]).
- Absence: $abs(T) = \prod_1^m q_i$, where $q_i = 1 - p_i$.

8.3.2 Impurity Types and Impurity Functions

Let $f_i \in E(T)$, and $E(T, f_i) = \{a_1, \ldots, a_t\}$. The attribute f_i divides the table T into subtables $T_1 = T(f_i, a_1), \ldots, T_t = T(f_i, a_t)$. For a fixed uncertainty measure U, we can define *impurity functions* $I(T, f_i)$ of three types which gives "impurity" of this partition:

- weighted max (*wm*): $I(T, f_i) = \max_{1 \leq j \leq t} U(T_j)N(T_j)$.
- weighted sum (*ws*): $I(T, f_i) = \sum_{j=1}^t U(T_j)N(T_j)$.
- multiplied weighted sum (*Mult_ws*): $I(T, f_i) = (\sum_{j=1}^t U(T_j)N(T_j)) \cdot \log_2 t$.

As a result, we have 12 (for each of three types and for each of four uncertainty measures) impurity functions.

8.3.3 Greedy Heuristics for Decision Tree Construction

For each impurity function I, we describe a greedy heuristic H_I which, for a given decision table T, constructs a decision tree $H_I(T)$ for the table T.

Table 8.1 Greedy heuristics

Name of heuristic	Uncertainty measure	Type of impurity function
wm_entML	entML	wm
wm_entSort	entSort	wm
wm_abs	abs	wm
wm_me	me	wm
ws_entML	entML	ws
ws_entSort	entSort	ws
ws_abs	abs	ws
ws_me	me	ws
Mult_ws_entML	entML	Mult_ws
Mult_ws_entSort	entSort	Mult_ws
Mult_ws_abs	abs	Mult_ws
Mult_ws_me	me	Mult_ws

Greedy heuristic H_I.
Input: A decision table with many-valued decisions T.
Output: Decision tree $H_I(T)$ for T.

1. Construct the tree G consisting of a single node labeled with the table T.
2. If no node of the tree G is labeled with a table then denote the tree G by $H_I(T)$, and return.
3. Otherwise, choose a node v in G which is labeled with a subtable T' of the table T.
4. Let U be the uncertainty measure for I. If $U(T') = 0$ then, instead of T', mark the node v with the common decision for T'. Return to step 2.
5. If $U(T') \neq 0$ then, for each $f_i \in E(T')$, we compute the value of the impurity function $I(T', f_i)$. Choose the attribute $f_{i_0} \in E(T')$, where i_0 is the minimum i for which $I(T', f_i)$ has the minimum value. Instead of T', mark the node v with the attribute f_{i_0}. For each $\delta \in E(T', f_i)$, add to the tree G the node v_δ and mark this node with the subtable $T'(f_{i_0}, \delta)$. Draw an edge from v to v_δ and mark this edge with δ. Return to step 2.

We consider 12 greedy heuristics (see Table 8.1). Each heuristic is specified by an uncertainty measure and a type of impurity function. The time complexities of these heuristics are polynomial depending on the size of the input decision table.

8.3.4 Greedy Heuristics for Inhibitory Tree Construction

The described heuristics can be extended to the construction of inhibitory trees. Let T be a decision table, H_I be one of the heuristics described in Table 8.1, and Γ be a decision tree for T^C constructed by the heuristic H_I. Then the tree Γ^- will be considered as an inhibitory tree for T constructed by the heuristic H_I.

8.4   Comparison of Greedy Heuristics

In this section, we compare 12 greedy heuristics for construction of decision and inhibitory trees as single-criterion and bi-criteria optimization algorithms. In the experiments, we use the same decision tables as in Chap. 7—see Table 7.3 in Sect. 7.4.1.

The first part of the experiments is the construction of the sets of Pareto optimal points (POPs) for different pairs of cost functions for both decision and inhibitory trees. We use the obtained results for comparison of heuristics as algorithms for single-criterion and bi-criteria optimization of decision and inhibitory trees.

In Table 8.2, we show the total number of POPs for each pair of cost functions, for each decision table, and for both decision and inhibitory trees. We can see that in

Table 8.2 Number of POPs

Decision table	Decision trees			Inhibitory trees		
	h, h_{avg}	h, L	h_{avg}, L	h, h_{avg}	h, L	h_{avg}, L
CARS-1	2	1	3	1	1	1
FLAGS-4	1	6	47	1	1	3
FLAGS-5	2	6	32	1	1	7
FLAGS-3	2	6	41	1	1	3
FLAGS-1	1	3	29	1	1	1
LYMPH-5	3	3	12	1	1	1
LYMPH-4	2	3	13	1	1	1
NURSERY-4	1	1	1	1	1	1
NURSERY-1	1	1	3	1	1	1
POKER-5A	1	1	2	1	2	2
POKER-5B	1	1	2	1	2	2
POKER-5C	1	1	1	1	1	1
ZOO-5	3	2	1	1	1	1
ZOO-4	2	1	2	1	1	1
ZOO-2	2	2	1	1	1	1

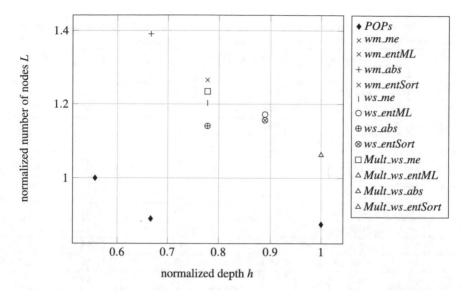

Fig. 8.1 Bi-criteria optimization of decision trees for LYMPH-5 relative to h and L—normalized POPs and normalized heuristic points

most cases there are more POPs for decision trees than for inhibitory trees. Probably, it is due to the fact that the complexity of decision trees is often larger than the complexity of inhibitory trees. Note that the number of POPs is equal to one if and only if there is a tree which is optimal relative to both cost functions simultaneously (totally optimal tree relative to these cost functions).

The second part of the experiments is the application of each heuristic to each decision table T and to its complimentary table T^C, and transformation the decision tree constructed for T^C into an inhibitory tree for T. For each of the constructed trees and for each of the cost functions, we find the cost of the tree. For each decision table, each pair of cost functions, and for each heuristic, we find a heuristic point whose coordinates are two costs of the constructed tree.

For any decision table and for any pair of cost functions, we can draw a two-dimensional graph to show the positions of POPs and corresponding heuristic points. As we might have different maximum values for the different coordinates of POPs, we normalize the coordinates of all points by the maximum values of POP coordinates. For example, we show in Fig. 8.1 the positions of normalized POPs and all 12 normalized heuristic points for the decision table LYMPH-5 for the depth and the number of nodes of decision trees. For each heuristic, we took the minimum distance from all POPs to the heuristic point. We call it "Min_Distance". Values of this parameter for the considered example can be found in Table 8.3. It is clear that the "Min_Distance" is smallest in the case of $Mult\_ws\_entML$, $Mult\_ws\_abs$, and $Mult\_ws\_entSort$ heuristics.

After getting all values of "Min_Distance" for all decision tables and pairs of cost functions, we can compare the heuristics as bi-criteria optimization algorithms

Table 8.3 The minimum distance from normalized POPs to each normalized heuristic point for decision table LYMPH-5 and cost functions depth and number of nodes

POPs		Heuristics	Heuristic points		Min_Distance
h	L		h	L	
0.56	1.00	wm_me	0.78	1.27	0.50
0.67	0.89	wm_entML	0.78	1.27	0.50
1.00	0.88	wm_abs	0.67	1.39	0.73
		wm_entSort	0.78	1.27	0.50
		ws_me	0.78	1.20	0.44
		ws_entML	0.89	1.17	0.31
		ws_abs	0.78	1.14	0.38
		ws_entSort	0.89	1.16	0.30
		Mult_ws_me	0.78	1.23	0.47
		Mult_ws_entML	1.00	1.06	0.14
		Mult_ws_abs	1.00	1.06	0.14
		Mult_ws_entSort	1.00	1.06	0.14

Table 8.4 Average of ranks of heuristics for decision trees

Heuristics	h	h_{avg}	L	h, h_{avg}	h, L	h_{avg}, L
wm_entML	**4.37**	6.33	9.13	6.50	9.17	9.07
wm_entSort	**4.80**	6.23	9.07	6.83	8.73	8.77
wm_abs	**4.80**	6.30	9.20	6.80	7.67	8.87
wm_me	5.27	7.33	9.63	6.13	9.03	8.90
ws_entML	7.30	**4.93**	**4.10**	**5.77**	**4.67**	**4.60**
ws_entSort	6.87	**4.50**	**4.33**	5.90	**4.63**	4.80
ws_abs	7.33	6.17	**3.70**	6.97	5.47	5.03
ws_me	**4.50**	**3.47**	6.47	**4.33**	6.63	6.30
Mult_ws_entML	8.23	8.13	5.03	8.77	5.37	5.70
Mult_ws_entSort	8.60	8.17	4.87	6.57	**4.83**	**4.50**
Mult_ws_abs	9.77	9.90	5.90	8.57	6.83	6.97
Mult_ws_me	6.17	6.53	6.57	**4.87**	4.97	**4.47**

by their average of ranks based on the "Min_Distance", i.e., we rank heuristics as Rank 1, Rank 2, and so on according to the value of the "Min_Distance". For single-criterion optimization, the ranking is based on the values of the cost functions for the constructed trees. After that, we consider the average of ranks among all 15 decision tables.

The results for decision and inhibitory trees are shown in Tables 8.4 and 8.5 where we list average of ranks for single- and bi-criteria optimization (total six possible combinations). The three top ranked heuristics are highlighted in bold.

Table 8.5 Average of ranks of heuristics for inhibitory trees

Heuristics	h	h_{avg}	L	h, h_{avg}	h, L	h_{avg}, L
wm_entML	6.33	7.10	8.10	7.10	8.10	8.10
wm_entSort	6.33	6.87	7.87	6.87	7.87	7.87
wm_abs	6.33	6.83	6.63	6.77	6.87	6.87
wm_me	6.33	6.87	7.87	6.87	7.87	7.87
ws_entML	**6.27**	**5.50**	**5.70**	**5.77**	**5.57**	5.97
ws_entSort	**5.93**	**5.33**	**5.53**	**5.33**	5.83	5.87
ws_abs	6.33	6.37	6.10	6.30	6.33	6.33
ws_me	6.33	6.00	6.43	6.00	6.63	6.67
Mult_ws_entML	7.37	7.23	**5.17**	7.23	6.23	**5.83**
Mult_ws_entSort	7.00	6.90	5.97	6.90	**5.23**	**5.17**
Mult_ws_abs	7.50	7.17	6.60	7.03	**5.23**	**5.30**
Mult_ws_me	**5.93**	**5.83**	6.03	**5.83**	6.23	6.17

Table 8.6 Average relative difference (ARD)

Heuristics	Decision trees			Inhibitory trees		
	h	h_{avg}	L	h	h_{avg}	L
wm_entML	**13.78**	13.42	88.99	1.67	9.29	79.96
wm_entSort	16.44	11.16	58.96	1.67	8.11	75.06
wm_abs	**15.67**	13.13	67.41	1.67	8.71	43.93
wm_me	18.44	13.72	66.41	1.67	8.11	75.06
ws_entML	31.56	**5.58**	**18.54**	2.22	**0.32**	**19.98**
ws_entSort	28.89	**5.39**	**19.20**	0.00	**0.12**	**18.80**
ws_abs	33.11	9.42	**17.29**	1.67	6.49	43.93
ws_me	**14.89**	**4.96**	31.34	1.67	4.19	52.89
Mult_ws_entML	68.33	42.00	24.09	31.11	17.09	**16.45**
Mult ws_entSort	57.44	40.50	26.23	13.33	11.59	21.94
Mult_ws_abs	82.44	53.56	33.74	17.22	14.31	33.34
Mult_ws_me	32.89	22.34	30.78	**0.00**	**0.63**	21.94

For single-criterion optimization relative to a cost function ψ, we consider also average relative difference (ARD) for each of 12 heuristics. For a heuristic and a decision table, the relative difference is $\frac{\psi_{greedy} - \psi_{opt}}{\psi_{opt}} \times 100\%$, where ψ_{greedy} is the cost of the tree constructed by the heuristic, and ψ_{opt} is the minimum cost of the tree for the considered table. ARD is the average value of relative differences among all the considered 15 decision tables. The results are shown in Table 8.6. The top three or two heuristics with minimum ARD are highlighted. For decision trees, we can see that some heuristics can give less than 5% ARD for the average depth. Interestingly, for inhibitory trees, we can see smaller or even zero ARD.

Table 8.7 Comparison of heuristics for decision trees

Heuristics	h	h'	h_{avg}	h'_{avg}	L	L'	h, h_{avg}	h, L	h_{avg}, L
wm_entML	1	1	7	7	10	12	6	12	12
wm_entSort	3.5	4	5	5	9	9	9	10	9
wm_abs	3.5	3	6	6	11	11	8	9	10
wm_me	5	5	9	8	12	10	5	11	11
ws_entML	8	7	3	3	2	2	3	2	3
ws_entSort	7	6	2	2	3	3	4	1	4
ws_abs	9	9	4	4	1	1	10	6	5
ws_me	2	2	1	1	7	7	1	7	7
Mult_ws_entML	10	11	10	11	5	4	12	5	6
Mult_ws_entSort	11	10	11	10	4	5	7	3	2
Mult_ws_abs	12	12	12	12	6	8	11	8	8
Mult_ws_me	6	8	8	9	8	6	2	4	1

Table 8.8 Comparison of heuristics for inhibitory trees

Heuristics	h	h'	h_{avg}	h'_{avg}	L	L'	h, h_{avg}	h, L	h_{avg}, L
wm_entML	6.5	5.5	10	9	12	12	11	12	12
wm_entSort	6.5	5.5	7.5	6.5	10.5	11	7.5	10.5	10.5
wm_abs	6.5	5.5	6	8	9	8	6	9	9
wm_me	6.5	5.5	7.5	6.5	10.5	10	7.5	10.5	10.5
ws_entML	3	9	2	2	3	3	2	3	5
ws_entSort	1.5	1.5	1	1	2	2	1	4	4
ws_abs	6.5	5.5	5	5	6	7	5	7	7
ws_me	6.5	5.5	4	4	7	9	4	8	8
Mult_ws_entML	11	12	12	12	1	1	12	5.5	3
Mult_ws_entSort	10	10	9	10	4	4.5	9	1.5	1
Mult_ws_abs	12	11	11	11	8	6	10	1.5	2
Mult_ws_me	1.5	1.5	3	3	5	4.5	3	5.5	6

To compare all heuristics, we make Tables 8.7 and 8.8 containing the positions of the heuristics based on the corresponding values from Tables 8.4, 8.5, and 8.6 (total nine possible combinations). In case of ties, average positions are assigned. Here, the columns "h", "h_{avg}", and "L" refer to the position based on average of ranks and "h'", "h'_{avg}", and "L'" refer to the position based on ARD.

The positions based on average of ranks are usually close to the positions based on ARD. There is no a heuristic which is good for all nine possible combinations both for decision and inhibitory trees. However, *ws_entSort* heuristic looks not bad. It is interesting that there exist heuristics which are good for bi-criteria optimization but bad for single-criterion optimization. An example is the heuristic $Mult\_ws\_me$ which is the best from the point of view of bi-criteria optimization of decision trees relative to the average depth and the number of nodes.

8.5 Decision Trees for Knowledge Representation

Decision trees are often used for knowledge representation. In this case, the usual goal is to minimize the number of nodes in the tree to make it more understandable. However, the depth and the average depth of the tree are also important: we need to understand conjunctions of conditions corresponding to paths in the tree from the root to terminal nodes.

To study tradeoffs number of nodes versus depth and number of nodes versus average depth, we can construct the sets of Pareto optimal points for corresponding bi-criteria optimization problems.

We consider two decision tables with many-valued decisions FLAGS-1 and LYMPH-5 (see Table 7.3 in Sect. 7.4.1). For these tables, we construct the sets of Pareto optimal points for bi-criteria optimization problems relative to the depth and number of nodes, and relative to the average depth and number of nodes.

For the table FLAGS-1 and bi-criteria optimization problem relative to the depth and number of nodes (see Fig. 8.2a), there are three Pareto optimal points (5,116), (6,113), and (7,112). The minimum depth of a decision tree with the minimum number of nodes 112 is equal to 7. Instead of such a tree, it is more reasonable to choose a decision tree which depth is equal to 5 and the number of nodes is equal to 116.

For the table FLAGS-1 and bi-criteria optimization problem relative to the average depth and number of nodes (see Fig. 8.2b), there are 29 Pareto optimal points. The minimum average depth of a decision tree with the minimum number of nodes 112 is equal to 3.9. Instead of such a tree, it is more reasonable to choose a decision tree which average depth is equal to 3.552632 and the number of nodes is equal to 113. There are other interesting possibilities. For example, a decision tree which average depth is equal to 2.878947 and the number of nodes is equal to 125. Note that the minimum possible average depth is equal to 2.768421 but, in this case, the minimum number of nodes is equal to 142.

For the table LYMPH-5 and bi-criteria optimization problem relative to the depth and number of nodes (see Fig. 8.2c), there are three Pareto optimal points (5,64), (6,57), and (9,56). The minimum depth of a decision tree with the minimum number of nodes 56 is equal to 9. Instead of such a tree, it is more reasonable to choose a decision tree which depth is equal to 6 and the number of nodes is equal to 57.

For the table LYMPH-5 and bi-criteria optimization problem relative to the average depth and number of nodes (see Fig. 8.2d), there are 12 Pareto optimal points. The minimum average depth of a decision tree with the minimum number of nodes 56 is equal to 4.721312. Instead of such a tree, it is more reasonable to choose a decision tree which average depth is equal to 4.172131 and the number of nodes is equal to

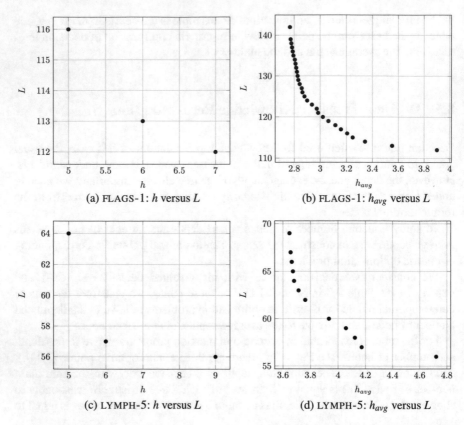

Fig. 8.2 Pareto optimal points for tables FLAGS-1 and LYMPH-5 for bi-criteria optimization problems relative to depth h and number of nodes L, and relative to average depth h_{avg} and number of nodes L

57. We even can use a decision tree with the minimum average depth 3.647541. For such a tree, the minimum number of nodes is equal to 69.

So, at the cost of a minor increase in the number of nodes, we can decrease essentially the depth or the average depth of decision trees.

References

1. AbouEisha, H., Amin, T., Chikalov, I., Hussain, S., Moshkov, M.: Extensions of Dynamic Programming for Combinatorial Optimization and Data Mining. Intelligent Systems Reference Library, vol. 146. Springer, Berlin (2019)
2. Alkhalid, A., Amin, T., Chikalov, I., Hussain, S., Moshkov, M., Zielosko, B.: Dagger: a tool for analysis and optimization of decision trees and rules. In: Ficarra, F.V.C., Kratky, A., Veltman, K.H., Ficarra, M.C., Nicol, E., Brie, M. (eds.) Computational Informatics, Social Factors and

New Information Technologies: Hypermedia Perspectives and Avant-Garde Experiencies in the Era of Communicability Expansion, pp. 29–39. Blue Herons (2011)

3. Alkhalid, A., Chikalov, I., Moshkov, M.: Comparison of greedy algorithms for decision tree construction. In: Filipe, J., Fred, A.L.N. (eds.) International Conference on Knowledge Discovery and Information Retrieval, KDIR 2011, Paris, France, 26–29 Oct 2011, pp. 438–443. SciTePress (2011)

4. Alkhalid, A., Chikalov, I., Moshkov, M.: Decision tree construction using greedy algorithms and dynamic programming – comparative study. In: Szczuka, M., Czaja, L., Skowron, A., Kacprzak, M. (eds.) 20th International Workshop on Concurrency, Specification and Programming, CS&P 2011, Pultusk, Poland, 28–30 Sept 2011, pp. 1–9. Białystok University of Technology (2011)

5. Alkhalid, A., Amin, T., Chikalov, I., Hussain, S., Moshkov, M., Zielosko, B.: Optimization and analysis of decision trees and rules: dynamic programming approach. Int. J. Gen. Syst. **42**(6), 614–634 (2013)

6. Azad, M., Moshkov, M.: Minimization of decision tree average depth for decision tables with many-valued decisions. In: Jedrzejowicz, P., Jain, L.C., Howlett, R.J., Czarnowski, I. (eds.) 18th International Conference in Knowledge Based and Intelligent Information and Engineering Systems, KES 2014, Gdynia, Poland, 15–17 Sept 2014. Procedia Computer Science, vol. 35, pp. 368–377. Elsevier (2014)

7. Azad, M., Moshkov, M.: Minimization of decision tree depth for multi-label decision tables. In: 2014 IEEE International Conference on Granular Computing, GrC 2014, Noboribetsu, Japan, 22–24 Oct 2014, pp. 7–12. IEEE Computer Society (2014)

8. Azad, M., Moshkov, M.: Minimizing size of decision trees for multi-label decision tables. In: Ganzha, M., Maciaszek, L.A., Paprzycki, M. (eds.) 2014 Federated Conference on Computer Science and Information Systems, FedCSIS 2014, Warsaw, Poland, 7–10 Sept 2014, pp. 67–74 (2014)

9. Azad, M., Moshkov, M.: Classification and optimization of decision trees for inconsistent decision tables represented as MVD tables. In: Ganzha, M., Maciaszek, L.A., Paprzycki, M. (eds.) 2015 Federated Conference on Computer Science and Information Systems, FedCSIS 2015, Lódz, Poland, 13–16 Sept 2015, pp. 31–38. IEEE (2015)

10. Chikalov, I., Hussain, S., Moshkov, M.: Relationships between average depth and number of nodes for decision trees. In: Sun, F., Li, T., Li, H. (eds.) Knowledge Engineering and Management, 7th International Conference on Intelligent Systems and Knowledge Engineering, ISKE 2012, Beijing, China, 15–17 Dec 2012. Advances in Intelligent Systems and Computing, vol. 214, pp. 519–529. Springer (2014)

11. Clare, A., King, R.D.: Knowledge discovery in multi-label phenotype data. In: Raedt, L.D., Siebes, A. (eds.) Principles of Data Mining and Knowledge Discovery, 5th European Conference, PKDD 2001, Freiburg, Germany, 3–5 Sept 2001. Lecture Notes in Computer Science, vol. 2168, pp. 42–53. Springer (2001)

12. Hüllermeier, E., Beringer, J.: Learning from ambiguously labeled examples. Intell. Data Anal. **10**(5), 419–439 (2006)

13. Hussain, S.: Greedy heuristics for minimization of number of terminal nodes in decision trees. In: 2014 IEEE International Conference on Granular Computing, GrC 2014, Noboribetsu, Japan, 22–24 Oct 2014, pp. 112–115. IEEE Computer Society (2014)

14. Hussain, S.: Relationships among various parameters for decision tree optimization. In: Faucher, C., Jain, L.C. (eds.) Innovations in Intelligent Machines-4 – Recent Advances in Knowledge Engineering. Studies in Computational Intelligence, vol. 514, pp. 393–410. Springer (2014)

Chapter 9
Bi-criteria Optimization Problem for Decision (Inhibitory) Trees: Cost Versus Uncertainty (Completeness)

In this chapter, we study bi-criteria optimization problems cost versus uncertainty for decision trees and cost versus completeness for inhibitory trees.

In Sect. 9.1, we consider an algorithm which constructs the sets of Pareto optimal points for bi-criteria optimization problems for decision trees relative to cost and uncertainty. We also show how the constructed set of Pareto optimal points can be transformed into the graphs of functions which describe the relationships between the considered cost function and uncertainty measure. Some of the initial results in this direction for decision tables with single-valued decisions were obtained in [1–3].

In Sect. 9.2, we generalize the obtained results to the case of inhibitory trees and, in Sect. 9.3, we consider illustrative examples. The created tools allow us to understand complexity versus accuracy trade-off for decision and inhibitory trees and to choose appropriate trees.

9.1 Bi-criteria Optimization Problem for Decision Trees: Cost Versus Uncertainty

In this section, we discuss bi-criteria optimization problem cost versus uncertainty for decision trees.

9.1.1 Pareto Optimal Points: Cost Versus Uncertainty

First, we consider an algorithm for construction of the set of Pareto optimal points.

Let ψ be an increasing cost function for decision trees, U be an uncertainty measure, T be a decision table with n conditional attributes $f_1, \ldots, f_n$, and $H \in \{\max, \text{sum}\}$ where $\max = \max(x, y)$ and $\text{sum} = x + y$.

© Springer Nature Switzerland AG 2020

F. Alsolami et al., *Decision and Inhibitory Trees and Rules for Decision
Tables with Many-valued Decisions*, Intelligent Systems Reference Library 156,
https://doi.org/10.1007/978-3-030-12854-8_9

For each node Θ of the graph $\Delta(T)$, we denote

$$t_{U,H,\psi}(\Theta) = \{(U^H(\Theta, \Gamma), \psi(\Theta, \Gamma)) : \Gamma \in Tree^*(\Delta(T), \Theta)\}$$

where $Tree^*(\Delta(T), \Theta)$ is, by Proposition 6.2, the set of decision trees for Θ. We denote by $Par(t_{U,H,\psi}(\Theta))$ the set of Pareto optimal points for $t_{U,H,\psi}(\Theta)$.

We now describe an algorithm $\mathscr{A}_7$ which constructs the set $Par(t_{U,H,\psi}(T))$. In fact, this algorithm constructs, for each node Θ of the graph $\Delta(T)$, the set $B(\Theta) = Par(t_{U,H,\psi}(\Theta))$.

Algorithm $\mathscr{A}_7$ (construction of POPs for decision trees, cost versus uncertainty).
Input: Increasing cost function for decision trees ψ given by triple of functions ψ^0, F, w, an uncertainty measure U, a function $H \in \{\max, \text{sum}\}$, a decision table T with n conditional attributes $f_1, \ldots, f_n$, and the graph $\Delta(T)$.
Output: The set $Par(t_{U,H,\psi}(T))$ of Pareto optimal points for the set of pairs $t_{U,H,\psi}(T) = \{(U^H(T, \Gamma), \psi(T, \Gamma)) : \Gamma \in Tree^*(\Delta(T), T)\}$.

1. If all nodes in $\Delta(T)$ are processed, then return the set $B(T)$. Otherwise, choose a node Θ in the graph $\Delta(T)$ which is not processed yet and which is either a terminal node of $\Delta(T)$ or a nonterminal node of $\Delta(T)$ such that, for any $f_i \in E(\Theta)$ and any $a_j \in E(\Theta, f_i)$, the node $\Theta(f_i, a_j)$ is already processed, i.e., the set $B(\Theta(f_i, a_j))$ is already constructed.
2. If Θ is a terminal node, then set $B(\Theta) = \{(U(\Theta), \psi^0(\Theta))\}$. Mark the node Θ as processed and proceed to step 1.
3. If Θ is a nonterminal node then, for each $f_i \in E(\Theta)$, apply the algorithm $\mathscr{A}_3$ to the functions H, F and the sets $B(\Theta(f_i, a_1)), \ldots, B(\Theta(f_i, a_t))$, where $\{a_1, \ldots, a_t\} = E(\Theta, f_i)$. Set $C(\Theta, f_i)$ the output of the algorithm $\mathscr{A}_3$ and

$$B(\Theta, f_i) = C(\Theta, f_i) \langle ++ \rangle \{(0, w(\Theta))\}$$
$$= \{(a, b + w(\Theta)) : (a, b) \in C(\Theta, f_i)\} .$$

4. Construct the multiset $A(\Theta) = \{(U(\Theta), \psi^0(\Theta))\} \cup \bigcup_{f_i \in E_G(\Theta)} B(\Theta, f_i)$ by simple transcription of elements from the sets $B(\Theta, f_i)$, $f_i \in E(\Theta)$. Apply to the obtained multiset $A(\Theta)$ the algorithm $\mathscr{A}_2$ which constructs the set $Par(A(\Theta))$. Set $B(\Theta) = Par(A(\Theta))$. Mark the node Θ as processed and proceed to step 1.

Proposition 9.1 *Let ψ be an increasing cost function for decision trees given by triple of functions ψ^0, F, w, U be an uncertainty measure, $H \in \{\max, \text{sum}\}$, and T be a decision table with n conditional attributes $f_1, \ldots, f_n$. Then, for each node Θ of the graph $\Delta(T)$, the algorithm $\mathscr{A}_7$ constructs the set $B(\Theta) = Par(t_{U,H,\psi}(\Theta))$.*

Proof We prove the considered statement by induction on nodes of $\Delta(T)$. Let Θ be a terminal node of $\Delta(T)$. Then $Tree^*(\Delta(T), \Theta) = \{tree(mcd(\Theta))\}$, $t_{U,H,\psi}(\Theta) = \{(U(\Theta), \psi^0(\Theta))\}$, and $B(\Theta) = Par(t_{U,H,\psi}(\Theta))$.

Let Θ be a nonterminal node of $\Delta(T)$ such that, for any $f_i \in E(\Theta)$ and any $a_j \in E(\Theta, f_i)$, the considered statement holds for the node $\Theta(f_i, a_j)$, i.e., $B(\Theta(f_i, a_j)) = Par(t_{U,H,\psi}(\Theta(f_i, a_j)))$.

Let $f_i \in E(\Theta)$ and $E(\Theta, f_i) = \{a_1, \ldots, a_t\}$. We denote

$$P(f_i) = \{(H(b_1, \ldots, b_t), F(c_1, \ldots, c_t) + w(\Theta))$$
$$: (b_j, c_j) \in t_{U,H,\psi}(\Theta(f_i, a_j)), j = 1, \ldots, t\}$$

and, for $j = 1, \ldots, t$, we denote $P_j = t_{U,H,\psi}(\Theta(f_i, a_j))$.

If we apply the algorithm $\mathscr{A}_3$ to the functions H, F and the sets

$$Par(P_1), \ldots, Par(P_t),$$

we obtain the set $Par(Q_t)$ where $Q_1 = P_1$, and, for $j = 2, \ldots, t$, $Q_j = Q_{j-1}\langle HF \rangle P_j$. It is not difficult to show that $P(f_i) = Q_t \langle ++ \rangle \{(0, w(\Theta))\} = \{(a, b + w(\Theta)) : (a, b) \in Q_t\}$ and $Par(P(f_i)) = Par(Q_t) \langle ++ \rangle \{(0, w(\Theta))\}$.

According to the induction hypothesis, $B(\Theta(f_i, a_j)) = Par(P_j)$ for $j = 1, \ldots, t$. Therefore $C(\Theta, f_i) = Par(Q_t)$ and $B(\Theta, f_i) = Par(P(f_i))$.

One can show that $t_{U,H,\psi}(\Theta) = \{(U(\Theta), \psi^0(\Theta))\} \cup \bigcup_{f_i \in E(\Theta)} P(f_i)$. By Lemma 5.7,

$$Par(t_{U,H,\psi}(\Theta)) = Par\left(\{(U(\Theta), \psi^0(\Theta))\} \cup \bigcup_{f_i \in E(\Theta)} P(f_i)\right)$$
$$\subseteq \{(U(\Theta), \psi^0(\Theta))\} \cup \bigcup_{f_i \in E(\Theta)} Par(P(f_i)).$$

Using Lemma 5.6 we obtain

$$Par(t_{U,H,\psi}(\Theta)) = Par\left(\{(U(\Theta), \psi^0(\Theta))\} \cup \bigcup_{f_i \in E(\Theta)} Par(P(f_i))\right).$$

Since $B(\Theta, f_i) = Par(P(f_i))$ for any $f_i \in E(\Theta)$, $Par(t_{U,H,\psi}(\Theta)) = Par(A(\Theta)) = B(\Theta)$. $\qquad\square$

We now analyze the number of elementary operations made by the algorithm $\mathscr{A}_7$ during the construction of the set $Par(t_{U,H,\psi}(T))$ for integral cost functions, for the cases when $F(x, y) = \max(x, y)$ (h is an example of such cost function) and when $F(x, y) = x + y$ (tpl, L, L_n, and L_t are examples of such cost functions).

Let us recall that $range(T) = \max\{|E(T, f_i)| : i = 1, \ldots, n\}$,

$$ub(\psi, T) = \max\{\psi(\Theta, \Gamma) : \Theta \in SEP(T), \Gamma \in DT(\Theta)\},$$

and $\psi(\Theta, \Gamma) \in \{0, 1, \ldots, ub(\psi, T)\}$ for any separable subtable Θ of T and for any decision tree Γ for Θ. Upper bounds on the value $ub(\psi, T)$ for $\psi \in \{h, tpl, L, L_n, L_t\}$ are given in Lemma 6.2: $ub(h, T) \le n$, $ub(tpl, T) \le nN(T)$, $ub(L, T) \le 2N(T)$, $ub(L_n, T) \le N(T)$, and $ub(L_t, T) \le N(T)$.

Proposition 9.2 *Let ψ be an increasing integral cost function for decision trees given by triple of functions $\psi^0, F, w, F \in \{\max(x, y), x + y\}$, U be an uncertainty measure, $H \in \{\max, \text{sum}\}$, and T be a decision table with n conditional attributes $f_1, \ldots, f_n$. Then, to construct the set $Par(t_{U,H,\psi}(T))$, the algorithm $\mathscr{A}_7$ makes*

$$O(L(\Delta(T))range(T)ub(\psi, T)^2 n \log(ub(\psi, T)n))$$

elementary operations (computations of ψ^0, F, w, H, U, comparisons and additions) if $F = \max(x, y)$, and

$$O(L(\Delta(T))range(T)^2 ub(\psi, T)^2 n \log(range(T)ub(\psi, T)n))$$

elementary operations (computations of ψ^0, F, w, H, U, comparisons and additions) if $F = x + y$.

Proof It is clear that, for any node Θ of the graph $\Delta(T)$,

$$t_{U,H,\psi}(\Theta)^{(2)} = \{b : (a, b) \in t_{U,H,\psi}(\Theta)\} \subseteq \{0, \ldots, ub(\psi, T)\} .$$

From Proposition 9.1 it follows that $B(\Theta) = Par(t_{U,H,\psi}(\Theta))$ for any node Θ of the graph $\Delta(T)$. Therefore, for any node Θ of the graph $\Delta(T)$, $B(\Theta)^{(2)} = \{b : (a, b) \in B(\Theta)\} \subseteq \{0, \ldots, ub(\psi, T)\}$.

Let $F(x, y) = \max(x, y)$. To process a terminal node Θ, the algorithm $\mathscr{A}_7$ makes two elementary operations (computations of ψ^0 and U).

We now consider the processing of a nonterminal node Θ (see description of the algorithm $\mathscr{A}_7$). We know that $B(\Theta(f_i, a_j))^{(2)} \subseteq \{0, \ldots, ub(\psi, T)\}$ for $j = 1, \ldots, t$, and $t \le range(T)$. From Proposition 5.7 it follows that $|C(\Theta, f_i)| \le ub(\psi, T) + 1$, and to construct the set $C(\Theta, f_i)$, the algorithm $\mathscr{A}_3$ makes

$$O(range(T)ub(\psi, T)^2 \log ub(\psi, T))$$

elementary operations (computations of F, H and comparisons). To construct the set $B(\Theta, f_i)$ from the set $C(\Theta, f_i)$, the algorithm makes a computation of w and at most $ub(\psi, T) + 1$ additions. From here it follows that, to construct the set $B(\Theta, f_i)$, the algorithm makes

$$O(range(T)ub(\psi, T)^2 \log ub(\psi, T))$$

elementary operations. It is clear that $|E(\Theta)| \le n$. Therefore, to construct the set $B(\Theta, f_i)$ for each $f_i \in E(\Theta)$, the algorithm makes

$$O(range(T)ub(\psi, T)^2 n \log ub(\psi, T))$$

elementary operations.

Since $|C(\Theta, f_i)| \le ub(\psi, T) + 1$, we have $|B(\Theta, f_i)| \le ub(\psi, T) + 1$. Since

$$|E(\Theta)| \le n ,$$

we have $|A(\Theta)| \le n(ub(\psi, T) + 1) + 1$ where

$$A(\Theta) = \{(U(\Theta), \psi^0(\Theta))\} \cup \bigcup_{f_i \in E(\Theta)} B(\Theta, f_i) .$$

From Proposition 5.5 it follows that, to construct the set $B(\Theta) = Par(A(\Theta))$, the algorithm $\mathscr{A}_2$ makes $O(ub(\psi, T)n \log(ub(\psi, T)n))$ comparisons. So, to process a nonterminal node Θ (to construct $B(\Theta) = Par(t_{U,H,\psi}(\Theta))$ if $B(\Theta(f_i, a_j)) = Par(t_{U,H,\psi}(\Theta(f_i, a_j)))$ is known for all $f_i \in E(\Theta)$ and $a_j \in E(\Theta, f_i)$), the algorithm $\mathscr{A}_7$ makes

$$O(range(T)ub(\psi, T)^2 n \log(ub(\psi, T)n))$$

elementary operations.

To construct the set $Par(t_{U,H,\psi}(T))$ for given decision table T with n conditional attributes and the graph $\Delta(T)$, it is enough to make

$$O(L(\Delta(T))range(T)ub(\psi, T)^2 n \log(ub(\psi, T)n))$$

elementary operations (computations of ψ^0, F, w, H, U, comparisons and additions).

Let $F(x, y) = x + y$. To process a terminal node Θ, the algorithm $\mathscr{A}_7$ makes two elementary operations (computations of ψ^0 and U).

We now consider the processing of a nonterminal node Θ (see description of the algorithm $\mathscr{A}_7$). We know that $B(\Theta(f_i, a_j))^{(2)} \subseteq \{0, \ldots, ub(\psi, T)\}$ for $j = 1, \ldots, t$, and $t \le range(T)$. From Proposition 5.7 it follows that $|C(\Theta, f_i)| \le t \times ub(\psi, T) + 1 \le range(T)ub(\psi, T) + 1$, and to construct the set $C(\Theta, f_i)$, the algorithm $\mathscr{A}_3$ makes

$$O(range(T)^2 ub(\psi, T)^2 \log(range(T)ub(\psi, T)))$$

elementary operations (computations of F, H and comparisons). To construct the set $B(\Theta, f_i)$ from the set $C(\Theta, f_i)$, the algorithm makes a computation of w and at most $range(T)ub(\psi, T) + 1$ additions. From here it follows that, to construct the set $B(\Theta, f_i)$, the algorithm makes

$$O(range(T)^2 ub(\psi, T)^2 \log(range(T)ub(\psi, T)))$$

elementary operations. It is clear that $|E(\Theta)| \le n$. Therefore, to construct the set $B(\Theta, f_i)$ for each $f_i \in E(\Theta)$, the algorithm makes

$$O(range(T)^2 ub(\psi, T)^2 n \log(range(T)ub(\psi, T)))$$

elementary operations.

Since $|C(\Theta, f_i)| \leq range(T)ub(\psi, T) + 1$, we have

$$|B(\Theta, f_i)| \leq range(T)ub(\psi, T) + 1 \, .$$

Since $|E(\Theta)| \leq n$, we have $|A(\Theta)| \leq (range(T)ub(\psi, T) + 1)n + 1$ where $A(\Theta)$ $= \{(U(\Theta), \psi^0(\Theta))\} \cup \bigcup_{f_i \in E(\Theta)} B(\Theta, f_i)$. From Proposition 5.5 it follows that, to construct the set $B(\Theta) = Par(A(\Theta))$, the algorithm $\mathscr{A}_2$ makes

$$O(range(T)ub(\psi, T)n \log(range(T)ub(\psi, T)n))$$

comparisons. So, to process a nonterminal node Θ (to construct

$$B(\Theta) = Par(t_{U,H,\psi}(\Theta))$$

if $B(\Theta(f_i, a_j)) = Par(t_{U,H,\psi}(\Theta(f_i, a_j)))$ is known for all $f_i \in E(\Theta)$ and $a_j \in E(\Theta, f_i)$), the algorithm $\mathscr{A}_7$ makes

$$O(range(T)^2 ub(\psi, T)^2 n \log(range(T)ub(\psi, T)n))$$

elementary operations.

To construct the set $Par(t_{U,H,\psi}(T))$ for given decision table T with n conditional attributes and the graph $\Delta(T)$, it is enough to make

$$O(L(\Delta(T))range(T)^2 ub(\psi, T)^2 n \log(range(T)ub(\psi, T)n))$$

elementary operations (computations of ψ^0, F, w, H, U, comparisons and additions). □

Proposition 9.3 *Let ψ be a cost function for decision trees from the set $\{h, tpl, L, L_n, L_t\}$ given by triple of functions ψ^0, F, w, U be an uncertainty measure from the set $\{me, rme, abs\}$, $H \in \{max, sum\}$, and $\mathscr{U}$ be a restricted information system. Then the algorithm $\mathscr{A}_7$ has polynomial time complexity for decision tables from $\mathscr{T}(\mathscr{U})$ depending on the number of conditional attributes in these tables.*

Proof Since $\psi \in \{h, tpl, L, L_n, L_t\}$, ψ^0 is a constant, the function F is either $max(x, y)$ or $x + y$, and the function w is either a constant or $N(T)$. From Proposition 9.2 and Lemma 6.2 it follows that, for the algorithm $\mathscr{A}_7$, the number of elementary operations (computations of ψ^0, F, w, H, U, comparisons and additions) is bounded from above by a polynomial depending on the size of input table T and on the number of separable subtables of T. All operations with numbers are basic ones. The computations of numerical parameters of decision tables used by the algorithm $\mathscr{A}_7$ (constants, $N(T)$, and $U(T)$) have polynomial time complexity depending on the size of decision tables.

According to Proposition 5.4, the algorithm $\mathscr{A}_7$ has polynomial time complexity for decision tables from $\mathscr{T}(\mathscr{U})$ depending on the number of conditional attributes in these tables. $\qquad\square$

We now consider the problem of construction of the set of Pareto optimal points for the sets $t_{U,\psi}^{\max}(T) = \{(U^{\max}(T,\Gamma), \psi(T,\Gamma)) : \Gamma \in DT_U^{\max}(T)\}$ and $t_{U,\psi}^{\text{sum}}(T) = \{(U^{\text{sum}}(T,\Gamma), \psi(T,\Gamma)) : \Gamma \in DT_U^{\text{sum}}(T)\}$. Let ψ be bounded and increasing cost function for decision trees (in particular, h, tpl, L, L_n, and L_t are bounded and increasing cost functions). Then, by Lemma 5.4 and Proposition 6.12, $Par(t_{U,\psi}^{\max}(T)) = Par(t_{U,\max,\psi}(T))$. By Lemma 5.4 and Proposition 6.13,

$$Par(t_{U,\psi}^{\text{sum}}(T)) = Par(t_{U,\text{sum},\psi}(T)) .$$

So to construct $Par(t_{U,\psi}^{\max}(T)$ and $Par(t_{U,\psi}^{\text{sum}}(T))$ we can use the algorithm $\mathscr{A}_7$.

9.1.2 Relationships Between Cost and Uncertainty

We now show how to construct graphs describing relationships between cost and uncertainty of decision trees.

Let ψ be bonded and increasing cost function for decision trees, U be an uncertainty measure, $H \in \{\max(x, y), \text{sum}(x, y)\}$, and T be a decision table.

To study relationships between cost function ψ and uncertainty U^H for decision trees on the set of decision trees $DT_U^H(T)$ we consider partial functions $\mathscr{T}_{T,H}^{U,\psi} : \mathbb{R} \to \mathbb{R}$ and $\mathscr{T}_{T,H}^{\psi,U} : \mathbb{R} \to \mathbb{R}$ defined as follows:

$$\mathscr{T}_{T,H}^{U,\psi}(x) = \min\{\psi(T,\Gamma) : \Gamma \in DT_U^H(T), U^H(T,\Gamma) \le x\} ,$$
$$\mathscr{T}_{T,H}^{\psi,U}(x) = \min\{U^H(T,\Gamma) : \Gamma \in DT_U^H(T), \psi(T,\Gamma) \le x\} .$$

Let $(a_1, b_1), \ldots, (a_k, b_k)$ be the normal representation of the set

$$Par(t_{U,\psi}^H(T)) = Par(t_{U,H,\psi}(T))$$

where $a_1 < \cdots < a_k$ and $b_1 > \cdots > b_k$. By Lemma 5.10 and Remark 5.4, for any $x \in \mathbb{R}$,

$$\mathscr{T}_{T,H}^{U,\psi}(x) = \begin{cases} undefined, & x < a_1 \\ b_1, & a_1 \le x < a_2 \\ \cdots & \cdots \\ b_{k-1}, & a_{k-1} \le x < a_k \\ b_k, & a_k \le x \end{cases} ,$$

$$\mathscr{T}_{T,H}^{\psi,U}(x) = \begin{cases} undefined, & x < b_k \\ a_k, & b_k \le x < b_{k-1} \\ \dots & \dots \\ a_2, & b_2 \le x < b_1 \\ a_1, & b_1 \le x \end{cases} .$$

9.2　Bi-criteria Optimization Problem for Inhibitory Trees: Cost Versus Completeness

In this section, we discuss bi-criteria optimization problem cost versus completeness for inhibitory trees.

Let T be a nondegenerate decision table with n conditional attributes $f_1, \dots, f_n$, T^C be the decision table complementary to T, U be an uncertainty measure, W be a completeness measure, U and W be dual, $\psi \in \{h, tpl, L, L_n, L_t\}$, and $H \in \{\max, \text{sum}\}$.

In Sect. 9.1, the algorithm $\mathscr{A}_7$ is described which constructs the set of Pareto optimal points $Par(t_{U,H,\psi}(T^C))$ for the set of points

$$t_{U,H,\psi}(T^C) = \{(U^H(T^C, \Gamma), \psi(T^C, \Gamma)) : \Gamma \in DT(T^C)\} .$$

We proved at the end of Sect. 9.1.1 that $Par(t_{U,H,\psi}(T^C)) = Par(t_{U,\psi}^H(T^C))$ where $t_{U,\psi}^H(T^C) = \{(U^H(T^C, \Gamma), \psi(T^C, \Gamma)) : \Gamma \in DT_U^H(T^C)\}$.

We denote $it_{W,H,\psi}(T) = \{(W^H(T, \Gamma), \psi(T, \Gamma)) : \Gamma \in IT(T)\}$ and $it_{W,\psi}^H(T) = \{(W^H(T, \Gamma), \psi(T, \Gamma)) : \Gamma \in IT_W^H(T)\}$. By Proposition 6.14, for any $\Gamma \in DT(T^C)$, $(U^H(T^C, \Gamma), \psi(T^C, \Gamma)) = (W^H(T, \Gamma^-), \psi(T, \Gamma^-))$. From Corollary 6.1 it follows that $DT(T^C)^- = IT(T)$ and $DT_U^H(T^C)^- = IT_W^H(T)$. Therefore

$$t_{U,H,\psi}(T^C) = it_{W,H,\psi}(T)$$

and $t_{U,\psi}^H(T^C) = it_{W,\psi}^H(T)$. Hence

$$Par(it_{W,H,\psi}(T)) = Par(it_{W,\psi}^H(T)) = Par(t_{U,H,\psi}(T^C)) .$$

To study relationships between cost function ψ and completeness W^H for inhibitory trees on the set of inhibitory trees $IT_W^H(T)$ we consider partial functions $\mathscr{I}\mathscr{T}_{T,H}^{W,\psi} : \mathbb{R} \to \mathbb{R}$ and $\mathscr{I}\mathscr{T}_{T,H}^{\psi,W} : \mathbb{R} \to \mathbb{R}$ defined as follows:

$$\mathscr{I}\mathscr{T}_{T,H}^{W,\psi}(x) = \min\{\psi(T, \Gamma) : \Gamma \in IT_W^H(T), W^H(T, \Gamma) \le x\} ,$$
$$\mathscr{I}\mathscr{T}_{T,H}^{\psi,W}(x) = \min\{W^H(T, \Gamma) : \Gamma \in IT_W^H(T), \psi(T, \Gamma) \le x\} .$$

Let $(a_1, b_1), \dots, (a_k, b_k)$ be the normal representation of the set

$$Par(it_{W,\psi}^{H}(T)) = Par(t_{U,H,\psi}(T^{C}))$$

where $a_1 < \cdots < a_k$ and $b_1 > \cdots > b_k$. By Lemma 5.10 and Remark 5.4, for any $x \in \mathbb{R}$,

$$\mathscr{I}\mathscr{T}_{T,H}^{W,\psi}(x) = \begin{cases} undefined, & x < a_1 \\ b_1, & a_1 \leq x < a_2 \\ \cdots & \cdots \\ b_{k-1}, & a_{k-1} \leq x < a_k \\ b_k, & a_k \leq x \end{cases},$$

$$\mathscr{I}\mathscr{T}_{T,H}^{\psi,W}(x) = \begin{cases} undefined, & x < b_k \\ a_k, & b_k \leq x < b_{k-1} \\ \cdots & \cdots \\ a_2, & b_2 \leq x < b_1 \\ a_1, & b_1 \leq x \end{cases}.$$

9.3 Illustrative Examples

To illustrate the use of tools developed in this chapter, we consider some experimental results for the decision table with many-valued decisions CARS-1 containing 432 rows and 5 conditional attributes (see Table 7.3). For this decision table, we describe how depth h, average depth h_{avg}, and number of nodes L of decision trees depend on the number of misclassifications me^{sum} which is a kind of decision tree uncertainty. We also describe how depth h, average depth h_{avg}, and number of nodes L of inhibitory trees depend on the number of misclassifications ime^{sum} which is a kind of inhibitory tree completeness. Corresponding relationships are shown in Figs. 9.1, 9.2, and 9.3. Filled circles in these figures are Pareto optimal points for the considered bi-criteria optimization problems cost versus uncertainty for decision trees and cost versus completeness for inhibitory trees.

Such relationships can be useful for the choice of decision and inhibitory trees which represent knowledge contained in the decision table: a tree with enough small number of nodes and reasonable uncertainty (completeness) can be considered as a good model of data. Similar situation is with the choice of decision and inhibitory trees that are considered as algorithms. In this case, we are interesting in finding of a tree with small depth or average depth and reasonable uncertainty (completeness). Similar approach can be useful in machine learning (see Chap. 10).

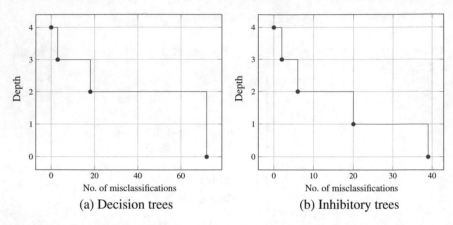

Fig. 9.1 Decision table CARS-1: depth versus no. of misclassifications

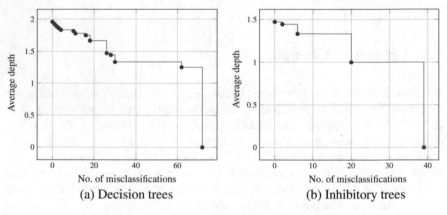

Fig. 9.2 Decision table CARS-1: average depth versus no. of misclassifications

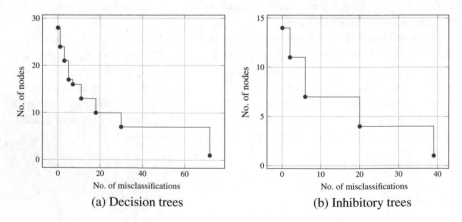

Fig. 9.3 Decision table CARS-1: no. of nodes versus no. of misclassifications

References

1. Chikalov, I., Hussain, S., Moshkov, M.: Average depth and number of misclassifications for decision trees. In: Popova-Zeugmann, L. (ed.) 21st International Workshop on Concurrency, Specification and Programming, CS&P 2012, Berlin, Germany, 26–28 Sept 2012. CEUR Workshop Proceedings, vol. 928, pp. 160–169. CEUR-WS.org (2012)
2. Chikalov, I., Hussain, S., Moshkov, M.: On cost and uncertainty of decision trees. In: Yao, J., Yang, Y., Slowinski, R., Greco, S., Li, H., Mitra, S., Polkowski, L. (eds.) Rough Sets and Current Trends in Computing – 8th International Conference, RSCTC 2012, Chengdu, China, 17–20 Aug 2012. Lecture Notes in Computer Science, vol. 7413, pp. 190–197. Springer, Berlin (2012)
3. Chikalov, I., Hussain, S., Moshkov, M.: Relationships between number of nodes and number of misclassifications for decision trees. In: Yao, J., Yang, Y., Slowinski, R., Greco, S., Li, H., Mitra, S., Polkowski, L. (eds.) Rough Sets and Current Trends in Computing – 8th International Conference, RSCTC 2012, Chengdu, China, 17–20 Aug 2012. Lecture Notes in Computer Science, vol. 7413, pp. 212–218. Springer, Berlin (2012)

Chapter 10
Multi-pruning and Restricted Multi-pruning of Decision Trees

In this chapter, we develop an approach to construct decision trees which can be used for knowledge representation and classification. This approach is applicable to decision tables with both categorical and numerical conditional attributes. It is based on the consideration of a large set of CART(Classification and Regression Trees)-like decision trees for the considered decision table. In particular, for the decision table BREAST-CANCER containing 266 rows and 10 conditional attributes, the number of CART-like decision trees is equal to 1.42×10^{193}. Such trees use binary splits created on the base of conditional attributes. In contrast with standard CART [3] which uses the best (from the point of view of impurity based on Gini index $gini$) splits among all attributes, CART-like trees use, additionally, the best splits for each attribute.

Created algorithms allow us to describe, for the considered set of CART-like decision trees given by a DAG, the set of Pareto optimal points for bi-criteria optimization problem relative to the number of nodes and the number of misclassifications. For each Pareto optimal point (a, b), we can derive the decision trees for which the number of misclassifications is equal to a and the number of nodes is equal to b.

Since the considered set of CART-like decision trees is closed under the operation of usual bottom up pruning, we call the created approach to construction of decision trees for knowledge representation and classification *multi-pruning* (MP). The initial study of multi-pruning approach was done in [1].

We consider two applications of multi-pruning approach. The first application is connected with knowledge representation. We use the initial decision table to build the DAG and the set of Pareto optimal points. Then we choose a suitable Pareto optimal point and derive the corresponding decision tree (we can choose, for example, a point with the minimum Euclidean distance from the origin point $(0, 0)$). If this tree has relatively small number of nodes and relatively small number of misclassifications, it can be considered as understandable and enough accurate model for the knowledge contained in the table. Results of experiments with decision

© Springer Nature Switzerland AG 2020

F. Alsolami et al., *Decision and Inhibitory Trees and Rules for Decision Tables with Many-valued Decisions*, Intelligent Systems Reference Library 156, https://doi.org/10.1007/978-3-030-12854-8_10

tables from the UCI ML Repository [5] show that we can construct such decision trees in many cases.

Another application is connected with machine learning. We divide the initial table into three subtables: training, validation, and testing. We use training subtable to build the DAG and the set of Pareto optimal points. We derive randomly a number of decision trees (five in our experiments) for each Pareto optimal point and find, based on the validation subtable, a decision tree with minimum number of misclassifications among all derived trees. We evaluate the accuracy of prediction for this tree using testing subtable. We compare this process with CART which continues to be one of the best algorithms for construction of classification trees. The classifiers constructed by the process often have better accuracy than the classifiers constructed by CART. The considered process is similar to the usual pruning of a decision tree but it is applied here to many decision trees since the set of CART-like decision trees is closed under the operation of usual bottom up pruning.

The multi-pruning approach is applicable to medium-sized decision tables and can be used as a research tool. To make this approach more scalable we consider *restricted multi-pruning approach* (RMP) where we use in each node of DAG only the best spits for a small number of attributes, instead of using the best splits for all attributes. The obtained DAGs with limited branching factor contain fewer nodes and edges and require less time for the construction and processing. However, the number of nodes and number of misclassifications of decision trees constructed by the restricted multi-pruning approach are comparable to the case when we use the best splits for all attributes. The prediction accuracy of decision trees constructed by RMP approach is comparable to the case when we use MP approach. For example, we did experiments with 15 decision tables from the UCI ML Repository [5]. For 11 tables, the decision trees constructed by multi-pruning approach which uses best splits for all attributes outperform the trees built by CART. The same situation is with decision trees constructed by restricted multi-pruning approach which uses only best splits for two attributes.

We extend multi-pruning and restricted multi-pruning approaches to the case of decision tables with many-valued decisions. We can not use *gini* uncertainty measure (Gini index) for the decision tables with many-valued decisions. Instead of *gini* we use the uncertainty measure *abs* which works for both decision tables with single- and many-valued decisions. The experimental results show that *abs* performs as good as *gini* for decision tables with single-valued decisions. We also compared the multi-pruning and restricted multi-pruning approaches in the case of decision tables with many-valued decisions, and found that restricted multi-pruning approach is comparable with multi-pruning approach.

We find a simplest variant of restricted multi-pruning approach which can work successfully for both knowledge representation and classification and for both decision tables with single-valued decisions and for decision tables with many-valued decisions. This variant of restricted multi-pruning approach uses only best splits for two attributes and uncertainty measure *abs*. The initial study of restricted multi-pruning approach was done in [2].

This chapter consists of eight sections. In Sects. 10.1–10.4, we consider decision tables, CART-like decision trees, DAGs, and sets of Pareto optimal points, respectively. We discuss MP and RMP approaches in Sect. 10.5. In Sect. 10.6, we present the decision tables that are used for experiments. In Sect. 10.7, we consider experimental results related to knowledge representation, and in Sect. 10.8 we discuss experimental results related to machine learning.

10.1 Decision Tables

A *decision table* consists of rows (objects) and columns (conditional attributes). Usually, each row is labeled with a single decision. We call such decision tables *decision tables with single-valued decisions*. There is another type of decision tables where each row is labeled with a set of decisions instead of single decision. We call such decision tables *decision tables with many-valued decisions*. Decision tables with single-valued decisions can be considered as a special case of decision tables with many-valued decisions.

In case of decision tables with single-valued decisions, the minimum decision which is attached to the maximum number of rows of the decision table T is called the *most common decision* for T. On the other hand, in the case of decision tables with many-valued decisions, the minimum decision which belongs to the maximum number of sets of decisions attached to rows of the table T is called the *most common decision* for T.

In the case of decision tables with single-valued decisions, if there is a decision attached to each row of T, then we call it the *common decision* for T. On the other hand, in the case of decision tables with many-valued decisions, if there is a decision which, for each row of T, belongs to the set of decisions attached to this row, then we call it a *common decision* for T.

In this chapter, an *uncertainty measure* U is a function from a set of nonempty decision tables to the set of real numbers such that $U(T) \geq 0$, and $U(T) = 0$ if and only if T has a common decision. We use two uncertainty measures in this chapter: *gini* and *abs*.

Let T be a decision table having $N = N(T)$ rows, and its rows be labeled with m different decisions $d_1, \ldots, d_m$ (in the case of decision tables with single-valued decisions) or sets containing m different decisions $d_1, \ldots, d_m$ (in the case of decision tables with many-valued decisions). For $i = 1, \ldots, m$, let N_i be the number of rows in T that are labeled with the decision d_i (in the case of decision tables with single-valued decisions) or with sets of decisions containing the decision d_i (in the case of decision tables with many-valued decisions), and $p_i = \frac{N_i}{N}$. Then

- $gini(T) = 1 - \sum_{i=1}^{m} p_i^2$ (this uncertainty measure, Gini index, is applicable to decision tables with single-valued decisions),
- $abs(T) = \prod_{i=1}^{m} q_i$ where $q_i = 1 - p_i$ (this uncertainty measure, absence, is applicable to all decision tables).

10.2 (m_1, m_2, U)-Decision Trees

In this section, we discuss the notion of (m_1, m_2, U)-decision trees. Such trees use the same type of binary splits as considered in CART [3]. We call such trees *CART-like decision trees*.

Let T be a decision table with n conditional attributes $f_1, \ldots, f_n$ which are categorical or numerical, and with a categorical decision attribute d that has either a single decision for each row in the case of decision table with single-valued decisions or a set of decisions for each row in the case of decision table with many-valued decisions.

Let Θ be a subtable of the table T obtained from T by removal of some rows. Instead of conditional attributes we use (as in CART) binary splits (binary attributes) each of which is based on a conditional attribute f_i. If f_i is a categorical attribute with the set of values B then we consider a partitioning of B into two nonempty subsets B_0 and B_1. The value of the corresponding binary split s is equal to 0 if the value of f_i belongs to B_0, and 1 otherwise. We consider splits for all possible partitionings of B. If f_i is a numerical attribute then we consider a real threshold α. The value of corresponding binary split s is equal to 0 if the value of f_i is less than α, and 1 otherwise. We consider splits for all possible thresholds α. Each binary split s divides the table Θ into two subtables $\Theta_{s=0}$ and $\Theta_{s=1}$ according to the values of s on rows of Θ.

Let U be an uncertainty measure: if T is a decision table with single-valued decisions then $U \in \{gini, abs\}$, and if T is a decision table with many-valued decisions, then $U = abs$. The impurity $I_U(\Theta, s)$ of the split s is equal to the weighted sum of uncertainties of subtables $\Theta_{s=0}$ and $\Theta_{s=1}$, where the weights are proportional to the number of rows in subtables $\Theta_{s=0}$ and $\Theta_{s=1}$, respectively. The impurity of splits is considered as a quality measure for splits where small impurity is better. A split s based on attribute f_i with minimum impurity $I_U(\Theta, s)$ is called the best split for U, Θ and f_i.

Let m_1 and m_2 be nonnegative integers such that $0 < m_1 + m_2 \leq n$. We now describe a way for construction of a set $S_{m_1, m_2, U}(\Theta)$ of admissible splits for the subtable Θ. Let $E(\Theta)$ be the set of all conditional attributes which are not constant on Θ, and $|E(\Theta)| = p$. For each attribute $f_i \in E(\Theta)$, we find a best split for U, Θ and the attribute f_i. Let $s_1, \ldots, s_p$ be the obtained splits in order from the best to the worst. If $m_1 \geq p$ then $S_{m_1, m_2, U}(\Theta) = \{s_1, \ldots, s_p\}$. Let $m_1 < p$. Then $S_{m_1, m_2, U}(\Theta)$ contains splits $s_1, \ldots, s_{m_1}$ and $\min(p - m_1, m_2)$ splits randomly chosen from the set $\{s_{m_1+1}, \ldots, s_p\}$.

We consider (m_1, m_2, U)-decision trees for T in which each terminal node is labeled with a decision (value of the decision attribute d), each nonterminal node is labeled with a binary split corresponding to one of the conditional attributes, and two outgoing edges from this node are labeled with 0 and 1, respectively. We correspond to each node v of a decision tree Γ a subtable $T(\Gamma, v)$ of T that contains all rows of T for which the computation of Γ passes through the node v. We assume that, for each nonterminal node v, the subtable $T(\Gamma, v)$ has no any common decisions, and the node v is labeled with a split from $S_{m_1, m_2, U}(T(\Gamma, v))$. We assume also that, for each terminal node v, the node v is labeled with the most common decision for $T(\Gamma, v)$.

10.3 DAG $G_{m_1,m_2,U}(T)$

In this section, we study a directed acyclic graph $G_{m_1,m_2,U}(T)$ which is used to describe the set of (m_1, m_2, U)-decision trees for T. Nodes of the graph $G_{m_1,m_2,U}(T)$ are some subtables of the table T. We now describe an algorithm for the construction of the directed acyclic graph $G_{m_1,m_2,U}(T)$ which can be considered as a definition of this DAG.

Algorithm $\mathscr{A}_{DAG}$ (construction of the DAG $G_{m_1,m_2,U}(T)$).

Input: A decision table T with n conditional attributes, an uncertainty measure U (if T is a decision table with single-valued decisions, then $U \in \{gini, abs\}$, and if T is a decision table with many-valued decisions then $U = abs$), and nonnegative integers m_1 and m_2 such that $0 < m_1 + m_2 \leq n$.

Output: The DAG $G_{m_1,m_2,U}(T)$.

1. Construct a graph which contains only one node T which is marked as not processed.
2. If all nodes of the graph are processed, then return it as $G_{m_1,m_2,U}(T)$ and finish. Otherwise, choose a node (subtable) Θ which is not processed yet.
3. If Θ has a common decision, mark Θ as processed and proceed to step 2.
4. Otherwise, construct the set $S_{m_1,m_2,U}(\Theta)$ of admissible splits for Θ and, for each split s from $S_{m_1,m_2,U}(\Theta)$, draw two edges from Θ to subtables $\Theta_{s=0}$ and $\Theta_{s=1}$, and label these edges with $s = 0$ and $s = 1$, respectively (this pair of edges is called an s-pair). If some of the subtables $\Theta_{s=0}$ and $\Theta_{s=1}$ are not in the graph, add them to the graph. Mark Θ as processed and proceed to step 2.

One can show that the time complexity of this algorithm is bounded from above by a polynomial in the size of decision table T and the number of nodes in the graph $G_{m_1,m_2,U}(T)$.

We correspond to each node Θ of the graph $G_{m_1,m_2,U}(T)$ a set of decision trees $DT_{m_1,m_2,U}(\Theta)$. We denote by $tree(\Theta)$ the decision tree with exactly one node labeled with the most common decision for Θ. If Θ has a common decision then $DT_{m_1,m_2,U}(\Theta)$ contains only one tree $tree(\Theta)$. Otherwise, $DT_{m_1,m_2,U}(\Theta)$ contains $tree(\Theta)$ and all trees of the following kind: the root of tree is labeled with a split s such that an s-pair of edges starts in Θ, two edges start in the root which are labeled with 0 and 1 and enter to the roots of decision trees from $DT_{m_1,m_2,U}(\Theta_{s=0})$ and $DT_{m_1,m_2,U}(\Theta_{s=1})$, respectively.

Note that the set $DT_{m_1,m_2,U}(T)$ is closed under the operation of usual bottom-up pruning of decision trees. One can prove that the set of decision trees $DT_{m_1,m_2,U}(T)$

coincides with the set of all (m_1, m_2, U)-decision trees for T if, for any subtable Θ of T, we consider the same set $S_{m_1,m_2,U}(\Theta)$ in the definition of the set $DT_{m_1,m_2,U}(T)$, and in the description of DAG $G_{m_1,m_2,U}(T)$.

10.4   Set of Pareto Optimal Points $POP_{m_1,m_2,U}(T)$

Let A be a finite set of points in two-dimensional Euclidean space. A point $(a, b) \in A$ is called a Pareto optimal point (POP) for A if there is no point $(c, d) \in A$ such that $(a, b) \neq (c, d)$, $c \le a$, and $d \le b$. We denote by $Par(A)$ the set of Pareto optimal points for A. It is easy to construct the set $Par(A)$ using $O(|A| \log |A|)$ comparisons (see the algorithm $\mathscr{A}_2$).

Let Γ be a decision tree from $DT_{m_1,m_2,U}(T)$. We denote by $L(\Gamma)$ the number of nodes in Γ and by $mc(T, \Gamma)$ the number of misclassifications of Γ on rows of T, i.e., the number of rows of T for which the work of Γ ends in a terminal node that is labeled with either a decision different from the decision attached to the row in the case of decision tables with single-valued decisions, or a decision which is not contained in the set of decisions attached to the row in the case of decision tables with many-valued decisions. We correspond to each decision tree $\Gamma \in DT_{m_1,m_2,U}(T)$ the point $(mc(T, \Gamma), L(\Gamma))$. As a result, we obtain the set of points $\{(mc(T, \Gamma), L(\Gamma)) : \Gamma \in DT_{m_1,m_2,U}(T)\}$. Our aim is to construct for this set the set of all Pareto optimal points $POP_{m_1,m_2,U}(T) = Par(\{(mc(T, \Gamma), L(\Gamma)) : \Gamma \in DT_{m_1,m_2,U}(T)\})$.

We describe now an algorithm which attaches to each node Θ of the DAG $G_{m_1,m_2,U}(T)$ the set

$$POP_{m_1,m_2,U}(\Theta) = Par(\{(mc(\Theta, \Gamma), L(\Gamma)) : \Gamma \in DT_{m_1,m_2,U}(\Theta)\}) .$$

This algorithm works in a bottom-up fashion beginning with subtables which have common decisions.

Algorithm $\mathscr{A}_{POPs}$ (construction of the set $POP_{m_1,m_2,U}(T)$).
Input: The DAG $G_{m_1,m_2,U}(T)$ for a decision table T.
Output: The set $POP_{m_1,m_2,U}(T)$.

1. If all nodes of $G_{m_1,m_2,U}(T)$ are processed then return the set $POP_{m_1,m_2,U}(T)$ attached to the node T and finish. Otherwise, choose a node Θ of $G_{m_1,m_2,U}(T)$ which is not processed yet and such that either Θ has a common decision or all children of Θ are already processed.
2. If Θ has a common decision then attach to Θ the set $POP_{m_1,m_2,U}(\Theta) = \{(0, 1)\}$, mark Θ as processed, and proceed to step 1.
3. If all children of Θ are already processed, and $S(\Theta)$ is the set of splits s such that an s-pair of edges starts in Θ then attach to Θ the set

$$POP_{m_1,m_2,U}(\Theta) = Par(\{(mc(\Theta, tree(\Theta)), 1)\} \cup \bigcup_{s \in S(\Theta)} \{(a+c, b+d+1)$$

$$: (a, b) \in POP_{m_1,m_2,U}(\Theta_{s=0}), (c, d) \in POP_{m_1,m_2,U}(\Theta_{s=1})\}) \,,$$

mark Θ as processed and proceed to step 1.

Let Θ be a node of the graph $G_{m_1,m_2,U}(T)$ and $N(\Theta)$ be the number of rows in Θ. It is clear that all points from $POP_{m_1,m_2,U}(\Theta)$ have pairwise different first coordinates which belong to the set $\{0, 1, \ldots, N(\Theta)\}$. Therefore the cardinality of the set $POP_{m_1,m_2,U}(\Theta)$ is bounded from above by the number $N(\Theta) + 1$. From here it follows that the time complexity of the considered algorithm is bounded from above by a polynomial in the size of the decision table T and the number of nodes in the graph $G_{m_1,m_2,U}(T)$.

For each Pareto optimal point from the set $POP_{m_1,m_2,U}(\Theta)$, we keep information about its construction: either it corresponds to the tree $tree(\Theta)$ or is obtained as a combination $(a + c, b + d + 1)$ of points $(a, b) \in POP_{m_1,m_2,U}(\Theta_{s=0})$ and $(c, d) \in POP_{m_1,m_2,U}(\Theta_{s=1})$ for some $s \in S(\Theta)$. In the last case, we keep all such combinations. This information allows us to derive, for each point (x, y) from $POP_{m_1,m_2,U}(T)$, decision trees from $DT_{m_1,m_2,U}(T)$, such that $(mc(T, \Gamma), L(\Gamma)) = (x, y)$ for each derived tree Γ.

10.5 Multi-pruning (MP) and Restricted Multi-pruning (RMP) Approaches

We know that the set $DT_{m_1,m_2,U}(T)$ is closed under the operation of usual bottom-up pruning of decision trees. In contrast with pruning of one decision tree, we consider here the pruning of many decision trees. When $m_1 = n$, where n is the number of conditional attributes in T, and $m_2 = 0$, we will call our approach to the study of decision trees, the *multi-pruning* (MP). Furthermore, in general case, when $0 < m_1 + m_2 \leq n$, we will call our approach, the *restricted multi-pruning* (RMP). In the case of MP, we can use optimal splits for all non-constant attributes, whereas, in the case of RMP, we can use optimal splits for at most $m_1 + m_2$ non-constant attributes. We decided to consider not only MP but also RMP due to time constraints.

10.6 Decision Tables Used in Experiments

We did experiments for two purposes: to represent important information from the decision table, and to classify unknown instances using decision trees. We used decision tables from the UCI ML Repository [5] containing categorical and/or numerical attributes.

Table 10.1 Decision tables with single-valued decisions from UCI ML repository

Decision table	Rows	Attr	Type of attributes
BALANCE-SCALE	625	4	Categorical
BANKNOTE	1372	4	Numerical
BREAST-CANCER	266	9	Categorical
CARS	1728	6	Categorical
GLASS	214	9	Numerical
HAYES-ROTH-DATA	69	4	Categorical
HOUSE-VOTES-84	279	16	Categorical
IRIS	150	4	Numerical
LYMPHOGRAPHY	148	18	Categorical
NURSERY	12960	8	Categorical
SOYBEAN-SMALL	47	35	Categorical
SPECT-TEST	169	22	Categorical
TIC-TAC-TOE	958	9	Categorical
WINE	178	12	Numerical
ZOO-DATA	59	16	Categorical

The first group of decision tables (Table 10.1) contains conventional decision tables with single-valued decisions. Before the experiment work, some preprocessing procedures were performed. An attribute is removed if it has unique value for each row. The missing value for an attribute is filled up with the most common value for that attribute. If there are duplicate rows with equal values of conditional attributes, then only one row is kept with the most common decision for that group of equal rows and others are removed. In Table 10.1, the first column "Decision table" refers to the name of the decision table from the UCI ML Repository, the column "Rows" refers to the number of rows, the column "Attr" refers to the number of conditional attributes, and the column "Types of attributes" refers to either categorical or numerical type of conditional attributes in the table.

The second group of decision tables (Table 10.2) contains decision tables with many-valued decisions. We took some decision tables from the UCI ML Repository and remove one or more conditional attributes from them. As a result, for each table, there are multiple rows that have equal values of conditional attributes but have different decisions which are then merged into a single row containing the set of decisions from the group of equal rows. Similar preprocessing steps as described above for the decision tables with single-valued decisions are performed as well.

In Table 10.2 (which is a part of Table 7.3), the first column "Decision table T" refers to the name of the new decision table T (that we get after removing attributes from the decision table from the UCI ML Repository), the second column "Original data set – removed attributes" refers to the name of the original decision table along with the indexes of attributes removed from the original decision table, the column "Rows" refers to the number of rows, the column "Attr" refers to the number of

Table 10.2 Decision tables with many-valued decisions based on tables from UCI ML repository

Decision table T	Original data set – removed attributes	Rows	Attr	$|D(T)|$	Spectrum #1, #2, #3, ...
CARS-1	CARS 1	432	5	4	258, 161, 13
LYMPH-5	LYMPHOGRAPHY 1, 13, 14, 15, 18	122	13	4	113, 9
LYMPH-4	LYMPHOGRAPHY 13, 14, 15, 18	136	14	4	132, 4
NURSERY-4	NURSERY 1, 5, 6, 7	240	4	5	97, 96, 47
NURSERY-1	NURSERY 1	4320	7	5	2858, 1460, 2
POKER-5B	POKER-HAND 2, 3, 4, 6, 8	3323	5	10	130, 1850, 1137, 199, 6, 1
POKER-5C	POKER-HAND 2, 4, 6, 8, 10	1024	5	10	0, 246, 444, 286, 44, 4
ZOO-5	ZOO-DATA 2, 6, 8, 9, 13	43	11	7	40, 1, 2
ZOO-4	ZOO-DATA 2, 9, 13, 14	44	12	7	40, 4
ZOO-2	ZOO-DATA 6, 13	46	14	7	44, 2

conditional attributes, the column "$|D(T)|$" refers to the total number of decisions in T, and the column "Spectrum" refers to a sequence #1, #2, #3, ..., where #i means the number of rows in T that are labeled with sets of decisions containing i decisions.

10.7 Experimental Results: Knowledge Representation

In this section, we consider the problem of knowledge representation. For a given decision table T, in the framework of MP or RMP approaches, we construct the set of Pareto-optimal points $POP_{m_1,m_2,U}(T)$ for two parameters of decision trees: number of nodes and number of misclassifications. We choose a POP with minimum Euclidean distance from the origin (see arrow sign in Fig. 10.1 where POPs are shown such that x axis corresponds to L, i.e., the number of nodes and y axis corresponds to mc, i.e., the number of misclassifications). In many cases, the chosen point corresponds to decision trees with reasonable number of nodes and number of misclassifications. For example, for HAYES-ROTH-DATA decision table, the number of nodes in the chosen tree is equal to 13, and the number of misclassifications is equal to 8 (the misclassification error rate is equal to 12%). Such trees can be used for knowledge representation. Also note that the user can choose other point and derive a decision tree with parameters corresponding to this point. For example, if the user chooses the

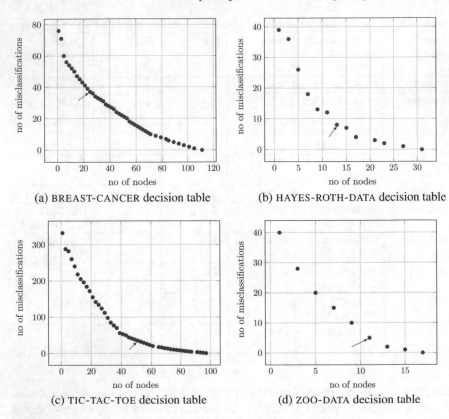

Fig. 10.1 Pareto-optimal points for decision tables with single-valued decisions and uncertainty measure *gini*

number of nodes equal to 9, then corresponding number of misclassifications will be 13 (the misclassification error rate is equal to 19%).

10.7.1   Decision Tables with Single-valued Decisions

We used 13 decision tables with single-valued decisions described in Table 10.1 (excluding GLASS and WINE). The results are shown for two uncertainty measures: *gini* and *abs*. For each decision table, we created the sets of POPs using both MP and RMP approaches. The aim of this experiment is to compare *gini* and *abs*, and to choose (m_1, m_2) RMP algorithm with small sum $m_1 + m_2$ and enough good results.

To compare the results among 35 RMP algorithms (we have 35 combinations of m_1 and m_2), we calculate the minimum distance of POPs from the origin. The parameter m_2 corresponds to randomly chosen attributes. To make the result stable, we repeated the experiment 10 times and took the average. Therefore we get the

Table 10.3 Average values of ranks for the average minimum distance from the origin to the Pareto optimal points for decision tables with single-valued decisions

$m_1 \backslash m_2$	0	1	2	3	4	5
(a) *gini* uncertainty measure						
0	N/A	33.69	32.54	30.23	25.42	24.04
1	24.85	23.77	20.65	18.73	17.88	15.58
2	22.15	18.81	17.12	16.27	14.58	13.42
3	17.08	15.73	14.96	14.19	13.88	13.62
4	15.96	15.38	14.69	13.81	13.31	**13.00**
5	15.54	14.96	14.31	13.46	**13.23**	**13.15**
(b) *abs* uncertainty measure						
0	N/A	33.69	32.54	30.00	25.27	23.81
1	24.77	23.38	21.50	18.35	17.23	15.38
2	22.27	19.85	16.54	15.42	13.65	13.12
3	18.77	16.65	15.54	14.19	13.58	13.23
4	16.38	15.54	14.69	13.96	13.15	**12.92**
5	15.96	14.92	14.42	13.38	**13.08**	**12.85**

average minimum distance of POPs from the origin. We name it by "avg_min_dist". We have results for each of 13 decision tables and for each algorithm. We can compare algorithms statistically over multiple decision tables by taking average of ranks [4]. For each decision table, we rank algorithms as Rank 1, Rank 2, and so on according to the value of "avg_min_dist". If there is a tie, we can break it by average of ranks. After that we take average of ranks over all decision tables. The average of ranks is shown in Table 10.3(a) for *gini*. We highlighted the top three algorithms that have the lowest ranks. Similar rankings are shown in Table 10.3(b) for *abs*.

We found $(5, 4)$ RMP algorithm (i.e., when $m_1 = 5$ and $m_2 = 4$), $(4, 5)$ RMP algorithm, and $(5, 5)$ RMP algorithm are the three top algorithms in terms of minimum average of ranks for both *gini* and *abs*. However, we are interested in finding of reasonable (m_1, m_2) RMP algorithms with smaller sum of m_1 and m_2.

We compare the minimum distance of POPs from the origin for MP approach with the best results for RMP approach in Table 10.4. For the case of RMP, we include the algorithm that produce the minimum distance (for the case of multiple algorithms, we took one with the minimum of sum of m_1 and m_2). It is clear that the results for RMP are very close to the results for MP, in fact they are the same for *gini* but slightly higher for *abs*. The last row shows the average value of the minimum distance.

The $(2, 0)$ RMP algorithm is closer to the MP algorithm for the maximum number of the decision tables. Therefore we compared the results for MP and $(2, 0)$ RMP algorithms in Table 10.5. One can see that, on average, the distance from the origin to the set of POPs for $(2, 0)$ RMP algorithm is very close to the distance for MP algorithm.

Table 10.4 Comparison of minimum distances from the origin to the Pareto optimal points for decision tables with single-valued decisions

Decision table	MP *gini*	Best RMP *gini*		MP *abs*	Best RMP *abs*	
	Distance	Distance	(m_1, m_2)	Distance	Distance	(m_1, m_2)
BALANCE-SCALE	97.05	97.05	(1, 1)	97.05	97.05	(3, 0)
BANKNOTE	19.21	19.21	(1, 1)	19.21	19.21	(1, 1)
BREAST-CANCER	44.65	44.65	(2, 4)	44.65	44.65	(2, 4)
CARS	76.06	76.06	(2, 1)	76.05	76.06	(4, 0)
HAYES-ROTH-DATA	15.26	15.26	(2, 0)	15.26	15.26	(2, 0)
HOUSE-VOTES-84	14.21	14.21	(2, 0)	14.21	14.21	(2, 0)
IRIS	7.81	7.81	(2, 0)	7.81	7.81	(2, 0)
LYMPHOGRAPHY	20.25	20.25	(4, 5)	20.25	20.52	(3, 3)
NURSERY	248.24	248.24	(2, 0)	248.24	248.24	(2, 0)
SOYBEAN-SMALL	7.00	7.00	(2, 0)	7.00	7.00	(2, 0)
SPECT-TEST	8.06	8.06	(2, 0)	8.06	8.06	(2, 0)
TIC-TAC-TOE	61.85	61.85	(1, 1)	61.85	61.85	(1, 1)
ZOO-DATA	12.08	12.08	(2, 0)	12.08	12.08	(2, 0)
Average:	48.60	48.60		48.60	48.62	

Table 10.5 Comparison of minimum distance from the origin to the Pareto optimal points for MP and (2, 0) RMP algorithms for decision tables with single-valued decisions

Decision table	MP *gini*	(2, 0) RMP *gini*	MP *abs*	(2, 0) RMP *abs*
BALANCE-SCALE	97.05	97.06	97.05	98.84
BANKNOTE	19.21	21.40	19.21	21.40
BREAST-CANCER	44.65	47.85	44.65	47.85
CARS	76.06	79.20	76.06	79.20
HAYES-ROTH-DATA	15.26	15.26	15.26	15.26
HOUSE-VOTES-84	14.21	14.21	14.21	14.21
IRIS	7.81	7.81	7.81	7.81
LYMPHOGRAPHY	20.25	21.93	20.25	21.93
NURSERY	248.24	248.24	248.24	248.24
SOYBEAN-SMALL	7.00	7.00	7.00	7.00
SPECT-TEST	8.06	8.06	8.06	8.06
TIC-TAC-TOE	61.85	64.82	61.85	64.82
ZOO-DATA	12.08	12.08	12.08	12.08
Average:	48.60	49.61	48.60	49.75

The obtained results show that, for knowledge representation for decision tables with single-valued decisions, instead of *gini* we can use *abs* and instead of MP algorithm we can use $(2, 0)$ RMP algorithm.

10.7.2 Decision Tables with Many-valued Decisions

The experiments are performed also with the decision tables with many-valued decisions. We took eight decision tables from Table 10.2 (excluding the POKER-HAND-decision tables). In these experiments, we only used the uncertainty measure *abs*. The aim of the experiments is to choose a (m_1, m_2) RMP algorithm with small sum of m_1 and m_2 which can be used instead of MP algorithm.

For each decision table, we created the set of POPs using both MP and RMP approaches. We compare the results in the same way as described in previous section. The average values of ranks of 35 RMP algorithms are shown in Table 10.6 in which the top three algorithms are highlighted.

For the case of decision tables with many-valued decisions, $(5, 5)$ RMP algorithm, $(5, 4)$ RMP algorithm, and $(5, 3)$ RMP algorithm are the top algorithms in terms of average of ranks. However, we are interested in finding of reasonable (m_1, m_2) RMP algorithms with smaller sum of m_1 and m_2.

We compared the minimum distance for MP and RMP algorithms in Table 10.7. For the case of RMP, we include the algorithm that gives the minimum distance (for the case of multiple algorithms, we took one with the minimum sum of m_1 and m_2). One can see that the minimum distance from the origin is the same for MP and RMP approaches.

Since $(2, 0)$ RMP algorithm is closer to the MP algorithm for the maximum number of decision tables, we compared the results for MP and $(2, 0)$ RMP algorithms in Table 10.8. On average, the distance from the origin to the set of POPs for $(2, 0)$ RMP algorithm is very close to the distance for MP algorithm.

Table 10.6 Average values of ranks for the average minimum distance from the origin to the Pareto optimal points for decision tables with many-valued decisions

$m_1 \backslash m_2$	0	1	2	3	4	5
0	N/A	35.00	34.00	33.00	28.12	26.75
1	21.56	20.56	20.19	19.31	18.44	16.62
2	19.88	17.94	17.31	16.88	14.5	13.38
3	19.88	18.31	16.88	14.88	14.31	13.81
4	16.38	15.50	13.69	13.19	13.06	12.69
5	12.81	12.56	12.44	**12.12**	**12.12**	**11.94**

Table 10.7 Comparison of minimum distances from the origin to the Pareto optimal points for decision tables with many-valued decisions

Decision table	MP *abs*	Best RMP *abs*	
	Distance	Distance	(m_1, m_2)
CARS-1	9.22	9.22	$(2, 0)$
LYMPH-5	19.24	19.24	$(4, 0)$
LYMPH-4	22.67	22.67	$(2, 2)$
NURSERY-4	5.00	5.00	$(2, 0)$
NURSERY-1	30.81	30.81	$(1, 1)$
ZOO-5	10.30	10.30	$(2, 0)$
ZOO-4	12.04	12.04	$(2, 0)$
ZOO-2	11.40	11.40	$(2, 0)$
Average:	15.08	15.08	

Table 10.8 Comparison of minimum distance from the origin to the Pareto optimal points for MP and $(2, 0)$ RMP algorithms for decision tables with many-valued decisions

Decision table	MP *abs*	$(2, 0)$ RMP *abs*
CARS-1	9.22	9.22
LYMPH-5	19.24	20.12
LYMPH-4	22.67	23.85
NURSERY-4	5.00	5.00
NURSERY-1	30.80	32.20
ZOO-5	10.30	10.30
ZOO-4	12.04	12.04
ZOO-2	11.40	11.40
Average:	15.08	15.52

The obtained results show that, for knowledge representation for decision tables with many-valued decisions, instead of MP algorithm we can use $(2, 0)$ RMP algorithm.

10.8   Experimental Results: Classification

The second group of experiments refers to the problem of classification when we should predict decisions for unknown instances. We use the misclassification error rate (the number of misclassifications divided by the whole number of rows) as a performance measure for classifiers. Let T be a decision table. We divide T into three parts: T_{train}, T_{val}, and T_{test}. The first part, T_{train}, is used to train the model, i.e., decision trees for the classification task. We construct the DAG

$G_{m_1,m_2,U}(T_{\text{train}})$ and based on this DAG we construct the set of Pareto optimal points $POP_{m_1,m_2,U}(T_{\text{train}})$. For each point $(a, b) \in POP_{m_1,m_2,U}(T_{\text{train}})$, we derive randomly k decision trees (in our experiments, $k = 5$) $\Gamma_1, \ldots, \Gamma_k$ from $DT_{m_1,m_2,U}(T_{\text{train}})$ such that $(a, b) = (mc(T_{\text{train}}, \Gamma_i), L(\Gamma_i))$ for $i = 1, \ldots, k$. Among such decision trees, we choose a decision tree Γ which has the minimum number of misclassifications on T_{val}. The third part of table, T_{test} is used to measure the performance of the final classifier Γ. We applied 2-fold cross-validation: 50% of data is used for training and the rest is for testing. Within 50% of data for training, 70% is for actual training and the rest 30% is for validation. We repeated the experiment five times and took the average misclassification error rate.

10.8.1 Decision Tables with Single-valued Decisions

We used 15 decision tables with single-valued decisions from the UCI ML Repository described in Table 10.1. The results are obtained using both *gini* and *abs*. For each of the uncertainty measures and each decision table, we have a classifier based on MP approach and 35 classifiers based on RMP approach. The aim of experiments is to compare *gini* and *abs*, and to choose (m_1, m_2) RMP algorithms with small sum $m_1 + m_2$ comparable with MP algorithm from the point of view of classifier accuracy and having less time complexity.

10.8.1.1 Uncertainty Measure *gini*

First, for *gini* uncertainty measure, we show the average of ranks of the 35 classifiers in Table 10.9(a) based on misclassification error rate where we rank Rank 1 the classifier which gives the minimum misclassification error rate. We highlighted the top three algorithms. We show also the average of ranks of the 35 classifiers in Table 10.9(b) based on the time of construction and usage of the classifiers where the lowest time consuming algorithm gets the Rank 1, and we also highlighted the top three algorithms.

From the results presented in Table 10.9 it follows that the classifiers based on $(2, 0)$ RMP algorithm and $(3, 0)$ RMP algorithm are the top ranked classifiers (note that classifiers based on $(5, 2)$ RMP algorithm are too time consuming and classifiers based on $(1, 0)$ RMP algorithm have bad misclassification error rate). Now, we compare the top ranked classifiers based on MP and RMP approaches with CART in Table 10.10 for misclassification error rate, and in Table 10.11 for the time of construction and usage.

From the point of view of the accuracy, the classifiers based on MP algorithm outperform CART classifiers for 11 out of 15 decision tables. Classifiers based on $(2, 0)$ RMP algorithm outperform CART classifiers in ten cases and have ties in one case, and classifiers based on $(3, 0)$ RMP algorithm outperform CART classifiers in ten cases. The average time for the construction and usage of the classifiers based on

Table 10.9 Average values of ranks of classifiers based on *gini* for decision tables with single-valued decisions

$m_1 \backslash m_2$	0	1	2	3	4	5
(a) Relative to misclassification error rate						
0	N/A	31.87	24.80	23.80	17.47	17.33
1	21.63	21.57	22.23	17.40	15.30	16.27
2	**15.23**	15.90	16.83	18.23	19.43	15.93
3	**15.13**	15.40	17.43	18.10	15.57	16.43
4	17.00	17.07	19.00	17.70	16.87	17.03
5	15.90	15.83	**14.33**	16.77	16.13	17.07
(b) Relative to time for construction and usage						
0	N/A	15.47	08.87	07.13	09.67	11.40
1	**04.53**	05.33	05.33	07.93	10.93	11.67
2	**04.33**	18.67	22.53	10.07	28.27	29.40
3	**03.40**	21.73	22.87	27.73	28.00	25.93
4	20.00	22.20	23.60	26.33	29.67	25.87
5	22.00	22.00	19.93	27.60	24.13	25.47

Table 10.10 Comparison of average misclassification error rates of classifiers based on *gini* for decision tables with single-valued decisions

Decision table	CART	MP	RMP	
			(3, 0)	(2, 0)
BALANCE-SCALE	23.81	23.68	24.70	25.83
BANKNOTE	3.20	1.91	1.96	2.14
BREAST-CANCER	29.17	29.32	29.77	28.12
CARS	5.16	5.08	5.03	4.86
GLASS	39.34	38.49	37.93	40.38
HAYES-ROTH-DATA	43.26	30.99	26.97	25.28
HOUSE-VOTES-84	6.81	6.67	6.60	7.03
IRIS	5.03	5.16	5.43	5.03
LYMPHOGRAPHY	27.97	24.86	28.92	24.86
NURSERY	1.40	1.44	1.39	1.28
SOYBEAN-SMALL	21.36	6.70	9.35	12.32
SPECT-TEST	5.21	4.74	4.74	4.74
TIC-TAC-TOE	10.21	7.81	5.97	7.52
WINE	12.47	11.80	11.35	11.57
ZOO-DATA	23.74	25.82	25.14	25.48
Average:	17.21	14.97	15.02	15.10

Table 10.11 Comparison of average time in seconds to construct and use classifiers based on *gini* for decision tables with single-valued decisions

Decision table	MP	RMP	
		(3, 0)	(2, 0)
BALANCE-SCALE	149.19	60.94	76.90
BANKNOTE	66.30	13.31	17.39
BREAST-CANCER	114.20	24.46	34.35
CARS	194.69	71.83	100.09
GLASS	1387.40	24.41	29.93
HAYES-ROTH-DATA	23.54	9.49	13.71
HOUSE-VOTES-84	24.92	9.70	9.77
IRIS	12.48	7.41	4.41
LYMPHOGRAPHY	125.06	16.28	13.07
NURSERY	1399.62	419.75	362.96
SOYBEAN-SMALL	13.44	4.21	3.32
SPECT-TEST	30.04	7.34	6.92
TIC-TAC-TOE	324.67	68.67	54.73
WINE	204.31	6.47	6.45
ZOO-DATA	19.06	8.28	7.01
Average:	272.59	50.17	49.40

RMP approach is more than five times less than for MP approach (note that CART time complexity is very small compared to any approach).

From the obtained results it follows that, in the case of classification problem, decision tables with single-valued decisions and *gini* uncertainty measure, we can use (2, 0) RMP and (3, 0) RMP algorithms instead of MP algorithm.

10.8.1.2 Uncertainty Measure *abs*

We did similar experiments with *abs*. In Table 10.12(a), the average of ranks relative to the misclassification error rate among 35 types of classifiers are shown and the top three types of classifiers are highlighted. Similar ranking is shown in Table 10.12(b) when we compare the time of construction and usage among 35 types of classifiers and the top three types of classifiers are highlighted also.

We can see that only classifiers based on (3, 0) RMP algorithm are among the top classifiers for both classification and time perspective. We also find that classifiers based on (2, 0) RMP algorithm are top based on the time and their rank based on classification accuracy is not far from the top ranked classifiers. Therefore, we include classifiers based on (3, 0) RMP and (2, 0) RMP algorithms for the comparison with CART as well as classifiers based on MP approach (see Tables 10.13 and 10.14).

Table 10.12 Average values of ranks of classifiers based on *abs* for decision tables with single-valued decisions

$m_1 \backslash m_2$	0	1	2	3	4	5
(a) Relative to misclassification error rate						
0	N/A	32.37	27.23	22.67	19.47	20.87
1	25.43	23.03	17.93	**14.57**	15.57	**14.40**
2	14.93	15.50	15.90	19.17	18.67	16.93
3	**13.73**	15.20	17.37	16.63	16.43	15.37
4	14.83	16.93	18.20	17.10	18.83	16.80
5	15.17	15.70	14.60	17.40	17.83	17.23
(b) Relative to time for construction and usage						
0	N/A	28.00	12.33	10.13	11.13	17.53
1	**3.80**	12.93	8.93	12.60	22.47	28.73
2	**2.00**	15.73	18.93	24.20	23.27	29.07
3	**7.13**	14.27	20.93	27.87	24.93	19.73
4	17.60	14.13	22.87	24.87	21.73	20.60
5	17.67	14.20	20.00	20.00	19.00	20.67

Table 10.13 Comparison of average misclassification error rates of classifiers based on *abs* for decision tables with single-valued decisions

Decision table	CART	MP	RMP	
			(3, 0)	(2, 0)
BALANCE-SCALE	23.81	24.26	24.35	24.38
BANKNOTE	3.20	1.91	1.96	2.14
BREAST-CANCER	29.17	29.85	30.6	27.89
CARS	5.16	5.20	5.02	5.02
GLASS	39.34	40.56	37.37	39.54
HAYES-ROTH-DATA	43.26	32.75	26.70	23.78
HOUSE-VOTES-84	6.81	6.67	6.60	7.10
IRIS	5.03	5.16	5.57	5.30
LYMPHOGRAPHY	27.97	25.68	27.70	24.46
NURSERY	1.40	1.43	1.40	1.33
SOYBEAN-SMALL	21.36	6.70	9.35	12.32
SPECT-TEST	5.21	4.74	4.74	4.74
TIC-TAC-TOE	10.21	7.72	5.95	7.77
WINE	12.47	11.46	11.24	11.57
ZOO-DATA	23.74	26.48	23.46	25.11
Average:	17.21	15.37	14.80	14.83

Table 10.14 Comparison of average time in seconds to construct and use classifiers based on *abs* for decision tables with single-valued decisions

Decision table	MP	RMP	
		(3, 0)	(2, 0)
BALANCE-SCALE	12.95	78.01	70.82
BANKNOTE	17.09	18.66	16.23
BREAST-CANCER	32.21	30.82	31.79
CARS	20.53	95.36	90.14
GLASS	5627.34	33.25	38.18
HAYES-ROTH-DATA	1.56	15.14	15.07
HOUSE-VOTES-84	4.35	15.24	12.57
IRIS	0.84	14.00	4.88
LYMPHOGRAPHY	51.97	31.00	14.74
NURSERY	483.39	959.63	410.19
SOYBEAN-SMALL	1.63	10.54	4.13
SPECT-TEST	7.32	18.51	8.17
TIC-TAC-TOE	117.89	141.16	64.91
WINE	179.55	14.51	8.18
ZOO-DATA	1.71	20.50	8.43
Average:	437.36	99.76	53.23

Table 10.13 shows that classifiers based on MP algorithm outperform CART classifiers for eight out of 15 decision tables. However, classifiers based on (3, 0) RMP algorithm outperform CART classifiers in 11 cases and have ties in one cases. Also classifiers based on (2, 0) RMP algorithm outperform CART classifiers in ten cases. The average time for the construction and usage of classifiers based on RMP approach is four times or eight times less than for MP classifiers (see Table 10.14).

The obtained results show that, in the case of classification problem, decision tables with single-valued decisions and *abs* uncertainty measure, we can use (2, 0) RMP and (3, 0) RMP algorithms instead of MP algorithm. The comparison of Tables 10.10 and 10.11 with Tables 10.13 and 10.14 shows also that instead of *gini* we can use *abs* uncertainty measure for classification problem and decision tables with single-valued decisions.

10.8.2 Decision Tables with Many-valued Decisions

We work also with classification problem for the decision tables with many-valued decisions. We took ten decision tables described in Table 10.2.

The experiment settings are similar to the case of decision tables with single-valued decisions, and we have results for both MP and RMP approaches. The aver-

Table 10.15 Average values of ranks of classifiers based on *abs* for decision tables with many-valued decisions

$m_1 \backslash m_2$	0	1	2	3	4	5
(a) Relative to the misclassification error rate						
0	N/A	29.90	28.85	21.45	15.85	17.85
1	25.25	25.50	22.75	21.50	20.85	19.55
2	**11.85**	**11.35**	13.40	17.30	20.10	17.90
3	15.70	13.60	15.10	14.65	17.15	17.35
4	**12.20**	18.00	15.30	15.50	17.05	16.20
5	13.50	17.25	19.20	18.10	16.65	16.30
(b) Relative to the time for construction and usage						
0	N/A	18.40	6.50	24.10	25.20	31.10
1	**2.90**	8.70	6.00	24.10	26.10	27.20
2	**1.80**	14.80	8.80	24.70	25.00	25.60
3	**5.10**	14.40	14.50	27.20	26.70	24.90
4	8.90	11.90	10.10	23.10	28.10	22.60
5	15.20	10.90	8.70	26.90	26.60	23.20

Table 10.16 Average misclassification error rates of classifiers based on *abs* for decision tables with many-valued decisions

Decision table	MP	RMP		
		(4, 0)	(3, 0)	(2,0)
CARS-1	1.53	1.57	1.57	1.62
LYMPH-5	28.20	26.89	27.38	26.07
LYMPH-4	25.29	26.76	28.38	28.97
NURSERY-4	1.17	1.17	1.17	1.17
NURSERY-1	0.69	0.63	0.63	0.57
POKER-5B	2.42	2.34	2.35	2.35
POKER-5C	0.27	0.27	0.27	0.27
ZOO-5	31.26	29.39	31.30	28.96
ZOO-4	38.64	39.55	35.91	37.27
ZOO-2	29.57	27.83	29.57	25.22
Average:	15.90	15.64	15.85	15.25

age of ranks based on misclassification error rate among classifiers constructed by 35 RMP algorithms are shown in Table 10.15(a) where the top three classifiers are highlighted. We have also the average of ranks based on the time for construction and usage of the classifiers in Table 10.15(b).

It is impossible to apply CART to decision tables with many-valued decisions. So we only compare accuracy of classifiers based on MP and RMP approaches

Table 10.17 Average time in seconds to construct and use of classifiers based on *abs* for decision tables with many-valued decisions

Decision table	MP	RMP		
		(4, 0)	(3, 0)	(2, 0)
CARS-1	1.23	1.54	1.32	1.28
LYMPH-5	2.83	3.24	2.89	2.32
LYMPH-4	4.58	3.46	2.91	2.55
NURSERY-4	0.79	0.80	0.90	0.73
NURSERY-1	26.13	17.90	9.02	5.31
POKER-5B	1072.20	1018.15	138.81	13.26
POKER-5C	1.10	1.14	0.99	0.84
ZOO-5	1.13	1.48	1.33	1.14
ZOO-4	1.08	1.61	1.29	1.10
ZOO-2	1.12	1.80	1.37	1.17
Average:	111.22	105.11	16.08	2.97

in Table 10.16. We took three types of classifiers according to the results from Table 10.15: classifiers based on $(2, 0)$ RMP algorithm are top classifiers relative to both accuracy and time, classifiers based on $(3, 0)$ RMP algorithm are top relative to the time, and classifiers based on $(4, 0)$ RMP algorithm are top relative to the misclassification error rate. The classifiers based on $(2, 0)$ RMP algorithm outperform the classifiers based on MP approach in six cases and have ties in two cases out of ten. We compared the time of classifier construction and usage in Table 10.17, and found that the classifiers based on $(2, 0)$ RMP algorithm require lowest time (on average, 37 times less than MP classifiers).

From the results presented in Tables 10.16 and 10.17 it follows that $(2, 0)$ RMP algorithm can be used instead of MP algorithm for the construction of classifiers for decision tables with many-valued decisions.

As a result, we have that $(2, 0)$ RMP algorithm based on *abs* uncertainty measure can be used both for decision tables with single-valued and many-valued decisions, and both for the problems of knowledge representation and classification instead of MP algorithm based on *gini* and *abs* uncertainty measures.

References

1. Azad, M., Chikalov, I., Hussain, S., Moshkov, M.: Multi-pruning of decision trees for knowledge representation and classification. In: 3rd IAPR Asian Conference on Pattern Recognition, ACPR 2015, Kuala Lumpur, Malaysia, 3–6 Nov 2015, pp. 604–608. IEEE (2015)
2. Azad, M., Chikalov, I., Hussain, S., Moshkov, M.: Restricted multi-pruning of decision trees. In: Liu, J., Lu, J., Xu, Y., Martinez, L., Kerre, E.E. (eds.) 13th International FLINS Conference on Data Science and Knowledge Engineering for Sensing Decision Support, FLINS 2018, Belfast,

Northern Ireland, UK, 21–24 Aug 2018, World Scientific Proceedings Series on Computer Engineering and Information Science, vol. 11, pp. 371–378 (2018). World Scientific (2018)

3. Breiman, L., Friedman, J.H., Olshen, R.A., Stone, C.J.: Classification and Regression Trees. Wadsworth and Brooks, Monterey (1984)

4. Demsar, J.: Statistical comparisons of classifiers over multiple data sets. J. Mach. Learn. Res. **7**, 1–30 (2006)

5. Lichman, M.: UCI Machine Learning Repository. University of California, Irvine, School of Information and Computer Sciences (2013). http://archive.ics.uci.edu/ml

Part III
Extensions of Dynamic Programming for Decision and Inhibitory Rules and Systems of Rules

This part is devoted to the development of extensions of dynamic programming for decision and inhibitory rules and rule systems for decision tables with many-valued decisions. It consists of four chapters.

In Chap. 11, we define various types of decision and inhibitory rules and systems of rules. We discuss the notion of cost function for rules, the notion of decision rule uncertainty, and the notion of inhibitory rule completeness. Similar notions are introduced for systems of decision and inhibitory rules.

In Chap. 12, we consider optimization of decision and inhibitory rules including multi-stage optimization relative to a sequence of cost functions. We discuss an algorithm for counting the number of optimal rules, and consider simulation of a greedy algorithm for construction of decision rule sets. We also discuss results of computer experiments with decision an inhibitory rules: existence of small systems of enough accurate decision rules that cover almost all rows, and existence of totally optimal decision and inhibitory rules that have minimum length and maximum coverage simultaneously.

In Chap. 13, we consider algorithms which construct the sets of Pareto optimal points for bi-criteria optimization problems for decision rules and rule systems relative to two cost functions. We show how the constructed set of Pareto optimal points can be transformed into the graphs of functions which describe the relationships between the considered cost functions. We compare 13 greedy heuristics for construction of decision rules from the point of view of single-criterion optimization (relative to the length or coverage) and bi-criteria optimization (relative to the length and coverage). At the end of the chapter, we generalize the obtained results to the case of inhibitory rules and systems of inhibitory rules.

In Chap. 14, we consider algorithms which construct the sets of Pareto optimal points for bi-criteria optimization problems for decision (inhibitory) rules and rule systems relative to a cost function and an uncertainty (completeness) measure. We show how the constructed set of Pareto optimal points can be transformed into the graphs of functions which describe the relationships between the considered cost function and uncertainty (completeness) measure. Computer experiments provide us with examples of trade-off between complexity and accuracy for decision and inhibitory rule systems.

Chapter 11
Decision and Inhibitory Rules and Systems of Rules

Decision rules are widely used in applications related to data mining and knowledge representation [5], and machine learning [3]. They are intensively studied in rough set theory [6, 7] and logical analysis of data [1, 2] (see also book [4]). Examples considered in Sect. 2.3 show that inhibitory rules can derive more knowledge from a decision table than decision rules, and can improve the quality of prediction in comparison with decision rules.

In this chapter, we consider various types of decision and inhibitory rules and systems of rules. We discuss the notion of cost function for rules, the notion of decision rule uncertainty, and the notion of inhibitory rule completeness. Similar notions are introduced for systems of decision and inhibitory rules.

11.1 Decision Rules and Systems of Rules

In this section, we discuss main notions related to decision rules and systems of decision rules.

11.1.1 Decision Rules

Let T be a decision table with n conditional attributes $f_1, \ldots, f_n$ and $r = (b_1, \ldots, b_n)$ be a row of T. A *decision rule over* T is an expression of the kind

$$(f_{i_1} = a_1) \wedge \ldots \wedge (f_{i_m} = a_m) \rightarrow t \tag{11.1}$$

where $f_{i_1}, \ldots, f_{i_m} \in \{f_1, \ldots, f_n\}$, and $a_1, \ldots, a_m, t$ are numbers from ω. It is possible that $m = 0$. For the considered rule, we denote $\beta_0 = \lambda$, and if $m > 0$ we denote

© Springer Nature Switzerland AG 2020
F. Alsolami et al., *Decision and Inhibitory Trees and Rules for Decision Tables with Many-valued Decisions*, Intelligent Systems Reference Library 156, https://doi.org/10.1007/978-3-030-12854-8_11

$\beta_j = (f_{i_1}, a_1) \dots (f_{i_j}, a_j)$ for $j = 1, \dots, m$. We will say that the decision rule (11.1) *covers* the row r if r belongs to $T\beta_m$, i.e., $b_{i_1} = a_1, \dots, b_{i_m} = a_m$.

A decision rule (11.1) over T is called a *decision rule for T* if $t = mcd(T\beta_m)$, and either $m = 0$, or $m > 0$ and, for $j = 1, \dots, m$, $T\beta_{j-1}$ is not degenerate, and $f_{i_j} \in E(T\beta_{j-1})$. A decision rule (11.1) for T is called a *decision rule for T and r* if it covers r.

We denote by $DR(T)$ the set of decision rules for T. By $DR(T, r)$ we denote the set of decision rules for T and r.

Let U be an uncertainty measure and $\alpha \in \mathbb{R}_+$. A decision rule (11.1) for T is called a *(U, α)-decision rule for T* if $U(T\beta_m) \le \alpha$ and, if $m > 0$, then $U(T\beta_j) > \alpha$ for $j = 0, \dots, m - 1$. A (U, α)-decision rule (11.1) for T is called a *(U, α)-decision rule for T and r* if it covers r.

We denote by $DR_{U,\alpha}(T)$ the set of (U, α)-decision rules for T, and we denote by $DR_{U,\alpha}(T, r)$ the set of (U, α) -decision rules for T and r.

We call a $(U, 0)$-decision rule for T and r an *exact decision rule for T and r*. The notion of an exact decision rule does not depend on the uncertainty measure U. Let U_1 and U_2 be uncertainty measures. Then the set of $(U_1, 0)$-decision rules for T and r coincides with the set of $(U_2, 0)$-decision rules for T and r.

A decision rule (11.1) for T is called a *U-decision rule for T* if there exists a nonnegative real number α such that (11.1) is a (U, α)-decision rule for T. A decision rule (11.1) for T and r is called a *U-decision rule for T and r* if there exists a nonnegative real number α such that (11.1) is a (U, α)-decision rule for T and r.

We denote by $DR_U(T)$ the set of U-decision rules for T. By $DR_U(T, r)$ we denote the set of U-decision rules for T and r.

We define *uncertainty $U(T, \rho)$ of a decision rule ρ for T relative to the table T* in the following way. Let ρ be equal to (11.1). Then $U(T, \rho) = U(T\beta_m)$.

We now consider a notion of *cost function for decision rules*. This is a function $\psi(T, \rho)$ which is defined on pairs T, ρ, where T is a nonempty decision table and ρ is a decision rule for T, and has values from the set $\mathbb{R}$ of real numbers. This cost function is given by pair of functions $\psi^0 : \mathscr{T}^+ \to \mathbb{R}$ and $F : \mathbb{R} \to \mathbb{R}$ where $\mathscr{T}^+$ is the set of nonempty decision tables. The value $\psi(T, \rho)$ is defined by induction:

- If ρ is equal to $\to mcd(T)$ then $\psi(T, \to mcd(T)) = \psi^0(T)$.
- If ρ is equal to $(f_i = a) \wedge \gamma \to t$ then

$$\psi(T, (f_i = a) \wedge \gamma \to t) = F(\psi(T(f_i, a), \gamma \to t)) .$$

The cost function ψ is called *strictly increasing* if $F(x_1) > F(x_2)$ for any $x_1, x_2 \in \mathbb{R}$ such that $x_1 > x_2$.

Let us consider examples of strictly increasing cost functions for decision rules:

- The *length $l(T, \rho) = l(\rho)$* for which $\psi^0(T) = 0$ and $F(x) = x + 1$. The length of the rule (11.1) is equal to m.
- The *coverage $c(T, \rho)$* for which $\psi^0(T) = N_{mcd(T)}(T)$ and $F(x) = x$. The coverage of the rule (11.1) for table T is equal to $N_{mcd(T\beta_m)}(T\beta_m)$.

- The *relative coverage* $rc(T, \rho)$ for which $\psi^0(T) = N_{mcd(T)}(T)/N(T)$ and $F(x) = x$. The relative coverage of the rule (11.1) for table T is equal to

$$N_{mcd(T^m)}(T\beta_m)/N(T\beta_m) .$$

- The *modified coverage* $c_M(T, \rho)$ for which $\psi^0(T) = N^M(T)$ and $F(x) = x$. Here M is a set of rows of T and, for any subtable Θ of T, $N^M(\Theta)$ is the number of rows of Θ which do not belong to M. The modified coverage of the rule (11.1) for table T is equal to $N^M(T\beta_m)$.
- The *miscoverage* $mc(T, \rho)$ for which $\psi^0(T) = N(T) - N_{mcd(T)}(T)$ and $F(x) = x$. The miscoverage of the rule (11.1) for table T is equal to

$$N(T\beta_m) - N_{mcd(T\beta_m)}(T\beta_m) .$$

- The *relative miscoverage* $rmc(T, \rho)$ for which

$$\psi^0(T) = (N(T) - N_{mcd(T)}(T))/N(T)$$

and $F(x) = x$. The relative miscoverage of the rule (11.1) for table T is equal to $(N(T\beta_m) - N_{mcd(T\beta_m)}(T\beta_m))/N(T\beta_m)$.

We need to minimize length, miscoverage and relative miscoverage, and maximize coverage, relative coverage, and modified coverage. However, we will consider only algorithms for the minimization of cost functions. Therefore, instead of maximization of coverage c we will minimize the negation of coverage $-c$ given by pair of functions $\psi^0(T) = -N_{mcd(T)}(T)$ and $F(x) = x$. Similarly, instead of maximization of relative coverage rc we will minimize the negation of relative coverage $-rc$ given by pair of functions $\psi^0(T) = -N_{mcd(T)}(T)/N(T)$ and $F(x) = x$. Instead of maximization of modified coverage c_M we will minimize the negation of modified coverage $-c_M$ given by pair of functions $\psi^0(T) = -N^M(T)$ and $F(x) = x$. The cost functions $-c$, $-rc$, and $-c_M$ are strictly increasing cost functions.

For a given cost function ψ and decision table T, we denote $Range_\psi(T) = \{\psi(\Theta, \rho) : \Theta \in SEP(T), \rho \in DR(\Theta)\}$. By $q_\psi(T)$ we denote the cardinality of the set $Range_\psi(T)$. It is easy to prove the following statement:

Lemma 11.1 *Let T be a decision table with n conditional attributes. Then*

- $Range_l(T) \subseteq \{0, 1, \ldots, n\}$, $q_l(T) \leq n + 1$.
- $Range_{-c}(T) \subseteq \{0, -1, \ldots, -N(T)\}$, $q_{-c}(T) \leq N(T) + 1$.
- $Range_{mc}(T) \subseteq \{0, 1, \ldots, N(T)\}$, $q_{mc}(T) \leq N(T) + 1$.
- $Range_{-rc}(T) \subseteq \{-a/b : a, b \in \{0, 1, \ldots, N(T)\}, b > 0\}$, $q_{-rc}(T) \leq N(T)(N(T) + 1)$.

- $Range_{-c_M}(T) \subseteq \{0, -1, \ldots, -N(T)\}$, $q_{-c_M}(T) \leq N(T) + 1$.
- $Range_{rmc}(T) \subseteq \{a/b : a, b \in \{0, 1, \ldots, N(T)\}, b > 0\}$, $q_{rmc}(T) \leq N(T)(N(T) + 1)$.

11.1.2 Systems of Decision Rules

Let T be a nonempty decision table with n conditional attributes $f_1, \ldots, f_n$ and $N(T)$ rows $r_1, \ldots, r_{N(T)}$, and U be an uncertainty measure.

A *system of decision rules for* T is an $N(T)$-tuple $S = (\rho_1, \ldots, \rho_{N(T)})$ where $\rho_1 \in DR(T, r_1), \ldots, \rho_{N(T)} \in DR(T, r_{N(T)})$. Let $\alpha \in \mathbb{R}_+$. The considered system is called a *(U, α)-system of decision rules for* T if $\rho_i \in DR_{U,\alpha}(T, r_i)$ for $i = 1, \ldots, N(T)$. This system is called a *U-system of decision rules for* T if $\rho_i \in DR_U(T, r_i)$ for $i = 1, \ldots, N(T)$.

We now consider a notion of *cost function for systems of decision rules*. This is a function $\psi_f(T, S)$ which is defined on pairs T, S, where T is a nonempty decision table and $S = (\rho_1, \ldots, \rho_{N(T)})$ is a system of decision rules for T, and has values from the set $\mathbb{R}$. This function is given by a cost function for decision rules ψ and a function $f : \mathbb{R}^2 \to \mathbb{R}$. The value $\psi_f(T, S)$ is equal to $f(\psi(T, \rho_1), \ldots, \psi(T, \rho_{N(T)}))$ where the value $f(x_1, \ldots, x_k)$, for any natural k, is defined by induction: $f(x_1) = x_1$ and, for $k > 2, f(x_1, \ldots, x_k) = f(f(x_1, \ldots, x_{k-1}), x_k)$.

The cost function for systems of decision rules ψ_f is called *strictly increasing* if ψ is strictly increasing cost function for decision rules and f is an increasing function from $\mathbb{R}^2$ to $\mathbb{R}$, i.e., $f(x_1, y_1) \leq f(x_2, y_2)$ for any $x_1, x_2, y_1, y_2 \in \mathbb{R}$ such that $x_1 \leq x_2$ and $y_1 \leq y_2$.

For example, if $\psi \in \{l, -c, -rc, -c_M, mc, rmc\}$ and

$$f \in \{\text{sum}(x, y) = x + y, \max(x, y)\},$$

then ψ_f is a strictly increasing cost function for systems of decision rules.

We now consider a notion of *uncertainty for systems of decision rules*. This is a function $U_g(T, S)$ which is defined on pairs T, S, where T is a nonempty decision table and $S = (\rho_1, \ldots, \rho_{N(T)})$ is a system of decision rules for T, and has values from the set $\mathbb{R}$. This function is given by an uncertainty measure U and a function $g : \mathbb{R}^2 \to \mathbb{R}$. The value $U_g(T, S)$ is equal to $g(U(T, \rho_1), \ldots, U(T, \rho_{N(T)}))$ where the value $g(x_1, \ldots, x_k)$, for any natural k, is defined by induction: $g(x_1) = x_1$ and, for $k > 2, g(x_1, \ldots, x_k) = g(g(x_1, \ldots, x_{k-1}), x_k)$.

11.2 Inhibitory Rules and Systems of Rules

In this section, we discuss main notions related to inhibitory rules and systems of inhibitory rules.

11.2.1 Inhibitory Rules

Let T be a nondegenerate decision table with n conditional attributes $f_1, \ldots, f_n$ and $r = (b_1, \ldots, b_n)$ be a row of T. An *inhibitory rule over* T is an expression of the kind

$$(f_{i_1} = a_1) \wedge \ldots \wedge (f_{i_m} = a_m) \to \neq t \tag{11.2}$$

where $f_{i_1}, \ldots, f_{i_m} \in \{f_1, \ldots, f_n\}$, and $a_1, \ldots, a_m, t$ are numbers from ω. It is possible that $m = 0$. For the considered rule, we denote $\beta_0 = \lambda$, and if $m > 0$ we denote $\beta_j = (f_{i_1}, a_1) \ldots (f_{i_j}, a_j)$ for $j = 1, \ldots, m$. We will say that the inhibitory rule (11.2) *covers* the row r if r belongs to $T\beta_m$, i.e., $b_{i_1} = a_1, \ldots, b_{i_m} = a_m$.

An inhibitory rule (11.2) over T is called an *inhibitory rule for* T if $t = lcd(T\beta_m)$, and either $m = 0$, or $m > 0$ and, for $j = 1, \ldots, m$, the subtable $T\beta_{j-1}$ is not incomplete relative to T, and $f_{i_j} \in E(T\beta_{j-1})$. An inhibitory rule (11.2) for T is called an *inhibitory rule for* T *and* r if it covers r.

We denote by $IR(T)$ the set of inhibitory rules for T. By $IR(T, r)$ we denote the set of inhibitory rules for T and r.

Let W be a completeness measure and $\alpha \in \mathbb{R}_+$. An inhibitory rule (11.2) for T is called a (W, α)-*inhibitory rule for* T if $W(T, T\beta_m) \leq \alpha$ and, if $m > 0$, then $W(T, T\beta_j) > \alpha$ for $j = 0, \ldots, m - 1$. A (W, α)-inhibitory rule (11.2) for T is called a (W, α)-*inhibitory rule for* T *and* r if it covers r.

We denote by $IR_{W,\alpha}(T)$ the set of (W, α)-inhibitory rules for T, and we denote by $IR_{W,\alpha}(T, r)$ the set of (W, α)-inhibitory rules for T and r.

We call a $(W, 0)$-inhibitory rule for T and r an *exact inhibitory rule for* T *and* r. The notion of an exact inhibitory rule does not depend on the completeness measure W. Let W_1 and W_2 be completeness measures. Then the set of $(W_1, 0)$-inhibitory rules for T and r coincides with the set of $(W_2, 0)$-inhibitory rules for T and r.

An inhibitory rule (11.2) for T is called a W-*inhibitory rule for* T if there exists a nonnegative real number α such that (11.2) is a (W, α)-inhibitory rule for T. An inhibitory rule (11.2) for T and r is called a W-*inhibitory rule for* T *and* r if there exists a nonnegative real number α such that (11.2) is a (W, α)-inhibitory rule for T and r.

We denote by $IR_W(T)$ the set of W-inhibitory rules for T. By $IR_W(T, r)$ we denote the set of W-inhibitory rules for T and r.

We define *completeness* $W(T, \rho)$ *of an inhibitory rule* ρ *for* T *relative to the table* T in the following way. Let ρ be equal to (11.2). Then $W(T, \rho) = W(T, T\beta_m)$.

We now consider a notion of *cost function for inhibitory rules*. This is a function $\psi(T, \rho)$ which is defined on pairs T, ρ, where T is a nonempty decision table and ρ is an inhibitory rule for T, and has values from the set $\mathbb{R}$ of real numbers. Let us consider examples of cost functions for inhibitory rules. Let ρ be the rule (11.2).

- The *length* $l(T, \rho) = l(\rho)$ which is equal to m.
- The *coverage* $c(T, \rho)$ which is equal to $N(T\beta_m) - N_{lcd(T, T\beta_m)}(T\beta_m)$.
- The *relative coverage* $rc(T, \rho)$ which is equal to

$$(N(T\beta_m) - N_{lcd(T,T\beta_m)}(T\beta_m))/N(T\beta_m) \ .$$

- The *miscoverage* $mc(T,\rho)$ which is equal to $N_{lcd(T,T\beta_m)}(T\beta_m)$.
- The *relative miscoverage* $rmc(T,\rho)$ which is equal to $N_{lcd(T,T\beta_m)}(T\beta_m)/N(T\beta_m)$.

We need to minimize length, miscoverage and relative miscoverage, and maximize coverage and relative coverage. However, we will consider only algorithms for the minimization of cost functions. Therefore, instead of maximization of coverage c we will minimize the negation of coverage $-c$. Similarly, instead of maximization of relative coverage rc we will minimize the negation of relative coverage $-rc$.

11.2.2 Systems of Inhibitory Rules

Let T be a nondegenerate decision table with n conditional attributes $f_1,\dots,f_n$ and $N(T)$ rows $r_1,\dots,r_{N(T)}$, and W be a completeness measure.

A *system of inhibitory rules for* T is an $N(T)$-tuple $S = (\rho_1,\dots,\rho_{N(T)})$ where $\rho_1 \in IR(T,r_1),\dots,\rho_{N(T)} \in IR(T,r_{N(T)})$. Let $\alpha \in \mathbb{R}_+$. The considered system is called a (W,α)-*system of inhibitory rules for* T if, for $i = 1,\dots,N(T)$,

$$\rho_i \in IR_{W,\alpha}(T,r_i) \ .$$

This system is called a W-*system of inhibitory rules for* T if $\rho_i \in IR_W(T,r_i)$ for $i = 1,\dots,N(T)$.

We now consider a notion of *cost function for systems of inhibitory rules*. This is a function $\psi_f(T,S)$ which is defined on pairs T,S, where T is a nonempty decision table and $S = (\rho_1,\dots,\rho_{N(T)})$ is a system of inhibitory rules for T, and has values from the set $\mathbb{R}$. This function is given by a cost function for inhibitory rules ψ and a function $f : \mathbb{R}^2 \to \mathbb{R}$. The value $\psi_f(T,S)$ is equal to $f(\psi(T,\rho_1),\dots,\psi(T,\rho_{N(T)}))$ where the value $f(x_1,\dots,x_k)$, for any natural k, is defined by induction: $f(x_1) = x_1$ and, for $k > 2$, $f(x_1,\dots,x_k) = f(f(x_1,\dots,x_{k-1}),x_k)$. Later we consider only cases when $\psi \in \{l,-c,-rc,mc,rmc\}$ and $f \in \{\mathrm{sum}(x,y) = x + y, \max(x,y)\}$.

We now consider a notion of *completeness for systems of inhibitory rules*. This is a function $W_g(T,S)$ which is defined on pairs T,S, where T is a nonempty decision table and $S = (\rho_1,\dots,\rho_{N(T)})$ is a system of inhibitory rules for T, and has values from the set $\mathbb{R}$. This function is given by a completeness measure W and a function $g : \mathbb{R}^2 \to \mathbb{R}$. The value $W_g(T,S)$ is equal to

$$g(W(T,\rho_1),\dots,W(T,\rho_{N(T)}))$$

where the value $g(x_1,\dots,x_k)$, for any natural k, is defined by induction: $g(x_1) = x_1$ and, for $k > 2$, $g(x_1,\dots,x_k) = g(g(x_1,\dots,x_{k-1}),x_k)$. Later we consider only cases when $g \in \{\mathrm{sum}(x,y) = x + y, \max(x,y)\}$.

Let ρ_1 be an inhibitory rule over T and ρ_2 be a decision rule over T^C. We denote by ρ_1^+ a decision rule over T^C obtained from ρ_1 by changing the right hand side of ρ_1: if the right hand side of ρ_1 is $\neq t$, then the right hand side of ρ_1^+ is t. We denote by ρ_2^- an inhibitory rule over T obtained from ρ_2 by changing the right hand side of ρ_2: if the right hand side of ρ_2 is t, then the right hand side of ρ_2^- is $\neq t$. It is clear that $(\rho_1^+)^- = \rho_1$ and $(\rho_2^-)^+ = \rho_2$. Let A be a set of inhibitory rules over T. We denote $A^+ = \{\rho^+ : \rho \in A\}$. Let B be a set of decision rules over T^C. We denote $B^- = \{\rho^- : \rho \in B\}$. It is clear that $(A^+)^- = A$ and $(B^-)^+ = B$.

Proposition 11.1 *Let T be a nondegenerate decision table with attributes $f_1, \ldots,$ f_n, r be a row of T, ρ be a decision rule over T^C, W be a completeness measure, U be an uncertainty measure, W and U are dual, and $\alpha \in \mathbb{R}_+$. Then*

1. *$\rho \in DR(T^C)$ if and only if $\rho^- \in IR(T)$.*
2. *$\rho \in DR_{U,\alpha}(T^C)$ if and only if $\rho^- \in IR_{W,\alpha}(T)$.*
3. *$\rho \in DR_U(T^C)$ if and only if $\rho^- \in IR_W(T)$.*
4. *$\rho \in DR(T^C, r)$ if and only if $\rho^- \in IR(T, r)$.*
5. *$\rho \in DR_{U,\alpha}(T^C, r)$ if and only if $\rho^- \in IR_{W,\alpha}(T, r)$.*
6. *$\rho \in DR_U(T^C, r)$ if and only if $\rho^- \in IR_W(T, r)$.*
7. *$W(T, \rho^-) = U(T^C, \rho)$.*
8. *If $\rho \in DR(T^C)$ then $\psi(T, \rho^-) = \psi(T^C, \rho)$ for any $\psi \in \{l, -c, -rc, mc, rmc\}$.*

Proof Let ρ^- be the rule (11.2). Since W and U are dual, $W(T, T\beta_j) = U(T^C\beta_j)$ for $j = 0, \ldots, m$. From Lemma 5.1 it follows that $T\beta_j$ is incomplete relative to T if and only if $T^C\beta_j$ is degenerate for $j = 0, \ldots, m$. It is clear that $E(T\beta_j) = E(T^C\beta_j)$ for $j = 0, \ldots, m$. By Lemma 5.1, $mcd(T^C\beta_m) = lcd(T, T\beta_m)$. It is clear that ρ covers r if and only if ρ^- covers r. Using these facts it is not difficult to show that the statements 1–7 of the considered proposition hold.

Let $\rho \in DR(T^C)$. It is clear that $l(T, \rho^-) = l(T^C, \rho)$. Set $\beta = \beta_m$ and $d = mcd(T^C\beta) = lcd(T, T\beta)$. By Lemma 5.1, $N(T\beta) = N(T^C\beta)$ and

$$N_d(T\beta) = N(T^C\beta) - N_d(T^C\beta) .$$

Therefore $-N(T\beta) + N_d(T\beta) = -N_d(T^C\beta)$. As a result, we have

$$-c(T, \rho^-) = -N(T\beta) + N_d(T\beta) = -N_d(T^C\beta) = -c(T^C, \rho) ,$$
$$-rc(T, \rho^-) = (-N(T\beta) + N_d(T\beta))/N(T\beta) = -N_d(T^C\beta)/N(T^C\beta)$$
$$= -rc(T^C, \rho) ,$$
$$mc(T, \rho^-) = N_d(T\beta) = N(T^C\beta) - N_d(T^C\beta) = mc(T^C, \rho) ,$$
$$rmc(T, \rho^-) = N_d(T\beta)/N(T\beta) = (N(T^C\beta) - N_d(T^C\beta))/N(T^C\beta)$$
$$= rmc(T^C, \rho) .$$

Therefore the statement 8 of the considered proposition holds. $\qquad\square$

Corollary 11.1 *Let T be a nondegenerate decision table, r be a row of T, W be a completeness measure, U be an uncertainty measure, W and U are dual, and $\alpha \in \mathbb{R}_+$. Then*

1. $IR(T) = DR(T^C)^-$.
2. $IR_{W,\alpha}(T) = DR_{U,\alpha}(T^C)^-$.
3. $IR_W(T) = DR_U(T^C)^-$.
4. $IR(T,r) = DR(T^C, r)^-$.
5. $IR_{W,\alpha}(T,r) = DR_{U,\alpha}(T^C, r)^-$.
6. $IR_W(T,r) = DR_U(T^C, r)^-$.

Proof Let $\rho_1 \in DR(T^C)$. Then, by the statement 1 of Proposition 11.1, $\rho_1^- \in IR(T)$. Therefore $DR(T^C)^- \subseteq IR(T)$. Let $\rho_2 \in IR(T)$. Then ρ_2^+ is a decision rule over T^C and $\rho_2 = (\rho_2^+)^-$. By the statement 1 of Proposition 11.1, $\rho_2^+ \in DR(T^C)$. Therefore $\rho_2 \in DR(T^C)^-$ and $DR(T^C)^- \supseteq IR(T)$. Hence the statement 1 of the corollary holds. The statements 2–6 can be proven in a similar way. $\qquad\square$

References

1. Boros, E., Hammer, P.L., Ibaraki, T., Kogan, A.: Logical analysis of numerical data. Math. Progr. **79**, 163–190 (1997)
2. Boros, E., Hammer, P.L., Ibaraki, T., Kogan, A., Mayoraz, E., Muchnik, I.: An implementation of logical analysis of data. IEEE Trans. Knowl. Data Eng. **12**, 292–306 (2000)
3. Carbonell, J.G., Michalski, R.S., Mitchell, T.M.: An overview of machine learning. In: Michalski, R.S., Carbonell, J.G., Mitchell, T.M. (eds.) Machine Learning, An Artificial Intelligence Approach, pp. 1–23. Tioga Publishing, Palo Alto, CA (1983)
4. Chikalov, I., Lozin, V.V., Lozina, I., Moshkov, M., Nguyen, H.S., Skowron, A., Zielosko, B.: Three Approaches to Data Analysis – Test Theory, Rough Sets and Logical Analysis of Data. Intelligent Systems Reference Library, vol. 41. Springer, Berlin (2013)
5. Fayyad, U., Piatetsky-Shapiro, G., Smyth, P.: From data mining to knowledge discovery in databases. AI Mag. **17**, 37–54 (1996)
6. Pawlak, Z.: Rough Sets – Theoretical Aspect of Reasoning About Data. Kluwer Academic Publishers, Dordrecht (1991)
7. Pawlak, Z., Skowron, A.: Rudiments of rough sets. Inf. Sci. **177**(1), 3–27 (2007)

Chapter 12
Multi-stage Optimization of Decision and Inhibitory Rules

In this chapter, we study optimization of decision rules for decision tables with many-valued decisions including multi-stage optimization relative to a sequence of cost functions. We discuss an algorithm for counting the number of optimal rules, and consider simulation of a greedy algorithm for construction of decision rule sets. We also discuss results of computer experiments with decision rules: existence of small systems of enough accurate decision rules that cover almost all rows, and existence of totally optimal decision rules that have minimum length and maximum coverage simultaneously. The considered algorithms are extensions of algorithms designed previously for decision tables with single-valued decisions [1].

We generalized the most part of created tools to the case of inhibitory rules for decision tables with many-valued decisions, and experimentally confirmed the existence of totally optimal inhibitory rules in many cases. Note that the dynamic programming algorithms for optimization of inhibitory rules for decision tables with single-valued decisions relative to the length, coverage, and number of misclassifications were studied in [4–6]. Multi-stage optimization of inhibitory rules for decision tables with single-valued decisions was considered in [3].

Some generalizations of the obtained results to the case of association rules can be found in [2].

12.1 Multi-stage Optimization of Decision Rules

In this section, we concentrate on optimization of decision rules.

12.1.1 Representation of the Set of (U, α)-Decision Rules

Let T be a nonempty decision table with n conditional attributes $f_1, \ldots, f_n$, U be an uncertainty measure, $\alpha \in \mathbb{R}_+$, and G be a bundle-preserving subgraph of $\Delta_{U,\alpha}(T)$ (it is possible that $G = \Delta_{U,\alpha}(T)$).

© Springer Nature Switzerland AG 2020
F. Alsolami et al., *Decision and Inhibitory Trees and Rules for Decision Tables with Many-valued Decisions*, Intelligent Systems Reference Library 156,
https://doi.org/10.1007/978-3-030-12854-8_12

Let τ be a directed path from a node Θ of G to a terminal node Θ' in which edges (in the order from Θ to Θ') are labeled with pairs $(f_{i_1}, c_{i_1}), \ldots, (f_{i_m}, c_{i_m})$, and $t = mcd(\Theta')$. We denote by $rule(\tau)$ the decision rule over T

$$(f_{i_1} = c_{i_1}) \wedge \cdots \wedge (f_{i_m} = c_{i_m}) \to t .$$

If $m = 0$ (if $\Theta = \Theta'$) then the rule $rule(\tau)$ is equal to $\to t$.

Let $r = (b_1, \ldots, b_n)$ be a row of T, and Θ be a node of G (subtable of T) containing the row r. We denote by $Rule(G, \Theta, r)$ the set of rules $rule(\tau)$ corresponding to all directed paths τ from Θ to terminal nodes Θ' containing r.

Proposition 12.1 *Let T be a nonempty decision table with n conditional attributes $f_1, \ldots, f_n$, $r = (b_1, \ldots, b_n)$ be a row of T, U be an uncertainty measure, $\alpha \in \mathbb{R}_+$, and Θ be a node of the graph $\Delta_{U,\alpha}(T)$ containing r. Then the set $Rule(\Delta_{U,\alpha}(T), \Theta, r)$ coincides with the set of all (U, α)-decision rules for Θ and r, i.e.,*

$$Rule(\Delta_{U,\alpha}(T), \Theta, r) = DR_{U,\alpha}(\Theta, r) .$$

Proof From the definition of the graph $\Delta_{U,\alpha}(T)$ it follows that each rule from $Rule(\Delta_{U,\alpha}, \Theta, r)$ is a (U, α)-decision rule for Θ and r.

Let us consider an arbitrary (U, α)-decision rule ρ for Θ and r:

$$(f_{i_1} = b_{i_1}) \wedge \cdots \wedge (f_{i_m} = b_{i_m}) \to t .$$

It is easy to show that there is a directed path $\Theta_0 = \Theta, \Theta_1, \ldots, \Theta_m$ in $\Delta_{U,\alpha}(T)$ such that, for $j = 1, \ldots, m$, $\Theta_j = \Theta(f_{i_1}, b_{i_1}) \ldots (f_{i_j}, b_{i_j})$, there is an edge from Θ_{j-1} to Θ_j labeled with (f_{i_j}, b_{i_j}), and Θ_m is a terminal node in $\Delta_{U,\alpha}(T)$. Therefore $\rho \in Rule(\Delta_{U,\alpha}, \Theta, r)$. $\qquad \square$

12.1.2 Procedure of Optimization

We describe now a procedure of optimization (minimization of cost) of rules for row $r = (b_1, \ldots, b_n)$ relative to a strictly increasing cost function ψ. We will move from terminal nodes of the graph G to the node T. We will attach to each node Θ of the graph G containing r the minimum cost $c(\Theta, r)$ of a rule from $Rule(G, \Theta, r)$ and, probably, we will remove some bundles of edges starting in nonterminal nodes. As a result, we obtain a bundle-preserving subgraph $G^\psi = G^\psi(r)$ of the graph G.

Algorithm $\mathscr{A}_8$ (procedure of decision rule optimization).

Input: A bundle-preserving subgraph G of the graph $\Delta_{U,\alpha}(T)$ for some decision table T with n conditional attributes $f_1, \ldots, f_n$, uncertainty measure U, and a number $\alpha \in \mathbb{R}_+$, a row $r = (b_1, \ldots, b_n)$ of T, and a strictly increasing cost function ψ for decision rules given by pair of functions ψ^0 and F.

Output: The bundle-preserving subgraph $G^\psi = G^\psi(r)$ of the graph G.

1. If all nodes of the graph G containing r are processed then return the obtained graph as G^ψ and finish the work of the algorithm. Otherwise, choose a node Θ of the graph G containing r which is not processed yet and which is either a terminal node of G or a nonterminal node of G for which all children containing r are processed.
2. If Θ is a terminal node then set $c(\Theta, r) = \psi^0(\Theta)$, mark node Θ as processed and proceed to step 1.
3. If Θ is a nonterminal node then, for each $f_i \in E_G(\Theta)$, compute the value $c(\Theta, r, f_i) = F(c(\Theta(f_i, b_i), r))$ and set $c(\Theta, r) = \min\{c(\Theta, r, f_i) : f_i \in E_G(\Theta)\}$. Remove all f_i-bundles of edges starting from Θ for which $c(\Theta, r) < c(\Theta, r, f_i)$. Mark the node Θ as processed and proceed to step 1.

Proposition 12.2 *Let G be a bundle-preserving subgraph of the graph $\Delta_{U,\alpha}(T)$ for some decision table T with n conditional attributes $f_1, \ldots, f_n$, uncertainty measure U, and a number $\alpha \in \mathbb{R}_+$, $r = (b_1, \ldots, b_n)$ be a row of T, and ψ be a strictly increasing cost function for decision rules given by pair of functions ψ^0 and F. Then, to construct the graph $G^\psi = G^\psi(r)$, the algorithm $\mathscr{A}_8$ makes*

$$O(nL(G))$$

elementary operations (computations of ψ^0, F, and comparisons).

Proof In each terminal node of the graph G, the algorithm $\mathscr{A}_8$ computes the value of ψ^0. In each nonterminal node of G, the algorithm $\mathscr{A}_8$ computes the value of F at most n times and makes at most $2n$ comparisons. Therefore the algorithm $\mathscr{A}_8$ makes

$$O(nL(G))$$

elementary operations.   □

Proposition 12.3 *Let $\mathscr{U}$ be a restricted information system and $\psi \in \{l, -c, -rc, -c_M, mc, rmc\}$. Then the algorithm $\mathscr{A}_8$ has polynomial time complexity for decision tables from $\mathscr{T}(\mathscr{U})$ depending on the number of conditional attributes in these tables.*

Proof Since $\psi \in \{l, -c, -rc, -c_M, mc, rmc\}$,

$$\psi^0 \in \{0, -N_{mcd(T)}(T), -N_{mcd(T)}(T)/N(T), -N^M(T), N(T) - N_{mcd(T)}(T),$$
$$(N(T) - N_{mcd(T)}(T))/N(T)\}$$

and F is either x or $x + 1$. Therefore the elementary operations used by the algorithm $\mathscr{A}_8$ are either basic numerical operations or computations of numerical parameters of decision tables which have polynomial time complexity depending on the size of decision tables. From Proposition 12.2 it follows that the number of elementary operations is bounded from above by a polynomial depending on the size of input table T and on the number of separable subtables of T.

According to Proposition 5.4, the algorithm $\mathscr{A}_8$ has polynomial time complexity for decision tables from $\mathscr{T}(\mathscr{U})$ depending on the number of conditional attributes in these tables. $\square$

Theorem 12.1 *Let T be a nonempty decision table with n conditional attributes $f_1, \ldots, f_n$, $r = (b_1, \ldots, b_n)$ be a row of T, U be an uncertainty measure, $\alpha \in \mathbb{R}_+$, G be a bundle-preserving subgraph of the graph $\Delta_{U,\alpha}(T)$, and ψ be a strictly increasing cost function given by pair of functions ψ^0 and F. Then, for any node Θ of the graph $G^\psi = G^\psi(r)$ containing the row r, $c(\Theta, r) = \min\{\psi(\Theta, \rho) : \rho \in Rule(G, \Theta, r)\}$ and the set $Rule(G^\psi, \Theta, r)$ coincides with set of rules from $Rule(G, \Theta, r)$ that have minimum cost relative to ψ.*

Proof We prove this theorem by induction on nodes of G^ψ containing r. If Θ is a terminal node containing r then $Rule(G^\psi, \Theta, r) = Rule(G, \Theta, r) = \{\to mcd(\Theta)\}$ and $c(\Theta, r) = \psi^0(\Theta) = \psi(\Theta, \to mcd(\Theta))$. Therefore the statement of theorem holds for Θ.

Let Θ be a nonterminal node containing r such that, for each child of Θ containing r, the statement of theorem holds. It is clear that

$$Rule(G, \Theta, r) = \bigcup_{f_i \in E_G(\Theta)} Rule(G, \Theta, r, f_i)$$

where, for $f_i \in E_G(\Theta)$,

$$Rule(G, \Theta, r, f_i) = \{(f_i = b_i) \wedge \gamma \to t : \gamma \to t \in Rule(G, \Theta(f_i, b_i), r)\} .$$

By the indiction hypothesis, for any $f_i \in E_G(\Theta)$, the minimum cost of rule from $Rule(G, \Theta(f_i, b_i), r)$ is equal to $c(\Theta(f_i, b_i), r)$. Since ψ is a strictly increasing cost function, the minimum cost of a rule from the set $Rule(G, \Theta, r, f_{i_j})$ is equal to $F(c(\Theta(f_i, b_i), r) = c(\Theta, r, f_i)$. Therefore $c(\Theta, r) = \min\{c(\Theta, r, f_i) : f_i \in E_G(\Theta)\}$ is the minimum cost of a rule from $Rule(G, \Theta, r)$. Set $q = c(\Theta, r)$.

Let $(f_i = b_i) \wedge \gamma \to t$ be a rule from $Rule(G, \Theta, r)$ which cost is equal to q. It is clear that G contains the node $\Theta(f_i, b_i)$ and the edge e which starts in Θ, enters $\Theta(f_i, b_i)$, and is labeled with (f_i, b_i). Let p be the minimum cost of a rule from $Rule(G, \Theta(f_i, b_i), r)$, i.e., $p = c(\Theta(f_i, b_i), r)$. The rule $\gamma \to t$ belongs to the set $Rule(G, \Theta(f_i, b_i), r)$ and, since ψ is strictly increasing, the cost of $\gamma \to t$ is equal to p (otherwise, the minimum cost of a rule from $Rule(G, \Theta, r)$ is less than q). Therefore $F(p) = q$, and the edge e belongs to the graph G^ψ. By the induction hypothesis, the set $Rule(G^\psi, \Theta(f_i, b_i), r)$ coincides with the set of rules from $Rule(G, \Theta(f_i, b_i), r)$ which cost is equal to p. From here it follows that $\gamma \to t$ belongs to $Rule(G^\psi, \Theta(f_i, b_i), r)$ and $(f_i = b_i) \wedge \gamma \to t$ belongs to $Rule(G^\psi, \Theta, r)$.

Let $(f_i = b_i) \wedge \gamma \to t$ belong to $Rule(G^\psi, \Theta, r)$. Then $\Theta(f_i, b_i)$ is a child of Θ in the graph G^ψ, $\gamma \to t$ belongs to $Rule(G^\psi, \Theta(f_i, b_i), r)$ and, by the induction hypothesis, the set $Rule(G^\psi, \Theta(f_i, b_i), r)$ coincides with the set of rules from $Rule(G, \Theta(f_i, b_i), r)$ which cost is equal to p – the minimum cost of a rule from $Rule(G, \Theta(f_i, b_i), r)$. From the description of the procedure of optimization and from the fact that $\Theta(f_i, b_i)$ is a child of Θ in the graph G^ψ it follows that $F(p) = q$. Therefore, the cost of rule $(f_i = b_i) \wedge \gamma \to t$ is equal to q.   □

We can make sequential optimization of (U, α)-rules for T and r relative to a sequence of strictly increasing cost functions $\psi_1, \psi_2, \ldots$ for decision rules. We begin from the graph $G = \Delta_{U,\alpha}(T)$ and apply to it the procedure of optimization (the algorithm $\mathscr{A}_8$) relative to the cost function ψ_1. As a result, we obtain a bundle-preserving subgraph $G^{\psi_1} = G^{\psi_1}(r)$ of the graph G. By Proposition 12.1, the set $Rule(G, T, r)$ is equal to the set of all (U, α)-rules for T and r. From here and from Theorem 12.1 it follows that the set $Rule(G^{\psi_1}, T, r)$ is equal to the set of all (U, α)-rules for T and r which have minimum cost relative to ψ_1. If we apply to the graph G^{ψ_1} the procedure of optimization relative to the cost function ψ_2 we obtain a bundle-preserving subgraph $G^{\psi_1, \psi_2} = G^{\psi_1, \psi_2}(r)$ of the graph G^{ψ_1}. The set $Rule(G^{\psi_1, \psi_2}, T, r)$ is equal to the set of all rules from the set $Rule(G^{\psi_1}, T, r)$ which have minimum cost relative to ψ_2, etc.

We described the work of optimization procedure for one row. If we would like to work with all rows in parallel, then instead of removal of bundles of edges we will change the list of bundles attached to row. We begin from the graph $G = \Delta_{U,\alpha}(T)$. In this graph, for each nonterminal node Θ, each row r of Θ is labeled with the set of attributes $E(G, \Theta, r) = E(\Theta)$. It means that, for the row r, we consider only f_i-bundles of edges starting from Θ such that $f_i \in E(G, \Theta, r)$. During the work of the procedure of optimization relative to a cost function ψ we will not change the "topology" of the graph G but will change sets $E(G, \Theta, r)$ attached to rows r of nonterminal nodes Θ. In a new graph G^ψ (we will say about this graph as about *labeled bundle-preserving subgraph* of the graph G), for each nonterminal node Θ, each row r of Θ is labeled with a subset $E(G^\psi, \Theta, r)$ of the set $E(G, \Theta, r)$ containing only attributes for which corresponding bundles were not removed during the optimization relative to ψ for the row r.

We can study also totally optimal decision rules relative to various combinations of cost functions. For a cost function ψ, we denote $\psi^{U,\alpha}(T, r) = \min\{\psi(T, \rho) : \rho \in DR_{U,\alpha}(T, r)\}$, i.e., $\psi^{U,\alpha}(T, r)$ is the minimum cost of a (U, α)-decision rule for T and r relative to the cost function ψ. Let $\psi_1, \ldots, \psi_m$ be cost functions and $m \geq 2$. A (U, α)-decision rule ρ for T and r is called a *totally optimal (U, α)-decision rule for T and r relative to the cost functions* $\psi_1, \ldots, \psi_m$ if $\psi_1(T, \rho) = \psi_1^{U,\alpha}(T, r), \ldots, \psi_m(T, \rho) = \psi_m^{U,\alpha}(T, r)$, i.e., ρ is optimal relative to $\psi_1, \ldots, \psi_m$ simultaneously.

Assume that $\psi_1, \ldots, \psi_m$ are strictly increasing cost functions for decision rules. We now describe how to recognize the existence of a (U, α)-decision rule for T and r which is a totally optimal (U, α)-decision rule for T and r relative to the cost functions $\psi_1, \ldots, \psi_m$.

First, we construct the graph $G = \Delta_{U,\alpha}(T)$ using the algorithm $\mathscr{A}_1$. For $i = 1, \ldots, m$, we apply to G and r the procedure of optimization relative to ψ_i (the algorithm $\mathscr{A}_8$). As a result, we obtain, for $i = 1, \ldots, m$, the graph $G^{\psi_i}(r)$ and the number $\psi_i^{U,\alpha}(T, r)$ attached to the node T of $G^{\psi_i}(r)$. Next, we apply to G sequentially the procedures of optimization relative to the cost functions $\psi_1, \ldots, \psi_m$. As a result, we obtain graphs $G^{\psi_1}(r), G^{\psi_1,\psi_2}(r), \ldots, G^{\psi_1,\ldots,\psi_m}(r)$ and numbers $\varphi_1, \varphi_2, \ldots, \varphi_m$ attached to the node T of these graphs. One can show that a totally optimal (U, α)-decision rule for T and r relative to the cost functions $\psi_1, \ldots, \psi_m$ exists if and only if $\varphi_i = \psi_i^{U,\alpha}(T, r)$ for $i = 1, \ldots, m$.

12.1.3 Experimental Results: Optimization of Decision Rules

In this section, we discuss results of experiments related to the optimization of decision rules. We consider decision tables with many-valued decisions derived from conventional decision tables from the UCI ML Repository [7] in the following way. Let T_1 be a decision table (data set) from [7]. We denote by T_2 the decision table obtained from T_1 by the removal of some conditional attributes. The table T_2 can contain groups of equal rows, possibly, with different decisions. We keep a single row from each group and label it with the set of all decisions attached to the rows in its group. As a result, we obtain a decision table T_3 with many-valued decisions. We remove from T_3 all rows r such that $D(r) = D(T_3)$, and denote by T_4 the obtained decision table. We work further with T_4 if it contains rows r with $|D(r)| \geq 2$, and does not contain rows r for which $D(r) = D(T_4)$.

We consider nine such decision tables with many-valued decisions. They are described in Table 12.1 which contains the name of decision table (the column "Table name"), the number or rows (the column "Rows"), the number of conditional attributes (the column "Attributes"), and positions of the removed conditional attributes in the original data set (the column "Removed attributes"). The name of each table consists of the name of the original data set and the number of attributes removed from this data set. For example, the "breast-cancer-5" decision table is obtained from the "breast-cancer" data set by the removal of five conditional attributes.

Table 12.1 Modified decision tables from UCI ML repository

Table name	Rows	Attributes	Removed attributes
breast-cancer-1	169	8	3
breast-cancer-5	58	4	4, 5, 6, 8, 9
cars-1	432	5	1
mushroom-5	4048	17	5, 8, 11, 13, 22
nursery-1	4320	7	1
nursery-4	240	4	1, 5, 6, 7
teeth-1	22	7	1
teeth-5	14	3	2, 3, 4, 5, 8
zoo-data-5	42	11	2, 9, 10, 13, 14

Table 12.2 Minimization of decision rule length

Table name	Length of rules		
	Min	Avg	Max
breast-cancer-1	1	2.80	5
breast-cancer-5	1	1.69	3
cars-1	1	1.37	4
mushroom-5	1	1.48	4
nursery-1	1	2.03	5
nursery-4	1	1.33	2
teeth-1	1	2.23	3
teeth-5	1	1.93	3
zoo-data-5	1	2.10	4

Table 12.3 Maximization of decision rule coverage

Table name	Coverage of rules		
	Min	Avg	Max
breast-cancer-1	1	7.50	15
breast-cancer-5	1	5.17	11
cars-1	4	110.19	144
mushroom-5	20	746.62	1080
nursery-1	6	620.84	1440
nursery-4	16	59.92	80
teeth-1	1	1.00	1
teeth-5	1	1.00	1
zoo-data-5	1	7.17	12

For each decision table T in Table 12.1, we construct by the algorithm $\mathscr{A}_1$ the directed acyclic graph $\Delta(T)$ (see Remark 5.1). Table 12.2 shows in the column "Length of rules" minimum, average and maximum length of exact decision rules among all rows of T obtained as a result of applying to $\Delta(T)$ the procedure of optimization of rules (the algorithm $\mathscr{A}_8$) relative to the length. One can see that we obtained short on average exact decision rules for each decision table. For example, "nursery-1" has seven conditional attributes whereas on average the system of constructed rules has two conditions per rule. In a similar way, we apply to $\Delta(T)$ the procedure of optimization relative to the coverage (really, the procedure of optimization relative to the minus coverage). Minimum, average and maximum coverage of obtained exact rules among all rows of T can be found in Table 12.3 (the column "Coverage of rules"). We should mention good results obtained for decision tables "cars-1", "mushroom-5", "nursery-1", and "nursery-4".

Table 12.4 Sequential optimization of decision rules

Table name	Coverage + Length		Length + Coverage		Rows tot	Rows
	Coverage	Length	Length	Coverage		
breast-cancer-1	7.50	3.64	2.80	5.16	71	169
breast-cancer-5	5.17	1.81	1.69	4.71	51	58
cars-1	110.19	1.37	1.37	110.19	432	432
mushroom-5	746.62	2.26	1.48	516.54	1388	4048
nursery-1	620.84	2.03	2.03	620.84	4320	4320
nursery-4	59.92	1.33	1.33	59.92	240	240
teeth-1	1.00	2.23	2.23	1.00	22	22
teeth-5	1.00	1.93	1.93	1.00	14	14
zoo-data-5	7.17	2.26	2.10	7.02	39	42

We applied to the directed acyclic graph $\Delta(T)$ sequentially the procedure of optimization relative to the coverage, and then the procedure of optimization relative to the length. Average length and coverage of obtained exact decision rules among all rows of T can be found in Table 12.4 (the column "Coverage + Length"). We also present in Table 12.4 results of sequential optimization relative to the length, and then relative to the coverage (the column "Length + Coverage"). We used the obtained results to count the number of rows for which there exist totally optimal decision rules relative to the length and coverage (the column "Rows tot"). This number can be compared with the number of all rows (the column "Rows"). For decision tables in bold in Table 12.4, the order of optimization relative to the length and coverage do not matter since a totally optimal decision rule exists for each row in those tables.

12.1.4 Number of Rules in Rule(G, Θ, r)

Let T be a nonempty decision table with n conditional attributes $f_1, \ldots, f_n$, $r = (b_1, \ldots, b_n)$ be a row of T, U be an uncertainty measure, $\alpha \in \mathbb{R}_+$, and G be a bundle-preserving subgraph of the graph $\Delta_{U,\alpha}(T)$. We describe now an algorithm which counts, for each node Θ of the graph G containing r, the cardinality $C(\Theta, r)$ of the set $Rule(G, \Theta, r)$, and returns the number $C(T, r) = |Rule(G, T, r)|$.

Algorithm $\mathscr{A}_9$ (counting the number of decision rules).
Input: A bundle-preserving subgraph G of the graph $\Delta_{U,\alpha}(T)$ for some decision table T with n conditional attributes $f_1, \ldots, f_n$, uncertainty measure U, number $\alpha \in \mathbb{R}_+$, and row $r = (b_1, \ldots, b_n)$ of the table T.

Output: The number $|Rule(G, T, r)|$.

1. If all nodes of the graph G containing r are processed then return the number $C(T, r)$ and finish the work of the algorithm. Otherwise, choose a node Θ of the graph G containing r which is not processed yet and which is either a terminal node of G or a nonterminal node of G such that, for each $f_i \in E_G(T)$, the node $\Theta(f_i, b_i)$ is processed.
2. If Θ is a terminal node then set $C(\Theta, r) = 1$, mark the node Θ as processed, and proceed to step 1.
3. If Θ is a nonterminal node then set

$$C(\Theta, r) = \sum_{f_i \in E_G(\Theta)} C(\Theta(f_i, b_i), r) \, ,$$

mark the node Θ as processed, and proceed to step 1.

Proposition 12.4 *Let U be an uncertainty measure, $\alpha \in \mathbb{R}_+$, T be a decision table with n attributes $f_1, \ldots, f_n$, G be a bundle-preserving subgraph of the graph $\Delta_{U,\alpha}(T)$, and $r = (b_1, \ldots, b_n)$ be a row of the table T. Then the algorithm $\mathscr{A}_9$ returns the number $|Rule(G, T, r)|$ and makes at most $nL(G)$ operations of addition.*

Proof We prove by induction on the nodes of G that $C(\Theta, r) = |Rule(G, \Theta, r)|$ for each node Θ of G containing r. Let Θ be a terminal node of G. Then $Rule(G, \Theta, r) = \{\to mcd(\Theta)\}$ and $|Rule(G, \Theta, r)| = 1$. Therefore the considered statement holds for Θ. Let now Θ be a nonterminal node of G such that the considered statement holds for its children containing r. It is clear that

$$Rule(G, \Theta, r) = \bigcup_{f_i \in E_G(\Theta)} Rule(G, \Theta, r, f_i)$$

where, for each $f_i \in E_G(\Theta)$,

$$Rule(G, \Theta, r, f_i) = \{(f_i = b_i) \wedge \gamma \to t : \gamma \to t \in Rule(G, \Theta(f_i, b_i), r)\}$$

and $\left|Rule(G, \Theta, r, f_{i_j})\right| = \left|Rule(G, \Theta(f_{i_j}, b_{i_j}), r)\right|$. Therefore

$$|Rule(G, \Theta, r)| = \sum_{f_i \in E_G(\Theta)} |Rule(G, \Theta(f_i, b_i), r)| \, .$$

By the induction hypothesis, $C(\Theta(f_i, b_i), r) = |Rule(G, \Theta(f_i, b_i), r)|$ for any $f_i \in E_G(\Theta)$. Therefore $C(\Theta, r) = |Rule(G, \Theta, r)|$. Hence, the considered statement holds. From here it follows that $C(T, r) = |Rule(G, T, r)|$, and the algorithm $\mathscr{A}_9$ returns the cardinality of the set $Rule(G, T, r)$.

It is easy to see that the considered algorithm makes at most $nL(G)$ operations of addition where $L(G)$ is the number of nodes in the graph G. $\qquad \square$

Proposition 12.5 *Let $\mathcal{U}$ be a restricted information system. Then the algorithm $\mathcal{A}_9$ has polynomial time complexity for decision tables from $\mathcal{T}(\mathcal{U})$ depending on the number of conditional attributes in these tables.*

Proof All operations made by the algorithm $\mathcal{A}_9$ are basic numerical operations (additions). From Proposition 12.4 it follows that the number of these operations is bounded from above by a polynomial depending on the size of input table T and on the number of separable subtables of T.

According to Proposition 5.4, the algorithm $\mathcal{A}_9$ has polynomial time complexity for decision tables from $\mathcal{T}(\mathcal{U})$ depending on the number of conditional attributes in these tables.   □

12.1.5 Simulation of Greedy Algorithm for Construction of Decision Rule Set

Let U be an uncertainty measure, $\alpha, \beta \in \mathbb{R}_+$, $0 \le \beta \le 1$, and T be a decision table with n conditional attributes $f_1, \ldots, f_n$. A set of (U, α)-decision rules for T which cover at least $(1 - \beta)N(T)$ rows of T is called a *β-system of (U, α)-decision rules for T*. We would like to construct a β-system of (U, α)-decision rules for T with minimum cardinality. Unfortunately, our approach does not allow us to do this. However, we can simulate the work of a greedy algorithm for the set cover problem (see the algorithm $\mathcal{A}_{10}$).

The algorithm $\mathcal{A}_{10}$ works with the graph $G = \Delta_{U,\alpha}(T)$. During each step, this algorithm constructs (based on the algorithm $\mathcal{A}_8$) a (U, α)-rule for T with minimum length among all (U, α)-rules for T which cover maximum number of uncovered previously rows. The algorithm finishes the work when at least $(1 - \beta)N(T)$ rows of T are covered. We will call the constructed β-system of (U, α)-decision rules for T a (U, α, β)-*greedy set of decision rules for T*. Using results of Slavík for the set cover problem [9, 10] we obtain that the cardinality of the constructed set of rules is less than $C_{\min}(\beta)(\ln \lceil (1 - \beta)N(T)\rceil - \ln \ln \lceil (1 - \beta)N(T)\rceil + 0.78)$ where $C_{\min}(\beta)$ is the minimum cardinality of a β-system of (U, α)-decision rules for T.

Let us recall that the cost function $-c_M$ is given by the pair of functions $\psi^0(T) = -N^M(T)$ and $F(x) = x$, and the cost function l is given by the pair of functions $\varphi^0(T) = 0$ and $H(x) = x + 1$.

Algorithm $\mathcal{A}_{10}$ (construction of (U, α, β)-greedy set of decision rules for a decision table).

Input: A decision table T with n conditional attributes $f_1, \ldots, f_n$, uncertainty measure U, numbers $\alpha, \beta \in \mathbb{R}_+$, $0 \le \beta \le 1$, and the graph $\Delta_{U,\alpha}(T)$.

Output: A (U, α, β)-greedy set of decision rules for T.

1. Set $M = \emptyset$ and $S = \emptyset$.
2. If $|M| \ge (1 - \beta)N(T)$ then return S and finish the algorithm.

3. Apply the algorithm $\mathscr{A}_8$ to each row of T two times: first, as the procedure of optimization of rules relative to the cost function $-c_M$, and after that, as the procedure of optimization of rules relative to the cost function l.
4. As a result, for each row r of T, we obtain two numbers $-c_M(r)$ which is the minimum cost of a rule from $Rule(G, T, r)$ relative to the cost function $-c_M$, and $l(r)$ which is the minimum length among rules from $Rule(G, T, r)$ which have minimum cost relative to the cost function $-c_M$.
5. Choose a row r of T for which the value of $-c_M(r)$ is minimum among all rows of T, and the value of $l(r)$ is minimum among all rows of T with minimum value of $-c_M(r)$. In the graph $G^{-c_M, l}$, which is the result of the bi-stage procedure of optimization of rules for the row r relative to the cost functions $-c_M$ and l, choose a directed path τ from T to a terminal node of $G^{-c_M, l}$ containing r.
6. Add the rule $rule(\tau)$ to the set S, and add all rows covered by $rule(\tau)$, which do not belong to M, to the set M. Proceed to step 2.

Proposition 12.6 *Let T be a decision table with n conditional attributes, U be an uncertainty measure, and $\alpha, \beta \in \mathbb{R}_+$, $0 \leq \beta \leq 1$. Then the algorithm $\mathscr{A}_{10}$ returns a (U, α, β)-greedy set of decision rules for T and makes*

$$O(N(T)^2 n L(G))$$

elementary operations (computations of ψ^0, F, φ^0, H, and comparisons).

Proof Let us analyze one iteration of the algorithm $\mathscr{A}_{10}$ (steps 3–6) and rule $rule(\tau)$ added at the step 6 to the set S. Using Proposition 12.1 and Theorem 12.1 we obtain that the rule $rule(\tau)$ is a (U, α)-rule for T which covers the maximum number of uncovered rows (rows which does not belong to M) and has minimum length among such rules. Therefore the algorithm $\mathscr{A}_{10}$ returns a (U, α, β)-greedy set of decision rules for T.

Let us analyze the number of elementary operations (computations of ψ^0, F, φ^0, H, and comparisons) which the algorithm $\mathscr{A}_{10}$ makes during one iteration. We know that the algorithm $\mathscr{A}_8$, under the bi-stage optimization of rules for one row, makes

$$O(nL(G))$$

elementary operations (computations of ψ^0, F, φ^0, H, and comparisons). The number of rows is equal to $N(T)$. To choose a row r of T for which the value of $-c_M(r)$ is minimum among all rows of T, and the value of $l(r)$ is minimum among all rows of T with minimum value of $-c_M(r)$, the algorithm $\mathscr{A}_{10}$ makes at most $2N(T)$ comparisons. Therefore the number of elementary operations which the algorithm $\mathscr{A}_{10}$ makes during one iteration is $O(N(T)nL(G))$.

The number of iterations is at most $N(T)$. Therefore, during the construction of a (U, α, β)-greedy set of decision rules for T, the algorithm $\mathscr{A}_{10}$ makes

$$O(N(T)^2 n L(G))$$

elementary operations (computations of ψ^0, F, φ^0, H, and comparisons).   □

Proposition 12.7 *Let $\mathcal{U}$ be a restricted information system. Then the algorithm $\mathscr{A}_{10}$ has polynomial time complexity for decision tables from $\mathscr{T}(\mathcal{U})$ depending on the number of conditional attributes in these tables.*

Proof From Proposition 12.6 it follows that, for the algorithm $\mathscr{A}_{10}$, the number of elementary operations (computations of ψ^0, F, φ^0, H, and comparisons) is bounded from above by a polynomial depending on the size of input table T and on the number of separable subtables of T. The computations of numerical parameters of decision tables used by the algorithm $\mathscr{A}_{10}$ (constant 0 and $-N^M(T)$) have polynomial time complexity depending on the size of decision tables. All operations with numbers are basic ones (x, $x + 1$, and comparisons).

According to Proposition 5.4, the algorithm $\mathscr{A}_{10}$ has polynomial time complexity for decision tables from $\mathscr{T}(\mathcal{U})$ depending on the number of conditional attributes in these tables.   □

Using information based on the work of the algorithm $\mathscr{A}_{10}$, we can obtain lower bound on the parameter $C_{\min}(U, \alpha, \beta, T)$ which is the minimum cardinality of a β-system of (U, α)-decision rules for T. During the construction of (U, α, β)-greedy set of decision rules for T, let the algorithm $\mathscr{A}_{10}$ choose consequently rules $\rho_1, \ldots, \rho_t$. Let $B_1, \ldots, B_t$ be sets of rows covered by rules $\rho_1, \ldots, \rho_t$, respectively. Set $B_0 = \emptyset$, $\delta_0 = 0$ and, for $i = 1, \ldots, t$, set $\delta_i = |B_i \setminus (B_0 \cup \ldots \cup B_{i-1})|$. The information derived from the algorithm's $\mathscr{A}_{10}$ work consists of the tuple $(\delta_1, \ldots, \delta_t)$ and the numbers $N(T)$ and β.

From the results obtained in [8] regarding the greedy algorithm for the set cover problem it follows that

$$C_{\min}(U, \alpha, \beta, T) \geq \max \left\{ \left\lceil \frac{\lceil (1 - \beta)N(T) \rceil - (\delta_0 + \cdots + \delta_i)}{\delta_{i+1}} \right\rceil : \\ i = 0, \ldots, t - 1 \right\}. \tag{12.1}$$

12.1.6 Experimental Results: Simulation of Greedy Algorithm

Let T be a decision table with many-valued decisions, $0 \leq \alpha \leq 1$, and $0 \leq \beta \leq 1$. We denote by $Rule_{\alpha}^{\beta}(T)$ the (rme, α, β)-greedy set of decision rules for T constructed by the algorithm $\mathscr{A}_{10}$.

Table 12.5 presents information regarding the set $Rule_0^0(T)$ for nine decision tables T described in Table 12.1. The column "Rows" contains the number of rows in T. The column "Bounds" contains the number of rules in $Rule_0^0(T)$ on the left and a lower bound on the cardinality of a 0-system of $(rme, 0)$-decision rules for T obtained

Table 12.5 Bounds on the cardinality of 0-systems of $(rme, 0)$-decision rules for decision tables from Table 12.1

Table name	Rows	Bounds	Max cov
breast-cancer-1	169	53/24	15
breast-cancer-5	58	27/15	11
cars-1	432	17/7	144
mushroom-5	4048	30/9	1080
nursery-1	4320	71/17	1440
nursery-4	240	9/4	80
teeth-1	22	22/22	1
teeth-5	14	14/14	1
zoo-data-5	42	9/5	12

Table 12.6 Bounds on the cardinality of β-system of (rme, α)-decision rules for "breast-cancer-1"

β	α		
	0	0.1	0.3
0	53/24	53/24	38/15
0.01	52/24	52/24	37/15
0.05	45/20	45/20	30/12
0.1	37/17	37/17	24/10
0.15	32/14	32/14	19/8
0.2	28/13	28/13	16/8

based on inequality (12.1) on the right. Finally, the column "Max cov" contains the maximum coverage of rules from $Rule_0^0(T)$.

We consider a threshold of 30 as a reasonable upper bound on the number of rules for a system of rules used for knowledge representation. In the cases where this threshold is exceeded (decision tables "breast-cancer-1" and "nursery-1"), we consider β-systems $Rule_\alpha^\beta(T)$ of (rme, α)-decision rules for $\beta \in \{0, 0.01, 0.05, 0.1, 0.15, 0.2\}$ and $\alpha \in \{0, 0.1, 0.3\}$. Results of experiments show that, for each of these two decision tables, we can find combinations of β and α such that the set of rules constructed by the algorithm $\mathscr{A}_{10}$ contains at most 30 rules. In Tables 12.6 and 12.7, we present, for each β and α, the cardinality of the set of rules constructed by the algorithm $\mathscr{A}_{10}$ and a lower bound on the cardinality of a β-system of (rme, α)-decision rules for "breast-cancer-1" and "nursery-1", respectively.

Table 12.7 Bounds on the cardinality of β-system of (rme, α)-decision rules for "nursery-1"

β	α		
	0	0.1	0.3
0	71/17	35/9	6/3
0.01	45/15	22/9	6/3
0.05	25/10	14/8	6/3
0.1	16/9	12/7	5/3
0.15	12/8	10/7	4/3
0.2	9/7	9/6	4/3

12.2 Multi-stage Optimization of Inhibitory Rules

In this section, we concentrate on optimization of inhibitory rules.

12.2.1 From Decision to Inhibitory Rules

Let T be a nondegenerate decision table with n conditional attributes $f_1, \ldots, f_n$, r be a row of T, T^C be the decision table complementary to T, U be an uncertainty measure, W be a completeness measure, U is dual to W, and $\alpha \in \mathbb{R}_+$.

Let G be a bundle-preserving subgraph of the graph $\Delta_{U,\alpha}(T^C)$. We correspond to the node T^C of G and row r a set $Rule(G, T^C, r)$ of (U, α)-decision rules for T^C and r. If $G = \Delta_{U,\alpha}(T^C)$ then, by Proposition 12.1, $Rule(G, T^C, r) = DR_{U,\alpha}(T^C, r)$. In general case, $Rule(G, T^C, r) \subseteq DR_{U,\alpha}(T^C, r)$.

Let us consider the set $Rule(G, T^C, r)^-$. By Corollary 11.1,

$$IR_{W,\alpha}(T, r) = DR_{U,\alpha}(T^C, r)^- .$$

Therefore, if $G = \Delta_{U,\alpha}(T^C)$ then $Rule(G, T^C, r)^- = IR_{W,\alpha}(T, r)$. In general case, $Rule(G, T^C, r)^- \subseteq DR_{U,\alpha}(T^C, r)^- = IR_{W,\alpha}(T, r)$.

In Sect. 12.1.2, the algorithm $\mathscr{A}_8$ is considered which, for a strictly increasing cost function ψ for decision rules, constructs the bundle-preserving subgraph $G^\psi = G^\psi(r)$ of the graph G. Let $\psi \in \{l, -c, -rc, mc, rmc\}$. Then, by Theorem 12.1, the set $Rule(G^\psi, T^C, r)$ is equal to the set of all rules from $Rule(G, T^C, r)$ which have minimum cost relative to ψ among all decision rules from the set $Rule(G, T^C, r)$. From Proposition 11.1 it follows that, for any $\rho \in Rule(G, T^C, r)$, $\psi(T^C, \rho) = \psi(T, \rho^-)$. Thus, the set $Rule(G^\psi, T^C, r)^-$ is equal to the set of all rules from $Rule(G, T^C, r)^-$ which have minimum cost relative to ψ among all inhibitory rules from the set $Rule(G, T^C, r)^-$.

We can make sequential optimization of (W, α)-inhibitory rules for T and r relative to a sequence of cost functions $\psi_1, \psi_2, \ldots$ from $\{l, -c, -rc, mc, rmc\}$. We begin from the graph $G = \Delta_{U,\alpha}(T^C)$ and apply to it the procedure of optimization (the algorithm $\mathscr{A}_8$) relative to the cost function ψ_1. As a result, we obtain a bundle-preserving subgraph $G^{\psi_1} = G^{\psi_1}(r)$ of the graph G. The set $Rule(G, T^C, r)^-$ is equal to the set of all (W, α)-inhibitory rules for T and r. The set $Rule(G^{\psi_1}, T^C, r)^-$ is equal to the set of all (W, α)-inhibitory rules for T and r which have minimum cost relative to ψ_1. If we apply to the graph G^{ψ_1} the procedure of optimization relative to the cost function ψ_2 we obtain a bundle-preserving subgraph $G^{\psi_1, \psi_2} = G^{\psi_1, \psi_2}(r)$ of the graph G^{ψ_1}. The set $Rule(G^{\psi_1, \psi_2}, T^C, r)^-$ is equal to the set of all rules from the set $Rule(G^{\psi_1}, T, r)^-$ which have minimum cost relative to ψ_2 among all inhibitory rules from $Rule(G^{\psi_1}, T, r)^-$, etc.

We can study also totally optimal inhibitory rules relative to various combinations of cost functions. For a cost function ψ, we denote $\psi^{W,\alpha}(T, r) = \min\{\psi(T, \rho) : \rho \in IR_{W,\alpha}(T, r)\}$, i.e., $\psi^{W,\alpha}(T, r)$ is the minimum cost of a (W, α)-inhibitory rule for T and r relative to the cost function ψ. Let $\psi_1, \ldots, \psi_m$ be cost functions and $m \geq 2$. A (W, α)-inhibitory rule ρ for T and r is called a *totally optimal (W, α)-inhibitory rule for T and r relative to the cost functions* $\psi_1, \ldots, \psi_m$ if $\psi_1(T, \rho) = \psi_1^{W,\alpha}(T, r), \ldots, \psi_m(T, \rho) = \psi_m^{W,\alpha}(T, r)$, i.e., ρ is optimal relative to $\psi_1, \ldots, \psi_m$ simultaneously.

Assume that $\psi_1, \ldots, \psi_m \in \{l, -c, -rc, mc, rmc\}$. We now describe how to recognize the existence of a (W, α)-inhibitory rule for T and r which is a totally optimal (W, α)-inhibitory rule for T and r relative to the cost functions $\psi_1, \ldots, \psi_m$.

First, we construct the graph $G = \Delta_{U,\alpha}(T^C)$ using the algorithm $\mathscr{A}_1$. For $i = 1, \ldots, m$, we apply to G and r the procedure of optimization relative to ψ_i (the algorithm $\mathscr{A}_8$). As a result, we obtain, for $i = 1, \ldots, m$, the graph $G^{\psi_i}(r)$ and the number $\psi_i^{U,\alpha}(T^C, r)$ attached to the node T^C of $G^{\psi_i}(r)$. Next, we apply to G sequentially the procedures of optimization relative to the cost functions $\psi_1, \ldots, \psi_m$. As a result, we obtain graphs $G^{\psi_1}(r), G^{\psi_1, \psi_2}(r), \ldots, G^{\psi_1, \ldots, \psi_m}(r)$ and numbers $\varphi_1, \varphi_2, \ldots, \varphi_m$ attached to the node T^C of these graphs. We know (see Sect. 12.1.2) that a totally optimal (U, α)-decision rule for T^C and r relative to the cost functions $\psi_1, \ldots, \psi_m$ exists if and only if $\varphi_i = \psi_i^{U,\alpha}(T^C, r)$ for $i = 1, \ldots, m$. Using Proposition 11.1 one can show that a totally optimal (W, α)-inhibitory rule for T and r relative to the cost functions $\psi_1, \ldots, \psi_m$ exists if and only if a totally optimal (U, α)-decision rule for T^C and r relative to the cost functions $\psi_1, \ldots, \psi_m$ exists.

Let G be a bundle-preserving subgraph of the graph $\Delta_{U,\alpha}(T^C)$ and r be a row of T^C. The algorithm $\mathscr{A}_9$ from Sect. 12.1.4 allows us to find the cardinality of the set $Rule(G, T^C, r)$ containing some (U, α)-decision rules for T^C and r, and in the same time, the cardinality of the set $Rule(G, T^C, r)^-$ containing some (W, α)-inhibitory rules for T and r. It can be, for example, the set of all (W, α)-inhibitory rules for T and r with minimum length, if $Rule(G, T^C, r)$ is the set of all (U, α)-decision rules for T^C and r with minimum length.

Table 12.8 Minimization of inhibitory rule length

Table name	Length of rules		
	Min	Avg	Max
breast-cancer-1	1	2.80	5
breast-cancer-5	1	1.69	3
cars-1	1	1.11	4
mushroom-5	1	1.48	4
nursery-1	1	1.00	1
nursery-4	1	1.00	1
teeth-1	1	1.00	1
teeth-5	1	1.00	1
zoo-data-5	1	1.00	1

12.2.2   Experimental Results: Optimization of Inhibitory Rules

In this section, we discuss results of experiments related to the optimization of inhibitory rules.

For each decision table T in Table 12.1, we converted T to the corresponding complementary decision table T^C by changing, for each row $r \in Row(T)$, the set $D(r)$ with the set $D(T) \setminus D(r)$. For each decision table T^C, we constructed by the algorithm $\mathscr{A}_1$ the directed acyclic graph $\Delta(T^C)$ (see Remark 5.1), and apply to $\Delta(T^C)$ the procedure of optimization of decision rules (the algorithm $\mathscr{A}_8$) relative to the length, relative to the coverage, relative to the coverage and then length, and relative to length and then coverage. The results obtained for exact decision rules for rows of T^C were transformed into results for exact inhibitory rules for rows of T.

Table 12.8 shows in the column "Length of rules" minimum, average, and maximum length of exact inhibitory rules with minimum length among all rows of T. One can see that we obtained short, on average, exact inhibitory rules for each decision table. Minimum, average and maximum coverage of obtained exact inhibitory rules with maximum coverage among all rows of T can be found in Table 12.9 (the column "Coverage of rules"). We should mention good results obtained for decision tables "cars-1", "mushroom-5", "nursery-1", and "nursery-4".

Average coverage and length of exact inhibitory rules obtained by multi-stage optimization relative to the coverage, and then relative to the length can be found in Table 12.10 (the column "Coverage + Length"). The output of multi-stage optimization relative to the length, and then relative to the coverage is shown in the column "Length + Coverage". We compared the number of rows for which there exist totally optimal inhibitory rules relative to the length and coverage (the column "Rows tot") with the number of all rows (the column "Rows"). For decision tables in bold, each row has a totally optimal inhibitory rule.

Table 12.9 Maximization of inhibitory rule coverage

Table name	Coverage of rules		
	Min	Avg	Max
breast-cancer-1	1	7.50	15
breast-cancer-5	1	5.17	11
cars-1	4	133.07	144
mushroom-5	20	746.62	1080
nursery-1	1440	1800.00	2160
nursery-4	80	80.00	80
teeth-1	13	15.18	16
teeth-5	4	7.36	8
zoo-data-5	22	33.79	35

Table 12.10 Sequential optimization of inhibitory rules

Table name	Coverage + Length		Length + Coverage		Rows tot	Rows
	Coverage	Length	Length	Coverage		
breast-cancer-1	7.50	3.64	2.80	5.16	71	169
breast-cancer-5	5.17	1.81	1.69	4.71	51	58
cars-1	133.07	1.11	1.11	133.07	432	432
mushroom-5	746.62	2.26	1.48	516.54	1388	4048
nursery-1	1800.00	1.00	1.00	1800.00	4320	4320
nursery-4	80.00	1.00	1.00	80.00	240	240
teeth-1	15.18	1.00	1.00	15.18	22	22
teeth-5	7.36	1.00	1.00	7.36	14	14
zoo-data-5	33.79	1.00	1.00	33.79	42	42

References

1. AbouEisha, H., Amin, T., Chikalov, I., Hussain, S., Moshkov, M.: Extensions of Dynamic Programming for Combinatorial Optimization and Data Mining. Intelligent Systems Reference Library, vol. 146. Springer, Berlin (2019)
2. Alsolami, F., Amin, T., Chikalov, I., Moshkov, M., Zielosko, B.: Dynamic programming approach for construction of association rule systems. Fundam. Inform. **147**(2–3), 159–171 (2016)
3. Alsolami, F., Chikalov, I., Moshkov, M.: Sequential optimization of approximate inhibitory rules relative to the length, coverage and number of misclassifications. In: Lingras, P., Wolski, M., Cornelis, C., Mitra, S., Wasilewski, P. (eds.) Rough Sets and Knowledge Technology – 8th International Conference, RSKT 2013, Halifax, NS, Canada, 11–14 Oct 2013. Lecture Notes in Computer Science, vol. 8171, pp. 154–165. Springer, Berlin (2013)
4. Alsolami, F., Chikalov, I., Moshkov, M., Zielosko, B.: Length and coverage of inhibitory decision rules. In: Nguyen, N.T., Hoang, K., Jedrzejowicz, P. (eds.) Computational Collective

Intelligence. Technologies and Applications – 4th International Conference, ICCCI 2012, Ho Chi Minh City, Vietnam, 28–30 Nov 2012, Part II. Lecture Notes in Computer Science, vol. 7654, pp. 325–334. Springer, Berlin (2012)

5. Alsolami, F., Chikalov, I., Moshkov, M., Zielosko, B.: Optimization of inhibitory decision rules relative to length and coverage. In: Li, T., Nguyen, H.S., Wang, G., Grzymala-Busse, J.W., Janicki, R., Hassanien, A.E., Yu, H. (eds.) Rough Sets and Knowledge Technology – 7th International Conference, RSKT 2012, Chengdu, China, 17–20 Aug 2012. Lecture Notes in Computer Science, vol. 7414, pp. 149–154. Springer, Berlin (2012)

6. Alsolami, F., Chikalov, I., Moshkov, M., Zielosko, B.: Optimization of approximate inhibitory rules relative to number of misclassifications. In: Watada, J., Jain, L.C., Howlett, R.J., Mukai, N., Asakura, K. (eds.) 17th International Conference in Knowledge Based and Intelligent Information and Engineering Systems, KES 2013, Kitakyushu, Japan, 9–11 Sept 2013. Procedia Computer Science, vol. 22, pp. 295–302. Elsevier, Amsterdam (2013)

7. Lichman, M.: UCI Machine Learning Repository. University of California, Irvine, School of Information and Computer Sciences (2013). http://archive.ics.uci.edu/ml

8. Moshkov, M., Piliszczuk, M., Zielosko, B.: Partial Covers, Reducts and Decision Rules in Rough Sets - Theory and Applications. Studies in Computational Intelligence, vol. 145. Springer, Heidelberg (2008)

9. Slavìk, P.: A tight analysis of the greedy algorithm for set cover. J. Algorithms 25(2), 237–254 (1997)

10. Slavìk, P.: Approximation algorithms for set cover and related problems. Ph.D. thesis, University of New York at Buffalo (1998)

Chapter 13
Bi-criteria Optimization Problem for Rules and Systems of Rules: Cost Versus Cost

In this chapter, we consider algorithms which construct the sets of Pareto optimal points for bi-criteria optimization problems for decision rules and rule systems relative to two cost functions. We show how the constructed set of Pareto optimal points can be transformed into the graphs of functions which describe the relationships between the considered cost functions. These results are extensions of the results obtained earlier for decision tables with single-valued decisions [1, 5].

We compare 13 greedy heuristics for construction of decision rules from the point of view of single-criterion optimization (relative to the length or coverage) and bi-criteria optimization (relative to the length and coverage). Note that five greedy heuristics for construction of decision rules for decision tables with many-valued decisions were considered in [3].

At the end of the chapter, we extend the obtained results to the case of inhibitory rules and systems of inhibitory rules. Note that some greedy heuristics for construction of inhibitory rules for decision tables with single-valued decisions were studied in [4, 6]. Note also that the paper [2] contains some results considered in this chapter.

13.1 Bi-criteria Optimization Problem for Decision Rules and Systems of Rules: Cost Versus Cost

In this section, we study bi-criteria cost versus cost optimization problem for decision rules and systems of decision rules.

© Springer Nature Switzerland AG 2020
F. Alsolami et al., *Decision and Inhibitory Trees and Rules for Decision Tables with Many-valued Decisions*, Intelligent Systems Reference Library 156, https://doi.org/10.1007/978-3-030-12854-8_13

13.1.1 Pareto Optimal Points for Decision Rules:
Cost Versus Cost

Let ψ and φ be strictly increasing cost functions for decision rules given by pairs
of functions ψ^0, F and φ^0, H, respectively. Let T be a nonempty decision table
with n conditional attributes $f_1, \ldots, f_n$, $r = (b_1, \ldots, b_n)$ be a row of T, U be an
uncertainty measure, $\alpha \in \mathbb{R}_+$, and G be a bundle-preserving subgraph of the graph
$\Delta_{U,\alpha}(T)$ (it is possible that $G = \Delta_{U,\alpha}(T)$).

For each node Θ of the graph G containing r, we denote $p_{\psi,\varphi}(G, \Theta, r) =$
$\{(\psi(\Theta, \rho), \varphi(\Theta, \rho)) : \rho \in Rule(G, \Theta, r)\}$. We denote by $Par(p_{\psi,\varphi}(G, \Theta, r))$ the
set of Pareto optimal points for $p_{\psi,\varphi}(G, \Theta, r)$. Note that, by Proposition 12.1, if
$G = \Delta_{U,\alpha}(T)$ then the set $Rule(G, \Theta, r)$ is equal to the set of (U, α)-decision rules
for Θ and r. Another interesting case is when G is the result of application of procedure of optimization of rules for r (the algorithm $\mathscr{A}_8$) relative to cost functions
different from ψ and φ to the graph $\Delta_{U,\alpha}(T)$.

We now describe an algorithm $\mathscr{A}_{11}$ constructing the set $Par(p_{\psi,\varphi}(G, T, r))$. In
fact, this algorithm constructs, for each node Θ of the graph G, the set $B(\Theta, r) =$
$Par(p_{\psi,\varphi}(G, \Theta, r))$. Let us remind that $A^{FH} = \{(F(a), H(b)) : (a, b) \in A\}$ for any
nonempty finite subset A of the set $\mathbb{R}^2$.

Algorithm $\mathscr{A}_{11}$ (construction of POPs for decision rules, cost versus cost).
Input: Strictly increasing cost functions ψ and φ for decision rules given by pairs
of functions ψ^0, F and φ^0, H, respectively, a nonempty decision table T with n conditional attributes $f_1, \ldots, f_n$, a row $r = (b_1, \ldots, b_n)$ of T, and a bundle-preserving
subgraph G of the graph $\Delta_{U,\alpha}(T)$ where U is an uncertainty measure and $\alpha \in \mathbb{R}_+$.
Output: The set $Par(p_{\psi,\varphi}(G, T, r))$ of Pareto optimal points for the set of pairs
$p_{\psi,\varphi}(G, T, r) = \{(\psi(T, \rho), \varphi(T, \rho)) : \rho \in Rule(G, T, r)\}$.

1. If all nodes in G containing r are processed, then return the set $B(T, r)$. Otherwise,
   choose in the graph G a node Θ containing r which is not processed yet and
   which is either a terminal node of G or a nonterminal node of G such that, for any
   $f_i \in E_G(\Theta)$, the node $\Theta(f_i, b_i)$ is already processed, i.e., the set $B(\Theta(f_i, b_i), r)$
   is already constructed.
2. If Θ is a terminal node, then set $B(\Theta, r) = \{(\psi^0(\Theta), \varphi^0(\Theta))\}$. Mark the node
   Θ as processed and proceed to step 1.
3. If Θ is a nonterminal node then, for each $f_i \in E_G(\Theta)$, construct the set

$$B(\Theta(f_i, b_i), r)^{FH} \, ,$$

and construct the multiset

$$A(\Theta, r) = \bigcup_{f_i \in E_G(\Theta)} B(\Theta(f_i, b_i), r)^{FH}$$

by simple transcription of elements from the sets $B(\Theta(f_i, b_i), r)^{FH}$, $f_i \in E_G(\Theta)$.

4. Apply to the multiset $A(\Theta, r)$ the algorithm $\mathscr{A}_2$ which constructs the set

$$Par(A(\Theta, r)) .$$

Set $B(\Theta, r) = Par(A(\Theta, r))$. Mark the node Θ as processed and proceed to step 1.

Proposition 13.1 *Let ψ and φ be strictly increasing cost functions for decision rules given by pairs of functions ψ^0, F and φ^0, H, respectively, T be a nonempty decision table with n conditional attributes $f_1, \ldots, f_n$, $r = (b_1, \ldots, b_n)$ be a row of T, U be an uncertainty measure, $\alpha \in \mathbb{R}_+$, and G be a bundle-preserving subgraph of the graph $\Delta_{U,\alpha}(T)$. Then, for each node Θ of the graph G containing r, the algorithm $\mathscr{A}_{11}$ constructs the set $B(\Theta, r) = Par(p_{\psi,\varphi}(G, \Theta, r))$.*

Proof We prove the considered statement by induction on nodes of G. Let Θ be a terminal node of G containing r. Then $Rule(G, \Theta, r) = \{\rightarrow mcd(\Theta)\}$,

$$p_{\psi,\varphi}(G, \Theta, r) = Par(p_{\psi,\varphi}(G, \Theta, r)) = \{(\psi^0(\Theta), \varphi^0(\Theta))\} ,$$

and $B(\Theta, r) = Par(p_{\psi,\varphi}(G, \Theta, r))$.

Let Θ be a nonterminal node of G containing r such that, for any $f_i \in E_G(\Theta)$, the considered statement holds for the node $\Theta(f_i, b_i)$, i.e.,

$$B(\Theta(f_i, b_i), r) = Par(p_{\psi,\varphi}(G, \Theta(f_i, b_i), r)) .$$

It is clear that

$$p_{\psi,\varphi}(G, \Theta, r) = \bigcup_{f_i \in E_G(\Theta)} p_{\psi,\varphi}(G, \Theta(f_i, b_i), r)^{FH} .$$

From Lemma 5.7 it follows that

$$Par(p_{\psi,\varphi}(G, \Theta, r)) \subseteq \bigcup_{f_i \in E_G(\Theta)} Par(p_{\psi,\varphi}(G, \Theta(f_i, b_i), r)^{FH}) .$$

By Lemma 5.9, $Par(p_{\psi,\varphi}(G, \Theta(f_i, b_i), r)^{FH}) = Par(p_{\psi,\varphi}(G, \Theta(f_i, b_i), r))^{FH}$ for any $f_i \in E_G(\Theta)$. Therefore

$$Par(p_{\psi,\varphi}(G, \Theta, r)) \subseteq \bigcup_{f_i \in E_G(\Theta)} Par(p_{\psi,\varphi}(G, \Theta(f_i, b_i), r))^{FH} \subseteq p_{\psi,\varphi}(G, \Theta, r) .$$

Using Lemma 5.6 we obtain

$$Par(p_{\psi,\varphi}(G, \Theta, r)) = Par \left(\bigcup_{f_i \in E_G(\Theta)} Par(p_{\psi,\varphi}(G, \Theta(f_i, b_i), r))^{FH} \right) .$$

Since $B(\Theta, r) = Par\left(\bigcup_{f_i \in E_G(\Theta)} B(\Theta(f_i, b_i), r)^{FH}\right)$ and

$$B(\Theta(f_i, b_i), r) = Par(p_{\psi, \varphi}(G, \Theta(f_i, b_i), r))$$

for any $f_i \in E_G(\Theta)$, we have $B(\Theta, r) = Par(p_{\psi, \varphi}(G, \Theta, r))$. $\square$

We now evaluate the number of elementary operations (computations of F, H, ψ^0, φ^0, and comparisons) made by the algorithm $\mathscr{A}_{11}$. Let us recall that, for a given cost function ψ for decision rules and decision table T,

$$q_\psi(T) = |\{\psi(\Theta, \rho) : \Theta \in SEP(T), \rho \in DR(\Theta)\}| .$$

In particular, by Lemma 11.1, $q_l(T) \leq n + 1$, $q_{-rc}(T) \leq N(T)(N(T) + 1)$, $q_{-c}(T) \leq N(T) + 1$, $q_{-c_M}(T) \leq N(T) + 1$, $q_{mc}(T) \leq N(T) + 1$, and $q_{rmc}(T) \leq N(T)(N(T) + 1)$.

Proposition 13.2 *Let ψ and φ be strictly increasing cost functions for decision rules given by pairs of functions ψ^0, F and φ^0, H, respectively, T be a nonempty decision table with n conditional attributes $f_1, \ldots, f_n$, $r = (b_1, \ldots, b_n)$ be a row of T, U be an uncertainty measure, $\alpha \in \mathbb{R}_+$, and G be a bundle-preserving subgraph of the graph $\Delta_{U, \alpha}(T)$. Then, to construct the set $Par(p_{\psi, \varphi}(G, T, r))$, the algorithm $\mathscr{A}_{11}$ makes*

$$O(L(G) \min(q_\psi(T), q_\varphi(T))n \log(\min(q_\psi(T), q_\varphi(T))n))$$

elementary operations (computations of F, H, ψ^0, φ^0, and comparisons).

Proof To process a terminal node, the algorithm $\mathscr{A}_{11}$ makes two elementary operations – computes ψ^0 and φ^0. We now evaluate the number of elementary operations under the processing of a nonterminal node Θ. From Lemma 5.5 it follows that $\left|Par(p_{\psi, \varphi}(G, \Theta(f_i, b_i), r))\right| \leq \min(q_\psi(T), q_\varphi(T))$ for any $f_i \in E_G(\Theta)$. It is clear that $|E_G(\Theta)| \leq n$, $\left|Par(p_{\psi, \varphi}(G, \Theta(f_i, b_i), r))^{FH}\right| = \left|Par(p_{\psi, \varphi}(G, \Theta(f_i, b_i), r))\right|$ for any $f_i \in E_G(\Theta)$. From Proposition 13.1 it follows that

$$B(\Theta(f_i, b_i), r) = Par(p_{\psi, \varphi}(G, \Theta(f_i, b_i), r))$$

and $B(\Theta(f_i, b_i), r)^{FH} = Par(p_{\psi, \varphi}(G, \Theta(f_i, b_i), r))^{FH}$ for any $f_i \in E_G(\Theta)$. Hence

$$|A(\Theta, r)| \leq \min(q_\psi(T), q_\varphi(T))n .$$

Therefore to construct the sets $B(\Theta(f_i, b_i), r)^{FH}$, $f_i \in E_G(\Theta)$, from the sets

$$B(\Theta(f_i, b_i), r) ,$$

$f_i \in E_G(\Theta)$, the algorithm $\mathscr{A}_{11}$ makes $O(\min(q_\psi(T), q_\varphi(T))n)$ computations of F and H, and to construct the set $Par(A(\Theta, r)) = Par(p_{\psi, \varphi}(G, \Theta, r))$ from the set $A(\Theta, r)$, the algorithm $\mathscr{A}_{11}$ makes

$$O(\min(q_\psi(T), q_\varphi(T))n \log(\min(q_\psi(T), q_\varphi(T))n))$$

comparisons (see Proposition 5.5). Hence, to process a nonterminal node Θ, the algorithm makes $O(\min(q_\psi(T), q_\varphi(T))n \log(\min(q_\psi(T), q_\varphi(T))n))$ computations of F, H, and comparisons.

To construct the set $Par(p_{\psi,\varphi}(G, T, r))$, the algorithm $\mathscr{A}_{11}$ makes

$$O(L(G) \min(q_\psi(T), q_\varphi(T))n \log(\min(q_\psi(T), q_\varphi(T))n))$$

elementary operations (computations of F, H, ψ^0, φ^0, and comparisons). $\qquad\square$

Proposition 13.3 *Let ψ and φ be strictly increasing cost functions for decision rules given by pairs of functions ψ^0, F and φ^0, H, respectively,*

$$\psi, \varphi \in \{l, -c, -rc, -c_M, mc, rmc\},$$

and $\mathscr{U}$ be a restricted information system. Then the algorithm $\mathscr{A}_{11}$ has polynomial time complexity for decision tables from $\mathscr{T}(\mathscr{U})$ depending on the number of conditional attributes in these tables.

Proof Since $\psi, \varphi \in \{l, -c, -rc, -c_M, mc, rmc\}$,

$$\psi^0, \varphi^0 \in \{0, -N_{mcd(T)}(T), -N_{mcd(T)}(T)/N(T), -N^M(T),$$
$$N(T) - N_{mcd(T)}(T), (N(T) - N_{mcd(T)}(T))/N(T)\},$$

and $F, H \in \{x, x + 1\}$. From Lemma 11.1 and Proposition 13.2 it follows that, for the algorithm $\mathscr{A}_{11}$, the number of elementary operations (computations of F, H, ψ^0, φ^0, and comparisons) is bounded from above by a polynomial depending on the size of input table T and on the number of separable subtables of T. All operations with numbers are basic ones. The computations of numerical parameters of decision tables used by the algorithm $\mathscr{A}_{11}$ ($0, -N_{mcd(T)}(T), -N_{mcd(T)}(T)/N(T), -N^M(T)$, $N(T) - N_{mcd(T)}(T)$, and $(N(T) - N_{mcd(T)}(T))/N(T)$) have polynomial time complexity depending on the size of decision tables.

According to Proposition 5.4, the algorithm $\mathscr{A}_{11}$ has polynomial time complexity for decision tables from $\mathscr{T}(\mathscr{U})$ depending on the number of conditional attributes in these tables. $\qquad\square$

13.1.2 Relationships for Decision Rules: Cost Versus Cost

Let ψ and φ be strictly increasing cost functions for decision rules, T be a nonempty decision table with n conditional attributes $f_1, \ldots, f_n$, $r = (b_1, \ldots, b_n)$ be a row of T, U be an uncertainty measure, $\alpha \in \mathbb{R}_+$, and G be a bundle-preserving subgraph of the graph $\Delta_{U,\alpha}(T)$ (it is possible that $G = \Delta_{U,\alpha}(T)$).

To study relationships between cost functions ψ and φ on the set of rules $Rule(G, T, r)$, we consider partial functions $\mathscr{R}_{G,T,r}^{\psi,\varphi} : \mathbb{R} \to \mathbb{R}$ and $\mathscr{R}_{G,T,r}^{\varphi,\psi} : \mathbb{R} \to \mathbb{R}$ defined in the following way:

$$\mathscr{R}_{G,T,r}^{\psi,\varphi}(x) = \min\{\varphi(T, \rho) : \rho \in Rule(G, T, r), \psi(T, \rho) \le x\},$$

$$\mathscr{R}_{G,T,r}^{\varphi,\psi}(x) = \min\{\psi(T, \rho) : \rho \in Rule(G, T, r), \varphi(T, \rho) \le x\}.$$

Let $p_{\psi,\varphi}(G, T, r) = \{(\psi(T, \rho), \varphi(T, \rho)) : \rho \in Rule(G, T, r)\}$ and

$$(a_1, b_1), \ldots, (a_k, b_k)$$

be the normal representation of the set $Par(p_{\psi,\varphi}(G, T, r))$ where $a_1 < \cdots < a_k$ and $b_1 > \cdots > b_k$. By Lemma 5.10 and Remark 5.4, for any $x \in \mathbb{R}$,

$$\mathscr{R}_{G,T,r}^{\psi,\varphi}(x) = \begin{cases} undefined, & x < a_1 \\ b_1, & a_1 \le x < a_2 \\ \ldots & \ldots \\ b_{k-1}, & a_{k-1} \le x < a_k \\ b_k, & a_k \le x \end{cases},$$

$$\mathscr{R}_{G,T,r}^{\varphi,\psi}(x) = \begin{cases} undefined, & x < b_k \\ a_k, & b_k \le x < b_{k-1} \\ \ldots & \ldots \\ a_2, & b_2 \le x < b_1 \\ a_1, & b_1 \le x \end{cases}.$$

13.1.3 Pareto Optimal Points for Systems of Decision Rules: Cost Versus Cost

Let T be a nonempty decision table with n conditional attributes $f_1, \ldots, f_n$ and $N(T)$ rows $r_1, \ldots, r_{N(T)}$, U be an uncertainty measure, $\alpha \in \mathbb{R}_+$, and $\mathbf{G} = (G_1, \ldots, G_{N(T)})$ be an $N(T)$-tuple of bundle-preserving subgraphs of the graph $\Delta_{U,\alpha}(T)$. Let $G = \Delta_{U,\alpha}(T)$ and ξ be a cost function for decision rules. Then two interesting examples of such $N(T)$-tuples are $(G, \ldots, G)$ and $(G^\xi(r_1), \ldots, G^\xi(r_{N(T)}))$.

We denote by $\mathscr{S}(\mathbf{G}, T)$ the set $Rule(G_1, T, r_1) \times \cdots \times Rule(G_{N(T)}, T, r_{N(T)})$ of (U, α)-systems of decision rules for T. Let ψ, φ be strictly increasing cost functions for decision rules, and f, g be increasing functions from $\mathbb{R}^2$ to $\mathbb{R}$. It is clear that ψ_f and φ_g are strictly increasing cost functions for systems of decision rules (see Sect. 11.1.2).

We describe now an algorithm which constructs the set of Pareto optimal points for the set of pairs $p_{\psi,\varphi}^{f,g}(\mathbf{G}, T) = \{(\psi_f(T, S), \varphi_g(T, S)) : S \in \mathscr{S}(\mathbf{G}, T)\}$.

Algorithm $\mathscr{A}_{12}$ (construction of POPs for decision rule systems, cost versus cost).

Input: Strictly increasing cost functions for decision rules ψ and φ given by pairs of functions ψ^0, F and φ^0, H, respectively, increasing functions f, g from $\mathbb{R}^2$ to $\mathbb{R}$, a nonempty decision table T with n conditional attributes $f_1, \ldots, f_n$ and $N(T)$ rows $r_1, \ldots, r_{N(T)}$, and an $N(T)$-tuple $\mathbf{G} = (G_1, \ldots, G_{N(T)})$ of bundle-preserving subgraphs of the graph $\Delta_{U,\alpha}(T)$ where U is an uncertainty measure and $\alpha \in \mathbb{R}_+$.

Output: The set $Par(p_{\psi,\varphi}^{f,g}(\mathbf{G}, T))$ of Pareto optimal points for the set of pairs $p_{\psi,\varphi}^{f,g}(\mathbf{G}, T) = \{((\psi_f(T, S), (\varphi_g(T, S)) : S \in \mathscr{S}(\mathbf{G}, T)\}$.

1. Using the algorithm $\mathscr{A}_{11}$ construct, for $i = 1, \ldots, N(T)$, the set $Par(P_i)$ where

$$P_i = p_{\psi,\varphi}(G_i, T, r_i) = \{(\psi(T, \rho), \varphi(T, \rho)) : \rho \in Rule(G_i, T, r_i)\}.$$

2. Apply the algorithm $\mathscr{A}_3$ to the functions f, g and the sets $Par(P_1), \ldots,$ $Par(P_{N(T)})$. Set $C(\mathbf{G}, T)$ the output of the algorithm $\mathscr{A}_3$ and return it.

Proposition 13.4 *Let* ψ, φ *be strictly increasing cost functions for decision rules,* f, g *be increasing functions from* $\mathbb{R}^2$ *to* $\mathbb{R}$, U *be an uncertainty measure,* $\alpha \in \mathbb{R}_+$, T *be a decision table with* n *conditional attributes* $f_1, \ldots, f_n$ *and* $N(T)$ *rows* $r_1, \ldots, r_{N(T)}$, *and* $\mathbf{G} = (G_1, \ldots, G_{N(T)})$ *be an* $N(T)$-*tuple of bundle-preserving subgraphs of the graph* $\Delta_{U,\alpha}(T)$. *Then the algorithm* $\mathscr{A}_{12}$ *constructs the set* $C(\mathbf{G}, T)=$ $Par(p_{\psi,\varphi}^{f,g}(\mathbf{G}, T))$.

Proof For $i = 1, \ldots, N(T)$, denote $P_i = p_{\psi,\varphi}(G_i, T, r_i)$. During the first step, the algorithm $\mathscr{A}_{12}$ constructs (using the algorithm $\mathscr{A}_{11}$) the sets $Par(P_1), \ldots,$ $Par(P_{N(T)})$ (see Proposition 13.1). During the second step, the algorithm $\mathscr{A}_{12}$ constructs (using the algorithm $\mathscr{A}_3$) the set $C(\mathbf{G}, T) = Par(Q_{N(T)})$ where $Q_1 = P_1$, and, for $i = 2, \ldots, N(T)$, $Q_i = Q_{i-1} \langle fg \rangle P_i$ (see Proposition 5.6). One can show that $Q_{N(T)} = p_{\psi,\varphi}^{f,g}(\mathbf{G}, T)$. Therefore $C(\mathbf{G}, T) = Par(p_{\psi,\varphi}^{f,g}(\mathbf{G}, T))$. $\square$

Let us recall that, for a given cost function ψ and a decision table T, $q_\psi(T) = |\{\psi(\Theta, \rho) : \Theta \in SEP(T), \rho \in DR(\Theta)\}|$. In particular, by Lemma 11.1, $q_l(T) \leq n + 1$, $q_{-c}(T) \leq N(T) + 1$, $q_{-rc}(T) \leq N(T)(N(T) + 1)$, $q_{-c_M}(T) \leq N(T) + 1$, $q_{mc}(T) \leq N(T) + 1$, and $q_{rmc}(T) \leq N(T)(N(T) + 1)$.

Let us recall also that, for a given cost function ψ for decision rules and a decision table T, $Range_\psi(T) = \{\psi(\Theta, \rho) : \Theta \in SEP(T), \rho \in DR(\Theta)\}$. By Lemma 11.1, $Range_l(T) \subseteq \{0, 1, \ldots, n\}$, $Range_{-c}(T) \subseteq \{0, -1, \ldots, -N(T)\}$, $Range_{-c_M}(T) \subseteq \{0, -1, \ldots, -N(T)\}$, and $Range_{mc}(T) \subseteq \{0, 1, \ldots, N(T)\}$. Let $t_l(T)=n$, $t_{-c}(T) = N(T)$, $t_{-c_M}(T) = N(T)$, and $t_{mc}(T) = N(T)$.

Proposition 13.5 *Let* ψ, φ *be strictly increasing cost functions for decision rules given by pairs of functions* ψ^0, F *and* φ^0, H, *respectively,* $\psi \in \{l, -c, -c_M, mc\}$, f, g *be increasing functions from* $\mathbb{R}^2$ *to* $\mathbb{R}$, $f \in \{x + y, \max(x, y)\}$, U *be an uncertainty measure,* $\alpha \in \mathbb{R}_+$, T *be a decision table with* n *conditional attributes* $f_1, \ldots, f_n$ *and* $N(T)$ *rows* $r_1, \ldots, r_{N(T)}$, $\mathbf{G} = (G_1, \ldots, G_{N(T)})$ *be an* $N(T)$-*tuple of bundle-preserving subgraphs of the graph* $\Delta_{U,\alpha}(T)$, *and* $L(\mathbf{G}) = L(\Delta_{U,\alpha}(T))$. *Then, to construct the set* $Par(p_{\psi,\varphi}^{f,g}(\mathbf{G}, T))$, *the algorithm* $\mathscr{A}_{12}$ *makes*

$$O(N(T)L(\mathbf{G}) \min(q_\psi(T), q_\varphi(T))n \log(\min(q_\psi(T), q_\varphi(T))n))$$
$$+ O(N(T)t_\psi(T)^2 \log(t_\psi(T)))$$

elementary operations (computations of F, H, ψ^0, φ^0, f, g and comparisons) if
$f = \max(x, y)$, *and*

$$O(N(T)L(\mathbf{G}) \min(q_\psi(T), q_\varphi(T))n \log(\min(q_\psi(T), q_\varphi(T))n))$$
$$+ O(N(T)^2 t_\psi(T)^2 \log(N(T)t_\psi(T)))$$

elementary operations (computations of F, H, ψ^0, φ^0, f, g and comparisons) if
$f = x + y$.

Proof To construct the sets $Par(P_i) = Par(p_{\psi,\varphi}(G_i, T, r_i)), i = 1, \ldots, N(T)$, the
algorithm $\mathscr{A}_{11}$ makes

$$O(N(T)L(\mathbf{G}) \min(q_\psi(T), q_\varphi(T))n \log(\min(q_\psi(T), q_\varphi(T))n))$$

elementary operations (computations of F, H, ψ^0, φ^0, and comparisons) – see Proposition 13.2.

We now evaluate the number of elementary operations (computations of f, g, and comparisons) made by the algorithm $\mathscr{A}_3$ during the construction of the set $C(\mathbf{G}, T) = Par(Q_{N(T)}) = Par(p_{\psi,\varphi}^{f,g}(\mathbf{G}, T))$. We know that $\psi \in \{l, -c, -c_M, mc\}$ and $f \in \{x + y, \max(x, y)\}$.

For $i = 1, \ldots, N(T)$, let $P_i^1 = \{a : (a, b) \in P_i\}$. Since $\psi \in \{l, -c, -c_M, mc\}$, we have $P_i^1 \subseteq \{0, 1, \ldots, t_\psi(T)\}$ for $i = 1, \ldots, N(T)$ or $P_i^1 \subseteq \{0, -1, \ldots, -t_\psi(T)\}$ for $i = 1, \ldots, N(T)$.

Using Proposition 5.7 we obtain the following.

If $f = x + y$, then to construct the set $Par(Q_{N(T)})$, the algorithm $\mathscr{A}_3$ makes

$$O(N(T)^2 t_\psi(T)^2 \log(N(T)t_\psi(T)))$$

elementary operations (computations of f, g, and comparisons).

If $f = \max(x, y)$, then to construct the sets $Par(Q_{N(T)})$, the algorithm $\mathscr{A}_3$ makes

$$O(N(T)t_\psi(T)^2 \log(t_\psi(T)))$$

elementary operations (computations of f, g, and comparisons). $\square$

Similar analysis can be done for the case when $\varphi \in \{l, -c, -c_M, mc\}$ and $g \in \{x + y, \max(x, y)\}$.

Proposition 13.6 *Let ψ, φ be strictly increasing cost functions for decision rules given by pairs of functions ψ^0, F and φ^0, H, respectively, f, g be increasing functions from $\mathbb{R}^2$ to $\mathbb{R}$, $\psi \in \{l, -c, -c_M, mc\}$, $\varphi \in \{l, -c, -rc, -c_M, mc, rmc\}$, f, g $\in \{x + y, \max(x, y)\}$, and $\mathscr{U}$ be a restricted information system. Then the algorithm $\mathscr{A}_{12}$*

has polynomial time complexity for decision tables from $\mathcal{T}(\mathcal{U})$ depending on the number of conditional attributes in these tables.

Proof Since $\psi \in \{l, -c, -c_M, mc\}$ and $\varphi \in \{l, -c, -rc, -c_M, mc, rmc\}$,

$$\psi^0, \varphi^0 \in \{0, -N_{mcd(T)}(T), -N_{mcd(T)}(T)/N(T), -N^M(T),$$
$$N(T) - N_{mcd(T)}(T), (N(T) - N_{mcd(T)}(T))/N(T)\},$$

and $F, H \in \{x, x + 1\}$. From Lemma 11.1 and Proposition 13.5 it follows that, for the algorithm $\mathscr{A}_{12}$, the number of elementary operations (computations of F, H, ψ^0, φ^0, f, g, and comparisons) is bounded from above by a polynomial depending on the size of input table T and on the number of separable subtables of T. All operations with numbers are basic ones. The computations of numerical parameters of decision tables used by the algorithm $\mathscr{A}_{12}$ $(0, -N_{mcd(T)}(T), -N_{mcd(T)}(T)/N(T),$ $-N^M(T), N(T) - N_{mcd(T)}(T),$ and $(N(T) - N_{mcd(T)}(T))/N(T))$ have polynomial time complexity depending on the size of decision tables.

According to Proposition 5.4, the algorithm $\mathscr{A}_{12}$ has polynomial time complexity for decision tables from $\mathcal{T}(\mathcal{U})$ depending on the number of conditional attributes in these tables. $\qquad\square$

13.1.4 Relationships for Systems of Decision Rules: Cost Versus Cost

Let ψ and φ be strictly increasing cost functions for decision rules, and f, g be increasing functions from $\mathbb{R}^2$ to $\mathbb{R}$. Let T be a nonempty decision table, U be an uncertainty measure, $\alpha \in \mathbb{R}_+$, and $\mathbf{G} = (G_1, \ldots, G_{N(T)})$ be an $N(T)$-tuple of bundle-preserving subgraphs of the graph $\Delta_{U,\alpha}(T)$.

To study relationships between cost functions ψ_f and φ_g on the set of systems of rules $\mathscr{S}(\mathbf{G}, T)$, we consider two partial functions $\mathscr{R}_{\mathbf{G},T}^{\psi,f,\varphi,g} : \mathbb{R} \to \mathbb{R}$ and $\mathscr{R}_{\mathbf{G},T}^{\varphi,g,\psi,f} : \mathbb{R} \to \mathbb{R}$ defined in the following way:

$$\mathscr{R}_{\mathbf{G},T}^{\psi,f,\varphi,g}(x) = \min\{\varphi_g(T, S) : S \in \mathscr{S}(\mathbf{G}, T), \psi_f(T, S) \le x\},$$
$$\mathscr{R}_{\mathbf{G},T}^{\varphi,g,\psi,f}(x) = \min\{\psi_f(T, S) : S \in \mathscr{S}(\mathbf{G}, T), \varphi_g(T, S) \le x\}.$$

Let $p_{\psi,\varphi}^{f,g}(\mathbf{G}, T) = \{(\psi_f(T, S), \varphi_g(T, S)) : S \in \mathscr{S}(\mathbf{G}, T)\}$, and $(a_1, b_1), \ldots,$ (a_k, b_k) be the normal representation of the set $Par(p_{\psi,\varphi}^{f,g}(\mathbf{G}, T))$ where $a_1 < \cdots < a_k$ and $b_1 > \cdots > b_k$. By Lemma 5.10 and Remark 5.4, for any $x \in \mathbb{R}$,

$$\mathscr{R}_{\mathbf{G},T}^{\psi,f,\varphi,g}(x) = \begin{cases} undefined, & x < a_1 \\ b_1, & a_1 \le x < a_2 \\ \cdots & \cdots \\ b_{k-1}, & a_{k-1} \le x < a_k \\ b_k, & a_k \le x \end{cases},$$

$$
\mathscr{R}_{G,T}^{\varphi, g, \psi, f}(x) = \begin{cases}
undefined, & x < b_k \\
a_k, & b_k \le x < b_{k-1} \\
\cdots & \cdots \\
a_2, & b_2 \le x < b_1 \\
a_1, & b_1 \le x
\end{cases}.
$$

13.2 Comparison of Heuristics for Decision Rule Construction

In this section, we compare 13 greedy heuristics for construction of decision rules from the point of view of single-criterion optimization (relative to length or coverage) and bi-criteria optimization (relative to length and coverage).

13.2.1 Greedy Heuristics

We consider greedy heuristics each of which, for a given decision table T, row r of T, and decision $t \in D(r)$, constructs an exact decision rule for T and r.

Each heuristic is described by the algorithm $\mathscr{A}_{\text{greedy}}$ which uses specific for this heuristic *attribute selection function* F. These functions are defined after the description of $\mathscr{A}_{\text{greedy}}$.

Algorithm $\mathscr{A}_{\text{greedy}}$ (greedy heuristic for decision rule construction).

Input: A decision table T with n conditional attributes $f_1, \ldots, f_n$, a row $r = (b_1, \ldots, b_n)$ from T, a decision t from $D(r)$, and an attribute selection function F.

Output: An exact decision rule for T and r.

1. Create a rule ρ_0 of the form $\to t$.
2. For $k \ge 0$, assume ρ_k is already defined as $(f_{i_1} = b_{i_1}) \wedge \cdots \wedge (f_{i_k} = b_{i_k}) \to t$ (step 1 defines ρ_k for $k = 0$).
3. Define a subtable T^k such that $T^k = T(f_{i_1}, b_{i_1}) \ldots (f_{i_k}, b_{i_k})$.
4. If T^k has a common decision t' (it is possible that $t \neq t'$), then replace the right-hand side of ρ_k with t' and end the algorithm with ρ_k as the constructed rule.
5. Select an attribute $f_{i_{k+1}} \in E(T^k)$ such that $F(T^k, r, t) = f_{i_{k+1}}$.
6. Define ρ_{k+1} by adding the term $f_{i_{k+1}} = b_{i_{k+1}}$ to the left-hand side of ρ_k.
7. Repeat from step 2 with $k = k + 1$.

Now we define the attribute selection functions $F(T, r, t)$ used by the algorithm $\mathscr{A}_{\text{greedy}}$. Given that T is a decision table with conditional attributes $f_1, \ldots, f_n$, $r = (b_1, \ldots, b_n)$ is a row T, and $t \in D(r)$, let $N_t(T)$ be the number of rows r' in T such that $t \in D(r')$. Let $M(T, t) = N(T) - N_t(T)$ and $P(T, t) = N_t(T)/N(T)$. For any attribute $f_j \in E(T)$, we define $a(T, r, f_j, t) = N_t(T) - N_t(T(f_j, b_j))$

and $b(T, r, f_j, t) = M(T, t) - M(T(f_j, b_j), t)$. By $mcd(T)$, we denote the *most common decision* for T which is the minimum decision t_0 from $D(T)$ such that $N_{t_0}(T) = \max\{N_t(T) : t \in D(T)\}$.

The following heuristics and corresponding attribute selection functions will be used in the experiments:

1. poly: $F(T, r, t) = f_j$ such that $f_j \in E(T)$ and $\frac{b(T,r,f_j,t)}{a(T,r,f_j,t)+1}$ is maximized.
2. log: $F(T, r, t) = f_j$ such that $f_j \in E(T)$ and $\frac{b(T,r,f_j,t)}{\log_2(a(T,r,f_j,t)+2)}$ is maximized.
3. maxCov: $F(T, r, t) = f_j$ such that $f_j \in E(T), b(T, r, f_j, t) > 0$ and $a(T, r, f_j, t)$ is minimized.
4. M: $F(T, r, t) = f_j$ such that $f_j \in E(T)$, $T' = T(f_j, b_j)$ and $M(T', t)$ is minimized.
5. RM: $F(T, r, t) = f_j$ such that $f_j \in E(T)$, $T' = T(f_j, b_j)$ and $\frac{M(T',t)}{N(T')}$ is minimized.
6. me: $F(T, r, t) = f_j$ such that $f_j \in E(T)$, $T' = T(f_j, b_j)$ and $M(T', mcd(T'))$ is minimized.
7. mep: $F(T, r, t) = f_j$ such that $f_j \in E(T)$, $T' = T(f_j, b_j)$ and $\frac{M(T',mcd(T'))}{N(T')}$ is minimized.
8. ABS: $F(T, r, t) = f_j$ such that $f_j \in E(T)$, $T' = T(f_j, b_j)$ and $\prod_{q \in D(T')}(1 - P(T', q))$ is minimized.

For the first five heuristics (poly, log, maxCov, M, and RM), we apply the algorithm $\mathscr{A}_{greedy}$ using each decision $t \in D(r)$. As a result, we obtain $|D(r)|$ rules. Depending on our aim, we either choose among these rules a single rule with minimum length (we denote this algorithm by ⟨name of heuristic⟩_L) or a single rule with maximum coverage (we denote this algorithm by ⟨name of heuristic⟩_C). For the last three heuristics (me, mep, and ABS), the algorithm $\mathscr{A}_{greedy}$ is only applied to each row r once using any given $t \in D(r)$ since the resulting rule will be the same regardless of the input decision. We apply each of the considered heuristics to each row of T. Overall, this results in 13 systems of decision rules for the decision table T with many-valued decisions (constructed by heuristics poly_C, poly_L, log_C, log_L, maxCov_C, maxCov_L, M_C, M_L, RM_C, RM_L, me, mep, and ABS).

13.2.2 Experimental Results

We carried out the experiments on nine decision tables with many-valued decisions described in Table 12.1. The number of rows and attributes in these tables can be found in the columns "Rows" and "Attributes", respectively, of Table 13.1. The last column "#POPs" of this table shows the number of Pareto optimal points constructed by the algorithm $\mathscr{A}_{12}$ for each decision table T. We apply the algorithm $\mathscr{A}_{12}$ to the cost functions l and $-c$, functions $f = g = \text{sum} = x + y$ (it means that we consider sum of lengths of rules from system and sum of negative coverages of rules from system), decision table T and $N(T)$-tuple $\mathbf{G} = (\Delta(T), \ldots, \Delta(T))$ where $\Delta(T) = \Delta_{U,0}(T)$

Table 13.1 Number of Pareto optimal points for bi-criteria optimization problem relative to l_{sum} and $-c_{sum}$ for systems of exact decision rules

Table name	Rows	Attributes	# POPs
breast-cancer-1	169	8	136
breast-cancer-5	58	4	8
cars-1	432	5	1
mushroom-5	4048	17	3154
nursery-1	4320	7	1
nursery-4	240	4	1
teeth-1	22	7	1
teeth-5	14	3	1
zoo-data-5	42	11	4

Table 13.2 Comparing heuristics for construction of exact decision rules

Heuristic name	Average rank			Overall rank		
	Cov	Len	Bi-crit	Cov	Len	Bi-crit
poly_C	2.78	7.11	5.67	**1**	8.5	5
poly_L	3.56	7.11	6.44	**2**	8.5	7
log_C	4.33	5.61	3.44	**3**	5.5	**1**
log_L	5.11	5.61	4.22	4	5.5	**3**
maxCov_C	8.22	12.39	11.22	7	12.5	12
maxCov_L	8.56	12.39	11.78	10	12.5	13
M_C	8.72	4.61	6.06	11	**1.5**	6
M_L	9.44	4.61	6.89	12	**1.5**	8
RM_C	6.06	4.67	4.17	5	**3.5**	2
RM_L	6.89	4.67	5.11	6	**3.5**	4
ABS	8.39	8.72	9.39	8	11	11
me	10.5	5.89	8.33	13	7	10
mep	8.44	7.61	8.28	9	10	9

for some uncertainty measure U (it means that we consider an exact decision rule for T and each row of T). More than half of the tables have only a single Pareto optimal point that means, for each row, there exists an exact decision rule which simultaneously has both minimum length and maximum coverage (totally optimal exact decision rule relative to the coverage and length).

We measured how well heuristics perform for single-criterion and bi-criteria optimizations for the length and coverage. For single-criterion optimization, we used ranking to avoid any bias. In the case of equal values for some heuristics, we break ties by averaging their ranks. Table 13.2 presents ranking of greedy heuristics averaged across all nine decision tables with many-valued decisions, as well as overall ranking of heuristics. The greedy heuristics that perform well for one cost function do not necessarily perform well for another cost function. By looking at Table 13.2, for example, we can see that poly_C and poly_L perform very well for the coverage, but quite badly for the length. Similarly, M_C and M_L are the best heuristics for the length but they do not give good results for the coverage.

For bi-criteria optimization, we utilized Pareto optimal points. We measured the Euclidean distance between a point corresponding to a rule system constructed by heuristic (*heuristic point*) and the nearest Pareto optimal point to get a good idea of heuristic performance. The coordinates of heuristic point are the sum of lengths of rules and the sum of negative coverages of rules from the system. For the convenience, we divide each coordinate of Pareto optimal point or heuristic point by the number of rows in the considered decision table. Moreover, to remove any bias we normalize the distance along each axis before calculating the distances: we divide each coordinate of Pareto optimal and heuristic points by the maximum absolute value of this coordinate among all Pareto optimal points. The heuristic that performed best for bi-criteria optimization is log_C. This heuristic is not the best for either length or coverage, but it gives reasonable results for both at the same time.

Figures 13.1, 13.2, 13.3, 13.4, and 13.5 depict the Pareto optimal points and heuristic points for five decision tables "cars-1", "nursery-4", "mushroom-5", "breast-cancer-5", and "zoo-data-5", respectively, before normalization.

The *average relative difference* (ARD) shows how close on average an approximate solution to the optimal solution for a single-criterion optimization problem. ARD for cost function ψ, decision table T, and heuristic H is equal to

$$\frac{\psi(T)^H - \psi(T)^{opt}}{\psi(T)^{opt}} \cdot 100\%$$

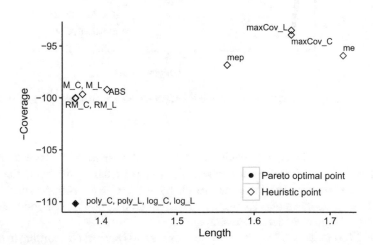

Fig. 13.1 Pareto optimal and heuristic points for the decision table "cars-1": case of decision rules

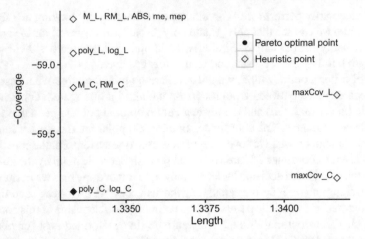

Fig. 13.2 Pareto optimal and heuristic points for the decision table "nursery-4": case of decision rules

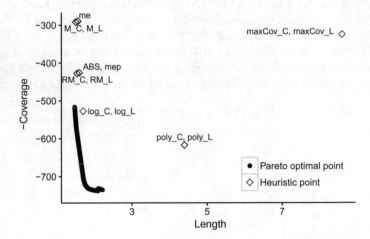

Fig. 13.3 Pareto optimal and heuristic points for the decision table "mushroom-5": case of decision rules

where $\psi(T)^{opt}$ is the cost of optimal relative to ψ system of decision rules and $\psi(T)^H$ is the cost of system of decision rules constructed by the heuristics H. If $\psi = -c$, we multiply the computed result by -1. ARD for cost function ψ and heuristic H is the mean of ARD for cost function ψ, decision table T, and heuristic H among all nine considered decision tables.

Table 13.3 contains ARD for cost functions $-c$ and l and best (according to overall ranking in Table 13.2) heuristics for the considered cost functions. The results seem to be promising: for the best heuristic poly_C for the coverage, ARD is at most 5% and, for the best heuristic M_C for the length, ARD is at most 2%.

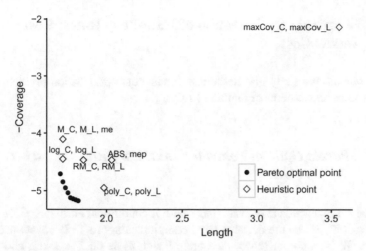

Fig. 13.4 Pareto optimal and heuristic points for the decision table "breast-cancer-5": case of decision rules

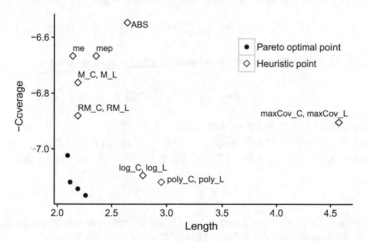

Fig. 13.5 Pareto optimal and heuristic points for the decision table "zoo-data-5": case of decision rules

Table 13.3 Average relative difference (ARD) for best heuristics for exact decision rule construction

Coverage		Length	
Heuristic	ARD (%)	Heuristic	ARD (%)
poly_C	5	M_C	2
poly_L	5	M_L	2
log_C	9	RM_L	4
		RM_C	4

13.3   Bi-criteria Optimization of Inhibitory Rules: Cost Versus Cost

In this section, we study bi-criteria cost versus cost optimization problem for inhibitory rules and systems of inhibitory rules.

13.3.1   Pareto Optimal Points for Inhibitory Rules: Cost Versus Cost

Let T be a nondegenerate decision table with n conditional attributes $f_1, \ldots, f_n, r$ be a row of T, T^C be the decision table complementary to T, U be an uncertainty measure, W be a completeness measure, U and W be dual, $\alpha \in \mathbb{R}_+$, and $\psi, \varphi \in \{l, -c, -rc, mc, rmc\}$.

Let G be a bundle-preserving subgraph of the graph $\Delta_{U,\alpha}(T^C)$ and

$$p_{\psi,\varphi}(G, T^C, r) = \{(\psi(T^C, \rho), \varphi(T^C, \rho)) : \rho \in Rule(G, T^C, r)\}.$$

In Sect. 13.1.1, the algorithm $\mathcal{A}_{11}$ is described which constructs the set

$$Par(p_{\psi,\varphi}(G, T^C, r))$$

of Pareto optimal points for the set of pairs $p_{\psi,\varphi}(G, T^C, r)$.

In Sect. 12.2, we show that $Rule(G, T^C, r)^- \subseteq IR_{W,\alpha}(T, r)$. Let $G = \Delta_{U,\alpha}(T^C)$. Then $Rule(G, T^C, r)^- = IR_{W,\alpha}(T, r)$. If $\eta \in \{l, -c, -rc, mc, rmc\}$ then

$$Rule(G^\eta(r), T^C, r)^-$$

is equal to the set of all rules from $IR_{W,\alpha}(T, r)$ which have minimum cost relative to η among all inhibitory rules from $IR_{W,\alpha}(T, r)$.

Denote $ip_{\psi,\varphi}(G, T, r) = \{(\psi(T, \rho^-), \varphi(T, \rho^-)) : \rho \in Rule(G, T^C, r)\}$. From Proposition 11.1 it follows that $(\psi(T, \rho^-), \varphi(T, \rho^-)) = (\psi(T^C, \rho), \varphi(T^C, \rho))$ for any $\rho \in Rule(G, T^C, r)$. Therefore $p_{\psi,\varphi}(G, T^C, r) = ip_{\psi,\varphi}(G, T, r)$ and

$$Par(p_{\psi,\varphi}(G, T^C, r)) = Par(ip_{\psi,\varphi}(G, T, r)).$$

To study relationships between cost functions ψ and φ on the set of inhibitory rules $Rule(G, T^C, r)^-$, we consider partial functions $\mathcal{IR}_{G,T,r}^{\psi,\varphi} : \mathbb{R} \to \mathbb{R}$ and

$$\mathcal{IR}_{G,T,r}^{\varphi,\psi} : \mathbb{R} \to \mathbb{R}$$

defined in the following way:

$$\mathscr{IR}^{\psi,\varphi}_{G,T,r}(x) = \min\{\varphi(T,\rho^-) : \rho^- \in Rule(G,T^C,r)^-, \psi(T,\rho^-) \le x\},$$

$$\mathscr{IR}^{\varphi,\psi}_{G,T,r}(x) = \min\{\psi(T,\rho)^- : \rho^- \in Rule(G,T^C,r)^-, \varphi(T,\rho^-) \le x\}.$$

Let $(a_1, b_1), \ldots, (a_k, b_k)$ be the normal representation of the set

$$Par(p_{\psi,\varphi}(G,T^C,r)) = Par(ip_{\psi,\varphi}(G,T,r))$$

where $a_1 < \cdots < a_k$ and $b_1 > \cdots > b_k$. By Lemma 5.10 and Remark 5.4, for any $x \in \mathbb{R}$,

$$\mathscr{IR}^{\psi,\varphi}_{G,T,r}(x) = \begin{cases} undefined, & x < a_1 \\ b_1, & a_1 \le x < a_2 \\ \cdots & \cdots \\ b_{k-1}, & a_{k-1} \le x < a_k \\ b_k, & a_k \le x \end{cases},$$

$$\mathscr{IR}^{\varphi,\psi}_{G,T,r}(x) = \begin{cases} undefined, & x < b_k \\ a_k, & b_k \le x < b_{k-1} \\ \cdots & \cdots \\ a_2, & b_2 \le x < b_1 \\ a_1, & b_1 \le x \end{cases}.$$

13.3.2 Pareto Optimal Points for Systems of Inhibitory Rules: Cost Versus Cost

Let T be a nondegenerate decision table with n conditional attributes $f_1, \ldots, f_n$ and $N(T)$ rows $r_1, \ldots, r_{N(T)}$, T^C be the decision table complementary to T, U be an uncertainty measure, W be a completeness measure, U and W be dual, $\alpha \in \mathbb{R}_+$, $\psi, \varphi \in \{l, -c, -rc, mc, rmc\}$, and $f, g \in \{sum(x,y), max(x,y)\}$.

Let $\mathbf{G} = (G_1, \ldots, G_{N(T)})$ be an $N(T)$-tuple of bundle-preserving subgraphs of the graph $\Delta_{U,\alpha}(T^C)$. Let $G = \Delta_{U,\alpha}(T^C)$ and $\xi \in \{l, -c, -rc, mc, rmc\} \setminus \{\psi, \varphi\}$. Then, as we mentioned, two interesting examples of such $N(T)$-tuples are $(G, \ldots, G)$ and $(G^\xi(r_1), \ldots, G^\xi(r_{N(T)}))$.

We denote by $\mathscr{S}(\mathbf{G}, T^C)$ the set

$$Rule(G_1, T^C, r_1) \times \cdots \times Rule(G_{N(T)}, T^C, r_{N(T)})$$

of (U,α)-systems of decision rules for T^C. It is clear that ψ_f and φ_g are strictly increasing cost functions for systems of decision rules. In the same time, ψ_f and φ_g are cost functions for systems of inhibitory rules.

In Sect. 13.1.3, we described the algorithm $\mathscr{A}_{12}$ which constructs the set of Pareto optimal points for the set of pairs

$$p_{\psi,\varphi}^{f,g}(\mathbf{G}, T^C) = \{(\psi_f(T^C, S), \varphi_g(T^C, S)) : S \in \mathscr{S}(\mathbf{G}, T^C)\} \ .$$

Let $S = (\rho_1, \dots, \rho_{N(T)}) \in \mathscr{S}(\mathbf{G}, T^C)$. We denote $S^- = (\rho_1^-, \dots, \rho_{N(T)}^-)$. From Proposition 11.1 it follows that S^- is a (W, α)-system of inhibitory rules for T, and $(\psi_f(T^C, S), \varphi_g(T^C, S)) = (\psi_f(T, S^-), \varphi_g(T, S^-))$.
We denote by $\mathscr{S}(\mathbf{G}, T^C)^-$ the set

$$Rule(G_1, T^C, r_1)^- \times \cdots \times Rule(G_{N(T)}, T^C, r_{N(T)})^- \ .$$

One can show that $\mathscr{S}(\mathbf{G}, T^C)^- = \{S^- : S \in \mathscr{S}(\mathbf{G}, T^C)\}$. Let

$$ip_{\psi,\varphi}^{f,g}(\mathbf{G}, T) = \{(\psi_f(T, S^-), \varphi_g(T, S^-)) : S^- \in \mathscr{S}(\mathbf{G}, T^C)^-\} \ .$$

It is clear that $p_{\psi,\varphi}^{f,g}(\mathbf{G}, T^C) = ip_{\psi,\varphi}^{f,g}(\mathbf{G}, T)$ and

$$Par(p_{\psi,\varphi}^{f,g}(\mathbf{G}, T^C)) = Par(ip_{\psi,\varphi}^{f,g}(\mathbf{G}, T)) \ .$$

From results obtained in Sect. 12.2 it follows that

$$Rule(G_i, T^C, r_i)^- \subseteq IR_{W,\alpha}(T, r_i) \ .$$

Let $G = \Delta_{U,\alpha}(T^C)$. If, for $i = 1, \dots, N(T)$, $G_i = G$ then

$$Rule(G_i, T^C, r_i)^- = IR_{W,\alpha}(T, r_i) \ .$$

If $\xi \in \{l, -c, -rc, mc, rmc\}$ and $G_i = G^\xi(r_i)$ for $i = 1, \dots, N(T)$, then

$$Rule(G_i, T^C, r_i)^-$$

is equal to the set of all rules from $IR_{W,\alpha}(T, r_i)$ which have minimum cost relative to ξ among all inhibitory rules from the set $IR_{W,\alpha}(T, r_i)$.

To study relationships between cost functions ψ_f and φ_g on the set of systems of inhibitory rules $\mathscr{S}(\mathbf{G}, T^C)^-$, we consider two partial functions $\mathscr{IR}_{\mathbf{G},T}^{\psi,f,\varphi,g} : \mathbb{R} \to \mathbb{R}$ and $\mathscr{IR}_{\mathbf{G},T}^{\varphi,g,\psi,f} : \mathbb{R} \to \mathbb{R}$ defined in the following way:

$$\mathscr{IR}_{\mathbf{G},T}^{\psi,f,\varphi,g}(x) = \min\{\varphi_g(T, S^-) : S^- \in \mathscr{S}(\mathbf{G}, T^C)^-, \psi_f(T, S^-) \le x\} \ ,$$
$$\mathscr{IR}_{\mathbf{G},T}^{\varphi,g,\psi,f}(x) = \min\{\psi_f(T, S^-) : S^- \in \mathscr{S}(\mathbf{G}, T^C)^-, \varphi_g(T, S^-) \le x\} \ .$$

Let $(a_1, b_1), \dots, (a_k, b_k)$ be the normal representation of the set

$$Par(p_{\psi,\varphi}^{f,g}(\mathbf{G}, T^C)) = Par(ip_{\psi,\varphi}^{f,g}(\mathbf{G}, T))$$

where $a_1 < \cdots < a_k$ and $b_1 > \cdots > b_k$. By Lemma 5.10 and Remark 5.4, for any $x \in \mathbb{R}$,

$$\mathscr{I}\mathscr{R}_{\mathbf{G},T}^{\psi,f,\varphi,g}(x) = \begin{cases} undefined, & x < a_1 \\ b_1, & a_1 \leq x < a_2 \\ \cdots & \cdots \\ b_{k-1}, & a_{k-1} \leq x < a_k \\ b_k, & a_k \leq x \end{cases},$$

$$\mathscr{I}\mathscr{R}_{\mathbf{G},T}^{\varphi,g,\psi,f}(x) = \begin{cases} undefined, & x < b_k \\ a_k, & b_k \leq x < b_{k-1} \\ \cdots & \cdots \\ a_2, & b_2 \leq x < b_1 \\ a_1, & b_1 \leq x \end{cases}.$$

13.3.3 Comparison of Greedy Heuristics for Inhibitory Rule Construction

In this section, we discuss possibilities to use greedy heuristics described in Sect. 13.2.1 for optimization of exact inhibitory rules. To work with inhibitory rules for a decision table T with many-valued decisions, we transform T to the corresponding complementary decision table T^C by changing, for each row $r \in Row(T)$, the set $D(r)$ with the set $D(T) \setminus D(r)$. Results obtained for decision rules for T^C are, in the same time, results for inhibitory rules for T. Let H be a greedy heuristic described in Sect. 13.2.1. We apply it to each row of T^C and, as a result, obtain a system $S = (\rho_1, \ldots, \rho_{N(T)})$ of exact decision rules for T^C. From Proposition 11.1 it follows that $S^- = (\rho_1^-, \ldots, \rho_{N(T)}^-)$ is a system of exact inhibitory rules for T, $-c_{\text{sum}}(T, S^-) = -c_{\text{sum}}(T^C, S)$, and $l_{\text{sum}}(T, S^-) = l_{\text{sum}}(T^C, S)$. We know also that the set of Pareto optimal points for bi-criteria optimization problem relative to $-c_{\text{sum}}$

Table 13.4 Number of Pareto optimal points for bi-criteria optimization problem relative to l_{sum} and $-c_{\text{sum}}$ for systems of exact inhibitory rules

Table name	# POPs
breast-cancer-1	136
breast-cancer-5	8
cars-1	1
mushroom-5	3154
nursery-1	1
nursery-4,	1
teeth-1	1
teeth-5	1
zoo-data-5	1

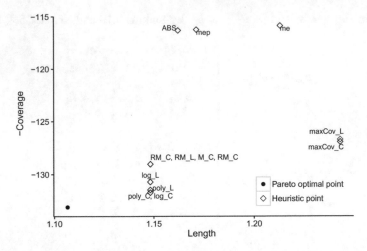

Fig. 13.6 Pareto optimal and heuristic points for the decision table "cars-1": case of inhibitory rules

Fig. 13.7 Pareto optimal and heuristic points for the decision table "zoo-data-5": case of inhibitory rules

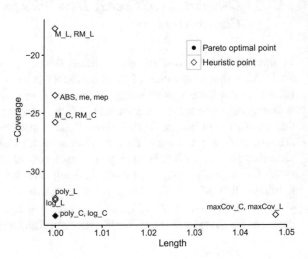

and l_{sum} for systems of exact decision rules for T^C is equal to the set of Pareto optimal points for bi-criteria optimization problem relative to $-c_{\text{sum}}$ and l_{sum} for systems of exact inhibitory rules for T.

Table 13.4 presents the number of Pareto optimal points for bi-criteria optimization problem relative to $-c_{\text{sum}}$ and l_{sum} for systems of exact inhibitory rules for each decision table described in Table 12.1. Figures 13.6 and 13.7 depict each the unique Pareto optimal point and points corresponding to systems of exact inhibitory rules constructed by 13 heuristics described in Sect. 13.2.1 (heuristic points) for decision tables "cars-1" and " zoo-data-5", respectively. Coordinates of each point are divided by the number of rows in decision table. We compared these heuristics as algorithms

Table 13.5 Comparing heuristics for construction of exact inhibitory rules

Heuristic name	Average rank			Overall rank		
	Cov	Len	Bi-crit	Cov	Len	Bi-crit
poly_C	1.94	7.44	4.17	**1**	8.5	**2**
poly_L	4.61	7.44	6.61	**3**	8.5	6
log_C	3.06	5.94	1.94	**2**	**3.5**	**1**
log_L	6.28	5.94	4.94	6	**3.5**	**3**
maxCov_C	5.72	10.67	7.06	4	12.5	7.5
maxCov_L	7.17	10.67	8.50	7	12.5	9
M_C	7.83	4.83	6.50	8.5	**1.5**	5
M_L	10.33	4.83	9.00	12	**1.5**	10
RM_C	5.78	6.00	5.00	5	5.5	4
RM_L	7.83	6.00	7.06	8.5	5.5	7.5
ABS	9.50	7.50	10.06	10	10	12
me	11.33	6.11	10.00	13	7	11
mep	9.61	7.61	10.17	11	11	13

Table 13.6 Average relative difference (ARD) for best heuristics for exact inhibitory rule construction

Coverage		Length	
Heuristic	ARD (%)	Heuristic	ARD (%)
poly_C	5	M_C	1
log_C	9	M_L	1
poly_L	16	log_C	3
		log_L	3

for optimization of exact inhibitory rules relative to $-c_{\mathrm{sum}}$, relative to l_{sum}, and relative to $-c_{\mathrm{sum}}$ and l_{sum} (see Table 13.5), and found ARD for the cost functions $-c$ and l, and best heuristics for these cost functions (see Table 13.6).

References

1. AbouEisha, H., Amin, T., Chikalov, I., Hussain, S., Moshkov, M.: Extensions of Dynamic Programming for Combinatorial Optimization and Data Mining. Intelligent Systems Reference Library, vol. 146. Springer, Berlin (2019)
2. Alsolami, F., Amin, T., Chikalov, I., Moshkov, M.: Bi-criteria optimization problems for decision rules. Ann. Oper. Res. **271**(2), 279–295 (2018)
3. Alsolami, F., Azad, M., Chikalov, I., Moshkov, M.: Decision rule classifiers for multi-label decision tables. In: Kryszkiewicz, M., Cornelis, C., Ciucci, D., Medina-Moreno, J., Motoda, H., Ras, Z.W. (eds.) Rough Sets and Intelligent Systems Paradigms – Second International

Conference, RSEISP 2014, held as part of JRS 2014, Granada and Madrid, Spain, 9–13 July 2014. Lecture Notes in Computer Science, vol. 8537, pp. 191–197. Springer, Berlin (2014)
4. Alsolami, F., Chikalov, I., Moshkov, M.: Comparison of heuristics for inhibitory rule optimization. In: Jedrzejowicz, P., Jain, L.C., Howlett, R.J., Czarnowski, I. (eds.) 18th International Conference in Knowledge Based and Intelligent Information and Engineering Systems, KES 2014, Gdynia, Poland, 15–17 Sept 2014. Procedia Computer Science, vol. 35, pp. 378–387. Elsevier, Amsterdam (2014)
5. Amin, T., Chikalov, I., Moshkov, M., Zielosko, B.: Relationships between length and coverage of exact decision rules. In: Popova-Zeugmann L. (ed.) 21th International Workshop on Concurrency, Specification and Programming, CS&P 2012, Berlin, Germany, 26–28 Sept 2012. CEUR Workshop Proceedings, vol. 928, pp. 1–12. CEUR-WS.org (2012)
6. Delimata, P., Moshkov, M., Skowron, A., Suraj, Z.: Inhibitory Rules in Data Analysis: A Rough Set Approach. Studies in Computational Intelligence, vol. 163. Springer, Berlin (2009)

Chapter 14
Bi-criteria Optimization Problem for Rules and Systems of Rules: Cost Versus Uncertainty (Completeness)

In this chapter, we consider algorithms which construct the sets of Pareto optimal points for bi-criteria optimization problems for decision rules and rule systems relative to a cost function and an uncertainty measure. These algorithms are extensions of tools created earlier for decision tables with single-valued decisions [1].

We generalize the obtained results to the case of bi-criteria optimization problems for inhibitory rules and rule systems relative to a cost function and a completeness measure. We also show how the constructed set of Pareto optimal points can be transformed into the graphs of functions which describe the relationships between the considered cost function and uncertainty (completeness) measure.

We made a number of computer experiments for decision tables with many-valued decisions derived from decision tables available in the UCI ML Repository [2]. These experiments provide us with examples of trade-off between complexity and accuracy for decision and inhibitory rule systems.

14.1 Bi-criteria Optimization Problem for Decision Rules and Systems of Rules: Cost Versus Uncertainty

In this section, we study bi-criteria cost versus uncertainty optimization problem for decision rules and systems of decision rules.

14.1.1 Pareto Optimal Points for Decision Rules: Cost Versus Uncertainty

Let ψ be a cost function for decision rules, U be an uncertainty measure, T be a nondegenerate decision table with n conditional attributes $f_1, \ldots, f_n$, and $r = (b_1, \ldots, b_n)$ be a row of T.

© Springer Nature Switzerland AG 2020
F. Alsolami et al., *Decision and Inhibitory Trees and Rules for Decision Tables with Many-valued Decisions*, Intelligent Systems Reference Library 156, https://doi.org/10.1007/978-3-030-12854-8_14

Let Θ be a node of the graph $\Delta(T) = \Delta_{U,0}(T)$ containing r. Let us recall that a rule over Θ is called a U-decision rule for Θ and r if there exists a nonnegative real number α such that the considered rule is a (U, α)-decision rule for Θ and r. Let $p_{\psi,U}(\Theta, r) = \{((\psi(\Theta, \rho), U(\Theta, \rho)) : \rho \in DR_U(\Theta, r)\}$ where $DR_U(\Theta, r)$ is the set of U-decision rules for Θ and r. Our aim is to find the set $Par(p_{\psi,U}(T, r))$. In fact, we will find the set $Par(p_{\psi,U}(\Theta, r))$ for each node Θ of the graph $\Delta(T)$ containing r.

To this end we consider an auxiliary set of rules $Path(\Theta, r)$ for each node Θ of $\Delta(T)$ containing r. Let τ be a directed path from the node Θ to a node Θ' of $\Delta(T)$ containing r in which edges are labeled with pairs $(f_{i_1}, b_{i_1}), \ldots, (f_{i_m}, b_{i_m})$. We denote by $rule(\tau)$ the decision rule over Θ

$$(f_{i_1} = b_{i_1}) \wedge \cdots \wedge (f_{i_m} = b_{i_m}) \to mcd(\Theta') .$$

If $m = 0$ (if $\Theta = \Theta'$) then the rule $rule(\tau)$ is equal to $\to mcd(\Theta)$. We denote by $Path(\Theta, r)$ the set of rules $rule(\tau)$ corresponding to all directed paths τ from Θ to a node Θ' of $\Delta(T)$ containing r (we consider also the case when $\Theta = \Theta'$). We correspond to the set of rules $Path(\Theta, r)$ the set of pairs $p_{\psi,U}^{path}(\Theta, r) = \{((\psi(\Theta, \rho), U(\Theta, \rho)) : \rho \in Path(\Theta, r)\}$.

Lemma 14.1 *Let U be an uncertainty measure, T be a nondegenerate decision table, r be a row of T, and Θ be a node of $\Delta(T)$ containing r. Then $DR_U(\Theta, r) \subseteq Path(\Theta, r)$.*

Proof Let $\alpha \in \mathbb{R}_+$. Then either $\Delta_{U,\alpha}(T) = \Delta_{U,0}(T) = \Delta(T)$ or the graph $\Delta_{U,\alpha}(T)$ is obtained from the graph $\Delta(T)$ by removal some nodes and edges. From here and from the definition of the set of rules $Rule(\Delta_{U,\alpha}(T), \Theta, r)$ it follows that this set is a subset of the set $Path(\Theta, r)$. From Proposition 12.1 it follows that the set $Rule(\Delta_{U,\alpha}(T), \Theta, r)$ coincides with the set of all (U, α)-decision rules for Θ and r. Since α is an arbitrary nonnegative real number, we have $DR_U(\Theta, r) \subseteq Path(\Theta, r)$. $\square$

Lemma 14.2 *Let $\psi \in \{l, -c, -c_M\}$, U be an uncertainty measure, T be a nondegenerate decision table with n conditional attributes $f_1, \ldots, f_n$, $r = (b_1, \ldots, b_n)$ be a row of T, and Θ be a node of $\Delta(T)$ containing r. Then $Par(p_{\psi,U}^{path}(\Theta, r)) = Par(p_{\psi,U}(\Theta, r))$.*

Proof From Lemma 14.1 it follows that $p_{\psi,U}(\Theta, r) \subseteq p_{\psi,U}^{path}(\Theta, r)$. Let us show that, for any pair $\alpha \in p_{\psi,U}^{path}(\Theta, r)$ there exists a pair $\beta \in p_{\psi,U}(\Theta, r)$ such that $\beta \leq \alpha$. Let $\alpha = (\psi(\Theta, \rho), U(\Theta, \rho))$ where $\rho \in Path(\Theta, r)$. If $\rho \in DR_U(\Theta, r)$ then $\alpha \in p_{\psi,U}(\Theta, r)$, and the considered statement holds.

Let $\rho \notin DR_U(\Theta, r)$ and $\rho = rule(\tau)$ where τ is a directed path from Θ to a node Θ' containing r which edges are labeled with pairs $(f_{i_1}, b_{i_1}), \ldots, (f_{i_m}, b_{i_m})$. Then $\rho = rule(\tau)$ is equal to

$$(f_{i_1} = b_{i_1}) \wedge \cdots \wedge (f_{i_m} = b_{i_m}) \to mcd(\Theta') .$$

Since $\rho \notin DR_U(\Theta, r)$, $m > 0$. Let

$$\Theta^0 = \Theta, \Theta^1 = \Theta(f_{i_1}, b_{i_1}), \ldots, \Theta^m = \Theta(f_{i_1}, b_{i_1}), \ldots, (f_{i_m}, b_{i_m}) = \Theta'.$$

By definition of the graph $\Delta(T)$, $f_{i_j} \in E(\Theta^{j-1})$ for $j = 1, \ldots, m$. Since

$$\rho \notin DR_U(\Theta, r),$$

there exists $j \in \{1, \ldots, m-1\}$ such that $U(\Theta^j) \leq U(\Theta^m)$. Let j_0 be the minimum number from $\{1, \ldots, m-1\}$ for which $U(\Theta^{j_0}) \leq U(\Theta^m)$. We denote by ρ' the rule

$$(f_{i_1} = b_{i_1}) \wedge \cdots \wedge (f_{i_{j_0}} = b_{i_{j_0}}) \rightarrow mcd(\Theta^{j_0}).$$

It is clear that ρ' is a $(U, U(\Theta^{j_0}))$-decision rule for Θ and r. Therefore $\rho' \in DR_U(\Theta, r)$ and $\beta = ((\psi(\Theta, \rho')), U(\Theta, \rho')) \in p_{\psi,U}(\Theta, r)$. We have $U(\Theta, \rho') = U(\Theta^{j_0}) \leq U(\Theta^m) = U(\Theta, \rho)$. If $\psi = l$ then $\psi(\Theta, \rho') = j_0 < m = \psi(\Theta, \rho)$. Let $\psi = -c$. Then

$$\psi(\Theta, \rho') = -N_{mcd(\Theta^{j_0})}(\Theta^{j_0}) \leq -N_{mcd(\Theta^m)}(\Theta^{j_0}) \leq -N_{mcd(\Theta^m)}(\Theta^m) = \psi(\Theta, \rho).$$

Let $\psi = -c_M$. Then $\psi(\Theta, \rho') = -N^M(\Theta^{j_0}) \leq -N^M(\Theta^m) = \psi(\Theta, \rho)$. Therefore $\beta \leq \alpha$. Using Lemma 5.4 we obtain $Par(p_{\psi,U}^{path}(\Theta, r)) = Par(p_{\psi,U}(\Theta, r))$. □

So in some cases, in particular, when $\psi \in \{l, -c, -c_M\}$, we have

$$Par(p_{\psi,U}(T, r)) = Par(p_{\psi,U}^{path}(T, r)).$$

In these cases we can concentrate on the construction of the set $Par(p_{\psi,U}^{path}(T, r))$.

Let ψ be a strictly increasing cost function for decision rules, U be an uncertainty measure, T be a decision table with n conditional attributes $f_1, \ldots, f_n$, and $r = (b_1, \ldots, b_n)$ be a row of T. We now describe an algorithm $\mathscr{A}_{13}$ which constructs the set $Par(p_{\psi,U}^{path}(T, r))$. In fact, this algorithm constructs, for each node Θ of the graph G, the set $B(\Theta, r) = Par(p_{\psi,U}^{path}(\Theta, r))$.

Algorithm $\mathscr{A}_{13}$ (construction of POPs for decision rules, cost versus uncertainty).
Input: A strictly increasing cost function ψ for decision rules given by pair of functions ψ^0, F, an uncertainty measure U, a nonempty decision table T with n conditional attributes $f_1, \ldots, f_n$, a row $r = (b_1, \ldots, b_n)$ of T, and the graph $\Delta(T)$.
Output: The set $Par(p_{\psi,U}^{path}(T, r))$ of Pareto optimal points for the set of pairs $p_{\psi,U}^{path}(T, r) = \{((\psi(T, \rho), U(T, \rho)) : \rho \in Path(T, r)\}$.

1. If all nodes in $\Delta(T)$ containing r are processed, then return the set $B(T, r)$. Otherwise, choose in the graph $\Delta(T)$ a node Θ containing r which is not processed yet and which is either a terminal node of $\Delta(T)$ or a nonterminal node of $\Delta(T)$

such that, for any $f_i \in E(\Theta)$, the node $\Theta(f_i, b_i)$ is already processed, i.e., the set $B(\Theta(f_i, b_i), r)$ is already constructed.

2. If Θ is a terminal node, then set $B(\Theta, r) = \{(\psi^0(\Theta), 0)\}$. Mark the node Θ as processed and proceed to step 1.

3. If Θ is a nonterminal node then construct $(\psi^0(\Theta), U(\Theta))$, for each $f_i \in E(\Theta)$, construct the set $B(\Theta(f_i, b_i), r)^{FH}$, where $H(x) = x$, and construct the multiset $A(\Theta, r) = \{(\psi^0(\Theta), U(\Theta))\} \cup \bigcup_{f_i \in E(\Theta)} B(\Theta(f_i, b_i), r)^{FH}$ by simple transcription of elements from the sets $B(\Theta(f_i, b_i), r)^{FH}$, $f_i \in E(\Theta)$, and $(\psi^0(\Theta), U(\Theta))$.

4. Apply to the multiset $A(\Theta, r)$ the algorithm $\mathscr{A}_2$ which constructs the set

$$Par(A(\Theta, r)) \,.$$

Set $B(\Theta, r) = Par(A(\Theta, r))$. Mark the node Θ as processed and proceed to step 1.

Proposition 14.1 *Let ψ be a strictly increasing cost function for decision rules given by pair of functions ψ^0, F, U be an uncertainty measure, T be a nonempty decision table with n conditional attributes $f_1, \ldots, f_n$, and $r = (b_1, \ldots, b_n)$ be a row of T. Then, for each node Θ of the graph $\Delta(T)$ containing r, the algorithm $\mathscr{A}_{13}$ constructs the set $B(\Theta, r) = Par(p_{\psi,U}^{path}(\Theta, r))$.*

Proof We prove the considered statement by induction on nodes of G. Let Θ be a terminal node of $\Delta(T)$ containing r. Then $U(\Theta) = 0$, $Path(\Theta, r) = \{\rightarrow mcd(\Theta)\}$, $p_{\psi,U}^{path}(\Theta, r) = Par(p_{\psi,U}^{path}(\Theta, r)) = \{(\psi^0(\Theta), 0)\}$, and $B(\Theta, r) = Par(p_{\psi,U}^{path}(\Theta, r))$.

Let Θ be a nonterminal node of $\Delta(T)$ containing r such that, for any $f_i \in E(\Theta)$, the considered statement holds for the node $\Theta(f_i, b_i)$, i.e.,

$$B(\Theta(f_i, b_i), r) = Par(p_{\psi,U}^{path}(\Theta(f_i, b_i), r)) \,.$$

One can show that

$$p_{\psi,\varphi}^{path}(\Theta, r) = \{(\psi^0(\Theta), U(\Theta))\} \cup \bigcup_{f_i \in E(\Theta)} p_{\psi,U}^{path}(\Theta(f_i, b_i), r)^{FH}$$

where H is a function from $\mathbb{R}$ to $\mathbb{R}$ such that $H(x) = x$. It is clear that

$$Par(\{(\psi^0(\Theta), U(\Theta))\}) = \{(\psi^0(\Theta), U(\Theta))\} \,.$$

From Lemma 5.7 it follows that

$$Par(p_{\psi,U}^{path}(\Theta, r)) \subseteq \{(\psi^0(\Theta), U(\Theta))\} \cup \bigcup_{f_i \in E(\Theta)} Par(p_{\psi,U}^{path}(\Theta(f_i, b_i), r)^{FH}) \,.$$

By Lemma 5.9, $Par(p_{\psi,U}^{path}(\Theta(f_i,b_i),r)^{FH}) = Par(p_{\psi,U}^{path}(\Theta(f_i,b_i),r))^{FH}$ for any $f_i \in E(\Theta)$. Therefore

$$Par(p_{\psi,U}^{path}(\Theta,r)) \subseteq \{(\psi^0(\Theta),U(\Theta))\}$$
$$\cup \bigcup_{f_i \in E(\Theta)} Par(p_{\psi,U}^{path}(\Theta(f_i,b_i),r))^{FH} \subseteq p_{\psi,U}^{path}(\Theta,r).$$

Using Lemma 5.6 we obtain

$$Par(p_{\psi,U}^{path}(\Theta,r))$$

$$= Par\left(\{(\psi^0(\Theta),U(\Theta))\} \cup \bigcup_{f_i \in E(\Theta)} Par(p_{\psi,U}^{path}(\Theta(f_i,b_i),r))^{FH}\right).$$

Since $B(\Theta,r) = Par\left(\{(\psi^0(\Theta),U(\Theta))\} \cup \bigcup_{f_i \in E(\Theta)} B(\Theta(f_i,b_i),r)^{FH}\right)$ and

$$B(\Theta(f_i,b_i),r) = Par(p_{\psi,U}^{path}(\Theta(f_i,b_i),r))$$

for any $f_i \in E(\Theta)$, we have $B(\Theta,r) = Par(p_{\psi,U}^{path}(\Theta,r))$.   □

We now evaluate the number of elementary operations (computations of F, H , ψ^0, U, and comparisons) made by the algorithm $\mathscr{A}_{13}$. Let us recall that, for a given cost function ψ for decision rules and decision table T,

$$q_\psi(T) = |\{\psi(\Theta,\rho) : \Theta \in SEP(T), \rho \in DR(\Theta)\}| .$$

In particular, by Lemma 11.1, $q_l(T) \le n+1$, $q_{-rc}(T) \le N(T)(N(T)+1)$, $q_{-c}(T) \le N(T)+1$, $q_{mc}(T) \le N(T)+1$, $q_{-c_M}(T) \le N(T)+1$, and $q_{rmc}(T) \le N(T)(N(T)+1)$.

Proposition 14.2 *Let ψ be a strictly increasing cost function for decision rules given by pair of functions ψ^0, F, H be a function from $\mathbb{R}$ to $\mathbb{R}$ such that $H(x) = x$, U be an uncertainty measure, T be a nonempty decision table with n conditional attributes $f_1, \ldots, f_n$, and $r = (b_1, \ldots, b_n)$ be a row of T. Then, to construct the set $Par(p_{\psi,U}^{path}(\Theta,r))$, the algorithm $\mathscr{A}_{13}$ makes*

$$O(L(\Delta(T))q_\psi(T)n \log(q_\psi(T)n))$$

elementary operations (computations of F, H, ψ^0, U, and comparisons).

Proof To process a terminal node, the algorithm $\mathscr{A}_{13}$ makes one elementary operation – computes ψ^0. We now evaluate the number of elementary operations under the processing of a nonterminal node Θ.

From Proposition 14.1 it follows that $B(\Theta, r) = Par(p_{\psi,U}^{path}(\Theta, r))$ and

$$B(\Theta(f_i, b_i), r) = Par(p_{\psi,U}^{path}(\Theta(f_i, b_i), r))$$

for any $f_i \in E(\Theta)$. From Lemma 5.5 it follows that $|B(\Theta(f_i, b_i), r)| \leq q_\psi(T)$ for any $f_i \in E(\Theta)$. It is clear that $|E(\Theta)| \leq n$,

$$\left| B(\Theta(f_i, b_i), r)^{FH} \right| = |B(\Theta(f_i, b_i), r)|$$

for any $f_i \in E(\Theta)$, and $|A(\Theta, r)| \leq q_\psi(T)n$. Therefore to construct the sets

$$B(\Theta(f_i, b_i), r)^{FH} \ ,$$

$f_i \in E(\Theta)$, from the sets $B(\Theta(f_i, b_i), r)$, $f_i \in E(\Theta)$, the algorithm $\mathscr{A}_{13}$ makes $O(q_\psi(T)n)$ computations of F and H. To construct the pair $(\psi^0(\Theta), U(\Theta))$, the algorithm $\mathscr{A}_{13}$ makes two operations – computes ψ^0 and U. To construct the set $Par(A(\Theta, r)) = B(\Theta, r)$ from the set $A(\Theta, r)$, the algorithm $\mathscr{A}_{13}$ makes

$$O(q_\psi(T)n \log(q_\psi(T)n))$$

comparisons (see Proposition 5.5). Hence, to process a nonterminal node Θ, the algorithm makes

$$O(q_\psi(T)n \log(q_\psi(T)n))$$

elementary operations.

To construct the set $Par(p_{\psi,U}^{path}(T, r))$, the algorithm $\mathscr{A}_{13}$ makes

$$O(L(\Delta(T))q_\psi(T)n \log(q_\psi(T)n))$$

elementary operations (computations of F, H, ψ^0, U, and comparisons). $\square$

Proposition 14.3 *Let ψ be a strictly increasing cost function for decision rules given by pair of functions ψ^0, F, H be a function from $\mathbb{R}$ to $\mathbb{R}$ such that $H(x) = x$, U be an uncertainty measure, $\psi \in \{l, -c, -rc, -c_M, mc, rmc\}$, $U \in \{me, rme, abs\}$, and $\mathscr{U}$ be a restricted information system. Then the algorithm $\mathscr{A}_{13}$ has polynomial time complexity for decision tables from $\mathscr{T}(\mathscr{U})$ depending on the number of conditional attributes in these tables.*

Proof Since $\psi \in \{l, -c, -rc, -c_M, mc, rmc\}$,

$$\psi^0 \in \{0, -N_{mcd(T)}(T), -N_{mcd(T)}(T)/N(T), -N^M(T), N(T) - N_{mcd(T)}(T),$$
$$(N(T) - N_{mcd(T)}(T))/N(T)\},$$

and $F, H \in \{x, x + 1\}$. From Lemma 11.1 and Proposition 14.2 it follows that, for the algorithm $\mathscr{A}_{13}$, the number of elementary operations (computations of F, H, ψ^0, U, and comparisons) is bounded from above by a polynomial depending on the size of input table T and on the number of separable subtables of T. All operations with numbers are basic ones. The computations of numerical parameters of decision tables used by the algorithm $\mathscr{A}_{13}$ $(0, -N_{mcd(T)}(T), -N_{mcd(T)}(T)/N(T), -N^M(T)$, $N(T) - N_{mcd(T)}(T), (N(T) - N_{mcd(T)}(T))/N(T)$ and $U \in \{me, rme, abs\})$ have polynomial time complexity depending on the size of decision tables.

According to Proposition 5.4, the algorithm $\mathscr{A}_{13}$ has polynomial time complexity for decision tables from $\mathscr{T}(\mathscr{U})$ depending on the number of conditional attributes in these tables. $\qquad\square$

14.1.2 Relationships for Decision Rules: Cost Versus Uncertainty

Let $\psi \in \{l, -c, -c_M\}$. Then the set $Par(p_{\psi,U}^{path}(T, r))$ constructed by the algorithm $\mathscr{A}_{13}$ is equal to the set $Par(p_{\psi,U}(T, r))$. Using this set we can construct two partial functions $\mathscr{R}_{T,r}^{\psi,U} : \mathbb{R} \to \mathbb{R}$ and $\mathscr{R}_{T,r}^{U,\psi} : \mathbb{R} \to \mathbb{R}$ which describe relationships between cost function ψ and uncertainty measure U on the set $DR_U(T, r)$ of U-decision rules for T and r, and are defined in the following way:

$$\mathscr{R}_{T,r}^{\psi,U}(x) = \min\{U(T, \rho) : \rho \in DR_U(T, r), \psi(T, \rho) \le x\},$$

$$\mathscr{R}_{T,r}^{U,\psi}(x) = \min\{\psi(T, \rho) : \rho \in DR_U(T, r), U(T, \rho) \le x\}.$$

Let $p_{\psi,U}(T, r) = \{(\psi(T, \rho), U(T, \rho)) : \rho \in DR_U(T, r)\}$ and $(a_1, b_1), \ldots,$ (a_k, b_k) be the normal representation of the set $Par(p_{\psi,U}(T, r))$ where $a_1 < \ldots < a_k$ and $b_1 > \cdots > b_k$. By Lemma 5.10 and Remark 5.4, for any $x \in \mathbb{R}$,

$$\mathscr{R}_{T,r}^{\psi,U}(x) = \begin{cases} undefined, & x < a_1 \\ b_1, & a_1 \le x < a_2 \\ \ldots & \ldots \\ b_{k-1}, & a_{k-1} \le x < a_k \\ b_k, & a_k \le x \end{cases},$$

$$\mathscr{R}_{T,r}^{U,\psi}(x) = \begin{cases} undefined, & x < b_k \\ a_k, & b_k \le x < b_{k-1} \\ \ldots & \ldots \\ a_2, & b_2 \le x < b_1 \\ a_1, & b_1 \le x \end{cases}.$$

14.1.3   Pareto Optimal Points for Systems of Decision Rules: Cost Versus Uncertainty

Let T be a nonempty decision table with n conditional attributes $f_1, \ldots, f_n$ and $N(T)$ rows $r_1, \ldots, r_{N(T)}$, and U be an uncertainty measure.

We denote by $\mathscr{I}_U(T)$ the set $DR_U(T, r_1) \times \cdots \times DR_U(T, r_{N(T)})$ of U-systems of decision rules for T. Let ψ be a strictly increasing cost function for decision rules, and f, g be increasing functions from $\mathbb{R}^2$ to $\mathbb{R}$. We consider two parameters of decision rule systems: cost $\psi_f(T, S)$ and uncertainty $U_g(T, S)$ which are defined on pairs T, S where $T \in \mathscr{T}^+$ and $S \in \mathscr{I}_U(T)$. Let $S = (\rho_1, \ldots, \rho_{N(T)})$. Then $\psi_f(T, S) = f(\psi(T, \rho_1), \ldots, \psi(T, \rho_{N(T)}))$ and $U_g(T, S) = g(U(T, \rho_1), \ldots, U(T, \rho_{N(T)}))$ where $f(x_1) = x_1$, $g(x_1) = x_1$, and, for $k > 2$, $f(x_1, \ldots, x_k) = f(f(x_1, \ldots, x_{k-1}), x_k)$ and $g(x_1, \ldots, x_k) = g(g(x_1, \ldots, x_{k-1}), x_k)$.

We assume that $\psi \in \{l, -c, -c_M\}$. According to Lemma 14.2, in this case

$$Par(p_{\psi,U}^{path}(T, r_i)) = Par(p_{\psi,U}(T, r_i))$$

for $i = 1, \ldots, N(T)$. We describe now an algorithm which constructs the set of Pareto optimal points for the set of pairs $p_{\psi,U}^{f,g}(T) = \{(\psi_f(T, S), U_g(T, S)) : S \in \mathscr{I}_U(T)\}$.

Algorithm $\mathscr{A}_{14}$ (construction of POPs for decision rule systems, cost versus uncertainty).

Input: A strictly increasing cost function for decision rules ψ given by pair of functions ψ^0, F, $\psi \in \{l, -c, -c_M\}$, an uncertainty measure U, increasing functions f, g from $\mathbb{R}^2$ to $\mathbb{R}$, a nonempty decision table T with n conditional attributes $f_1, \ldots, f_n$ and $N(T)$ rows $r_1, \ldots, r_{N(T)}$, and the graph $\Delta(T)$.

Output: The set $Par(p_{\psi,U}^{f,g}(T))$ of Pareto optimal points for the set of pairs $p_{\psi,U}^{f,g}(T) = \{(\psi_f(T, S), U_g(T, S)) : S \in \mathscr{I}_U(T)\}$.

1. Using the algorithm $\mathscr{A}_{13}$ construct, for $i = 1, \ldots, N(T)$, the set $Par(P_i)$ where

$$P_i = p_{\psi,U}^{path}(T, r) = \{((\psi(T, \rho), U(T, \rho)) : \rho \in Path(T, r)\} \, .$$

2. Apply the algorithm $\mathscr{A}_3$ to the functions f, g and the sets $Par(P_1), \ldots, Par(P_{N(T)})$. Set $C(T)$ the output of the algorithm $\mathscr{A}_3$ and return it.

Proposition 14.4 *Let ψ be a strictly increasing cost function for decision rules given by pair of functions ψ^0, F, $\psi \in \{l, -c, -c_M\}$, U be an uncertainty measure, f, g be increasing functions from $\mathbb{R}^2$ to $\mathbb{R}$, and T be a nonempty decision table with n conditional attributes $f_1, \ldots, f_n$ and $N(T)$ rows $r_1, \ldots, r_{N(T)}$. Then the algorithm $\mathscr{A}_{14}$ constructs the set $C(T) = Par(p_{\psi,U}^{f,g}(T))$.*

Proof For $i = 1, \ldots, N(T)$, denote $P_i = p^{path}_{\psi,U}(T, r_i)$ and $R_i = p_{\psi,U}(T, r_i)$. During the first step, the algorithm $\mathscr{A}_{14}$ constructs (using the algorithm $\mathscr{A}_{13}$) the sets $Par(P_1), \ldots, Par(P_{N(T)})$ (see Proposition 13.1). From Lemma 14.2 it follows that $Par(P_1) = Par(R_1), \ldots, Par(P_{N(T)}) = Par(R_{N(T)})$. During the second step of the algorithm $\mathscr{A}_{14}$ we apply the algorithm $\mathscr{A}_3$ to the functions f, g and the sets $Par(R_1), \ldots, Par(R_{N(T)})$. The algorithm $\mathscr{A}_3$ constructs the set $C(T) = Par(Q_{N(T)})$ where $Q_1 = R_1$, and, for $i = 2, \ldots, N(T)$, $Q_i = Q_{i-1} \langle fg \rangle R_i$ (see Proposition 5.6). One can show that $Q_{N(T)} = p^{f,g}_{\psi,U}(T)$). Therefore $C(T) = Par(p^{f,g}_{\psi,U}(T))$. $\qquad\square$

Let us recall that, for a given cost function ψ and a decision table T, $q_\psi(T) = |\{\psi(\Theta, \rho) : \Theta \in SEP(T), \rho \in DR(\Theta)\}|$. In particular, by Lemma 11.1, $q_l(T) \le n + 1$, $q_{-c}(T) \le N(T) + 1$, and $q_{-c_M}(T) \le N(T) + 1$.

Let us recall also that, for a given cost function ψ for decision rules and a decision table T, $Range_\psi(T) = \{\psi(\Theta, \rho) : \Theta \in SEP(T), \rho \in DR(\Theta)\}$. By Lemma 11.1, $Range_l(T) \subseteq \{0, 1, \ldots, n\}$, $Range_{-c}(T) \subseteq \{0, -1, \ldots, -N(T)\}$, and $Range_{-c_M}(T) \subseteq \{0, -1, \ldots, -N(T)\}$. Let $t_l(T) = n$, $t_{-c}(T) = N(T)$, and $t_{-c_M}(T) = N(T)$.

Proposition 14.5 *Let ψ be a strictly increasing cost function for decision rules given by pair of functions ψ^0, F, $\psi \in \{l, -c, -c_M\}$, U be an uncertainty measure, f, g be increasing functions from $\mathbb{R}^2$ to $\mathbb{R}$, $f \in \{x + y, \max(x, y)\}$, H be a function from $\mathbb{R}$ to $\mathbb{R}$ such that $H(x) = x$, and T be a nonempty decision table with n conditional attributes $f_1, \ldots, f_n$ and $N(T)$ rows $r_1, \ldots, r_{N(T)}$. Then, to construct the set $Par(p^{f,g}_{\psi,U}(T))$, the algorithm $\mathscr{A}_{14}$ makes*

$$O(N(T)L(\Delta(T))q_\psi(T)n \log(q_\psi(T)n)) + O(N(T)t_\psi(T)^2 \log(t_\psi(T)))$$

elementary operations (computations of F, H, ψ^0, U, f, g and comparisons) if $f = \max(x, y)$, and

$$O(N(T)L(\Delta(T))q_\psi(T)n \log(q_\psi(T)n)) + O(N(T)^2 t_\psi(T)^2 \log(N(T)t_\psi(T)))$$

elementary operations (computations of F, H, ψ^0, U, f, g and comparisons) if $f = x + y$.

Proof For $i = 1, \ldots, N(T)$, denote $P_i = p^{path}_{\psi,U}(T, r)$ and $R_i = p_{\psi,U}(T, r)$. To construct the sets $Par(P_i) = Par(R_i)$, $i = 1, \ldots, N(T)$, the algorithm $\mathscr{A}_{13}$ makes

$$O(N(T)L(\Delta(T))q_\psi(T)n \log(q_\psi(T)n))$$

elementary operations (computations of F, H, ψ^0, U, and comparisons) – see Proposition 14.2.

We now evaluate the number of elementary operations (computations of f, g, and comparisons) made by the algorithm $\mathscr{A}_3$ during the construction of the set

$C(T) = Par(p_{\psi,U}^{f,g}(T))$ from the sets $Par(R_i)$, $i = 1, \ldots, N(T)$. We know that $\psi \in \{l, -c, -c_M\}$ and $f \in \{x + y, \max(x, y)\}$.

For $i = 1, \ldots, N(T)$, let $R_i^1 = \{a : (a, b) \in R_i\}$. Since $\psi \in \{l, -c, -c_M\}$, we have $R_i^1 \subseteq \{0, 1, \ldots, t_\psi(T)\}$ for $i = 1, \ldots, N(T)$ or $R_i^1 \subseteq \{0, -1, \ldots, -t_\psi(T)\}$ for $i = 1, \ldots, N(T)$.

Using Proposition 5.7 we obtain the following.

If $f = x + y$, then to construct the set $C(T)$, the algorithm $\mathscr{A}_3$ makes

$$O(N(T)^2 t_\psi(T)^2 \log(N(T) t_\psi(T)))$$

elementary operations (computations of f, g, and comparisons).

If $f = \max(x, y)$, then to construct the set $C(T)$, the algorithm $\mathscr{A}_3$ makes

$$O(N(T) t_\psi(T)^2 \log(t_\psi(T)))$$

elementary operations (computations of f, g, and comparisons). $\square$

Proposition 14.6 *Let ψ be a strictly increasing cost function for decision rules given by pair of functions ψ^0, F, $\psi \in \{l, -c, -c_M\}$, U be an uncertainty measure, $U \in \{me, rme, abs\}$, $f, g \in \{\max(x, y), x + y\}$, H be a function from $\mathbb{R}$ to $\mathbb{R}$ such that $H(x) = x$, and $\mathscr{U}$ be a restricted information system. Then the algorithm $\mathscr{A}_{14}$ has polynomial time complexity for decision tables from $\mathscr{T}(\mathscr{U})$ depending on the number of conditional attributes in these tables.*

Proof We have $\psi^0 \in \{0, -N_{mcd(T)}(T), -N^M(T)\}$, and $F, H \in \{x, x + 1\}$. From Lemma 11.1 and Proposition 14.5 it follows that, for the algorithm $\mathscr{A}_{14}$, the number of elementary operations (computations of F, H, ψ^0, U, f, g, and comparisons) is bounded from above by a polynomial depending on the size of input table T and on the number of separable subtables of T. All operations with numbers are basic ones. The computations of numerical parameters of decision tables used by the algorithm $\mathscr{A}_{14}$ ($0, -N_{mcd(T)}(T), -N^M(T)$, and $U \in \{me, rme, abs\}$) have polynomial time complexity depending on the size of decision tables.

According to Proposition 5.4, the algorithm $\mathscr{A}_{14}$ has polynomial time complexity for decision tables from $\mathscr{T}(\mathscr{U})$ depending on the number of conditional attributes in these tables. $\square$

14.1.4 Relationships for Systems of Decision Rules: Cost Versus Uncertainty

Let ψ be a strictly increasing cost function for decision rules, U be an uncertainty measure, f, g be increasing functions from $\mathbb{R}^2$ to $\mathbb{R}$, and T be a nonempty decision table.

To study relationships between functions ψ_f and U_g which characterize cost and uncertainty of systems of rules from $\mathscr{I}_U(T)$ we can consider two partial functions from $\mathbb{R}$ to $\mathbb{R}$:

$$\mathscr{R}_T^{\psi,f,U,g}(x) = \min\{U_g(T, S) : S \in \mathscr{I}_U(T), \psi_f(T, S) \le x\},$$

$$\mathscr{R}_T^{U,g,\psi,f}(x) = \min\{\psi_f(T, S) : S \in \mathscr{I}_U(T), U_g(T, S) \le x\}.$$

Let $p_{\psi,U}^{f,g}(T) = \{(\psi_f(T, S), U_g(T, S)) : S \in \mathscr{I}_U(T)\}$, and $(a_1, b_1), \ldots, (a_k, b_k)$ be the normal representation of the set $Par(p_{\psi,U}^{f,g}(T))$ where $a_1 < \cdots < a_k$ and $b_1 > \cdots > b_k$. By Lemma 5.10 and Remark 5.4 , for any $x \in \mathbb{R}$,

$$\mathscr{R}_T^{\psi,f,U,g}(x) = \begin{cases} undefined, & x < a_1 \\ b_1, & a_1 \le x < a_2 \\ \ldots & \ldots \\ b_{k-1}, & a_{k-1} \le x < a_k \\ b_k, & a_k \le x \end{cases},$$

$$\mathscr{R}_T^{U,g,\psi,f}(x) = \begin{cases} undefined, & x < b_k \\ a_k, & b_k \le x < b_{k-1} \\ \ldots & \ldots \\ a_2, & b_2 \le x < b_1 \\ a_1, & b_1 \le x \end{cases}.$$

14.1.5 Experimental Results for Relationships Cost Versus Uncertainty

In this section, we consider some experimental results on relationships of cost and uncertainty for systems of decision rules for seven of nine decision tables described in Table 12.1. Using the algorithm $\mathscr{A}_{14}$ we constructed, for each of the considered decision tables, the set of Pareto optimal points for bi-criteria optimization problem for decision rule systems relative to $(l_{\text{sum}}, rme_{\text{max}})$, and the set of Pareto optimal points for bi-criteria optimization problem for decision rule systems relative to $(-c_{\text{sum}}, rme_{\text{max}})$. The number of Pareto optimal points in these sets can be found in the column "# POPs" of Table 14.1.

The constructed sets of Pareto optimal points can be transformed into the graphs of functions which describe relationships between cost and uncertainty of rule systems. For the convenience, we divide the coordinate of points corresponding to the cost of rule system by the number of rows in the decision table. As a result, we obtain relationships between average length or average negative coverage and maximum value

Table 14.1 Number of Pareto optimal points for bi-criteria optimization problems relative to (l_{sum}, rme_{max}) and relative to $(-c_{sum}, rme_{max})$ for systems of decision rules

Table name	Rows	Attributes	# POPs	
			(l_{sum}, rme_{max})	$(-c_{sum}, rme_{max})$
breast-cancer-1	169	8	475	7203
breast-cancer-5	58	4	99	266
cars-1	432	5	591	2533
nursery-4	240	4	321	401
teeth-1	22	7	50	1
teeth-5	14	3	28	1
zoo-data-5	42	11	89	35

Fig. 14.1 Relationship between average length (l) and maximum value of rme (rme) for systems of decision rules for decision table "cars-1"

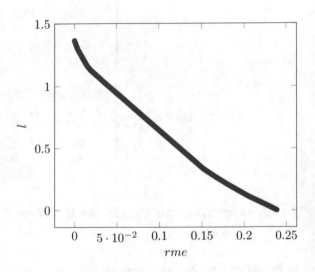

of rme for systems of decision rules. Figures 14.1 and 14.2 depict such relationships for the decision table "cars-1".

14.2 Bi-criteria Optimization of Inhibitory Rules: Cost Versus Completeness

In this section, we study bi-criteria cost versus completeness optimization problem for inhibitory rules and systems of inhibitory rules.

Fig. 14.2 Relationship between average negative coverage $(-c)$ and maximum value of rme (rme) for systems of decision rules for decision table "cars-1"

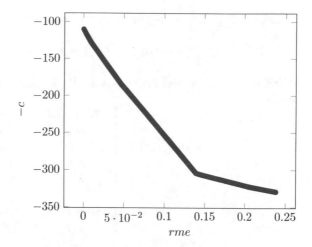

14.2.1 Pareto Optimal Points for Inhibitory Rules: Cost Versus Completeness

Let T be a nondegenerate decision table with n conditional attributes $f_1, \ldots, f_n$, r be a row of T, T^C be the decision table complementary to T, U be an uncertainty measure, W be a completeness measure, U and W be dual, and $\psi \in \{l, -c\}$. In Sect. 14.1.1, the algorithm $\mathscr{A}_{13}$ is described which constructs the set $Par(p_{\psi,U}(T^C, r))$ of Pareto optimal points for the set of points $p_{\psi,U}(T^C, r) = \{(\psi(T^C, \rho), U(T^C, \rho)) : \rho \in DR_U(T^C, r)\}$ (see Lemma 14.2).

We denote $ip_{\psi,W}(T, r) = \{(\psi(T, \rho), W(T, \rho)) : \rho \in IR_W(T, r)\}$. From Corollary 11.1 it follows that $DR_U(T^C, r)^- = IR_W(T, r)$. From Proposition 11.1 it follows that, for any $\rho \in DR_U(T^C, r)$,

$$(\psi(T^C, \rho), U(T^C, \rho)) = (\psi(T, \rho^-), W(T, \rho^-)) .$$

Therefore $p_{\psi,U}(T^C, r) = ip_{\psi,W}(T, r)$. Hence

$$Par(p_{\psi,U}(T^C, r)) = Par(ip_{\psi,W}(T, r)) .$$

To study relationships between ψ and W on the set $IR_W(T, r)$ of inhibitory rules for T and r we will consider two partial functions $\mathscr{IR}_{T,r}^{\psi,W} : \mathbb{R} \to \mathbb{R}$ and $\mathscr{IR}_{T,r}^{W,\psi} : \mathbb{R} \to \mathbb{R}$ which are defined in the following way:

$$\mathscr{IR}_{T,r}^{\psi,W}(x) = \min\{W(T, \rho) : \rho \in IR_W(T, r), \psi(T, \rho) \le x\} ,$$
$$\mathscr{IR}_{T,r}^{W,\psi}(x) = \min\{\psi(T, \rho) : \rho \in IR_W(T, r), W(T, \rho) \le x\} .$$

Let $(a_1, b_1), \ldots, (a_k, b_k)$ be the normal representation of the set

$$Par(p_{\psi, U}(T^C, r)) = Par(ip_{\psi, w}(T, r))$$

where $a_1 < \cdots < a_k$ and $b_1 > \cdots > b_k$. By Lemma 5.10 and Remark 5.4, for any $x \in \mathbb{R}$,

$$\mathscr{IR}_{T,r}^{\psi, W}(x) = \begin{cases} undefined, & x < a_1 \\ b_1, & a_1 \le x < a_2 \\ \ldots & \ldots \\ b_{k-1}, & a_{k-1} \le x < a_k \\ b_k, & a_k \le x \end{cases},$$

$$\mathscr{IR}_{T,r}^{W, \psi}(x) = \begin{cases} undefined, & x < b_k \\ a_k, & b_k \le x < b_{k-1} \\ \ldots & \ldots \\ a_2, & b_2 \le x < b_1 \\ a_1, & b_1 \le x \end{cases}.$$

14.2.2 Pareto Optimal Points for Systems of Inhibitory Rules: Cost Versus Completeness

Let T be a nondegenerate decision table with n conditional attributes $f_1, \ldots, f_n$ and $N(T)$ rows $r_1, \ldots, r_{N(T)}$, T^C be the decision table complementary to T, U be an uncertainty measure, W be a completeness measure, U and W be dual, $\psi \in \{l, -c\}$, and $f, g \in \{\text{sum}(x, y), \max(x, y)\}$.

We denote by $\mathscr{I}_U(T^C)$ the set $DR_U(T^C, r_1) \times \cdots \times DR_U(T^C, r_{N(T)})$ of U-systems of decision rules for T^C, and consider two parameters of decision rule systems: cost $\psi_f(T, S)$ and uncertainty $U_g(T, S)$.

In Sect. 14.1.3, the algorithm $\mathscr{A}_{14}$ is described which constructs the set of Pareto optimal points for the set of pairs $p_{\psi, U}^{f, g}(T^C) = \{(\psi_f(T^C, S), U_g(T^C, S)) : S \in \mathscr{I}_U(T^C)\}$.

Let $S = (\rho_1, \ldots, \rho_{N(T)}) \in \mathscr{I}_U(T^C)$. We denote $S^- = (\rho_1^-, \ldots, \rho_{N(T)}^-)$. From Proposition 11.1 it follows that S^- is a W-system of inhibitory rules for T, and

$$(\psi_f(T^C, S), U_g(T^C, S)) = (\psi_f(T, S^-), W_g(T, S^-)) .$$

We denote by $\mathscr{I}_U(T^C)^-$ the set $DR_U(T^C, r_1)^- \times \cdots \times DR_U(T^C, r_{N(T)})^-$. One can show that $\mathscr{I}_U(T^C)^- = \{S^- : S \in \mathscr{I}_U(T^C)\}$. Let

$$ip_{\psi, W}^{f, g}(T) = \{(\psi_f(T, S^-), W_g(T, S^-)) : S^- \in \mathscr{I}_U(T^C)^-\} .$$

It is clear that $p_{\psi,U}^{f,g}(T^C) = ip_{\psi,W}^{f,g}(T)$ and $Par(p_{\psi,U}^{f,g}(T^C)) = Par(ip_{\psi,W}^{f,g}(T))$.

From Corollary 11.1 it follows that $IR_W(T, r_i) = DR_U(T^C, r_i)^-$ for $i = 1, \ldots,$ $N(T)$. Therefore $\mathscr{S}_U(T^C)^- = IR_W(T, r_1) \times \cdots \times IR_W(T, r_{N(T)})$.

To study relationships between functions ψ_f and W_g on the set $\mathscr{S}_U(T^C)^-$ of W-systems of inhibitory rules for T we can consider two partial functions from $\mathbb{R}$ to $\mathbb{R}$:

$$\mathscr{IR}_T^{\psi,f,W,g}(x) = \min\{W_g(T, S^-) : S^- \in \mathscr{S}_U(T)^-, \psi_f(T, S^-) \le x\},$$
$$\mathscr{IR}_T^{W,g,\psi,f}(x) = \min\{\psi_f(T, S^-) : S^- \in \mathscr{S}_U(T)^-, W_g(T, S^-) \le x\}.$$

Let $(a_1, b_1), \ldots, (a_k, b_k)$ be the normal representation of the set

$$Par(p_{\psi,U}^{f,g}(T^C)) = Par(ip_{\psi,W}^{f,g}(T))$$

where $a_1 < \cdots < a_k$ and $b_1 > \cdots > b_k$. By Lemma 5.10 and Remark 5.4, for any $x \in \mathbb{R}$,

$$\mathscr{IR}_T^{\psi,f,W,g}(x) = \begin{cases} undefined, & x < a_1 \\ b_1, & a_1 \le x < a_2 \\ \ldots & \ldots \\ b_{k-1}, & a_{k-1} \le x < a_k \\ b_k, & a_k \le x \end{cases},$$

$$\mathscr{IR}_T^{W,g,\psi,f}(x) = \begin{cases} undefined, & x < b_k \\ a_k, & b_k \le x < b_{k-1} \\ \ldots & \ldots \\ a_2, & b_2 \le x < b_1 \\ a_1, & b_1 \le x \end{cases}.$$

14.2.3 Experimental Results for Relationships Cost Versus Completeness

To work with inhibitory rules for a decision table T with many-valued decisions, we transform T to the corresponding complementary decision table T^C. Let $\psi \in \{-c, l\}$. We know that the set of Pareto optimal points for bi-criteria optimization problem relative to $(\psi_{\text{sum}}, rme_{\text{max}})$ for systems of decision rules for T^C is equal to the set of Pareto optimal points for bi-criteria optimization problem relative to $(\psi_{\text{sum}}, irme_{\text{max}})$ for systems of inhibitory rules for T.

Table 14.2 presents the number of Pareto optimal points for bi-criteria optimization problems relative to $(l_{\text{sum}}, irme_{\text{max}})$ and relative to $(-c_{\text{sum}}, irme_{\text{max}})$ for systems of

Table 14.2 Number of Pareto optimal points for bi-criteria optimization problems relative to $(l_{sum}, irme_{max})$ and relative to $(-c_{sum}, irme_{max})$ for systems of inhibitory rules

Table name	Rows	Attributes	# POPs	
			$(l_{sum}, irme_{max})$	$(-c_{sum}, irme_{max})$
breast-cancer-1	169	8	475	7203
breast-cancer-5	58	4	99	266
cars-1	432	5	479	3236
nursery-4	240	4	241	241
teeth-1	22	7	23	23
teeth-5	14	3	15	15
zoo-data-5	42	11	43	53

Fig. 14.3 Relationship between average negative coverage ($-c$) and maximum value of $irme$ ($irme$) for systems of inhibitory rules for decision table "zoo-data-5"

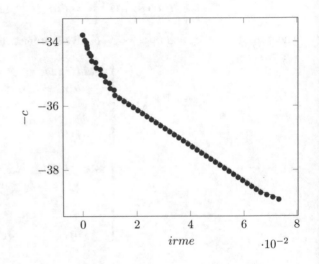

Fig. 14.4 Relationship between average length (l) and maximum value of $irme$ ($irme$) for systems of inhibitory rules for decision table "zoo-data-5"

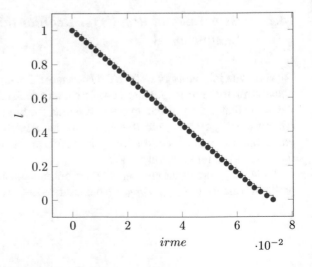

inhibitory rules for seven of nine decision tables described in Table 12.1. Figures 14.3 and 14.4 depict the relationships between average cost and maximum value of $irme$ for systems of inhibitory rules for the decision table "zoo-data-5".

References

1. AbouEisha, H., Amin, T., Chikalov, I., Hussain, S., Moshkov, M.: Extensions of Dynamic Programming for Combinatorial Optimization and Data Mining. Intelligent Systems Reference Library, vol. 146. Springer, Berlin (2019)
2. Lichman, M.: UCI Machine Learning Repository. University of California, Irvine, School of Information and Computer Sciences (2013). http://archive.ics.uci.edu/ml

Part IV
Study of Decision and Inhibitory Trees and Rule Systems Over Arbitrary Information Systems

This part is devoted to the study of time complexity of decision and inhibitory trees and rule systems over arbitrary sets of attributes represented by information systems. It consists of Chap. 15.

In Chap. 15, we consider the notion of an information system $U = (A, B, F)$ containing a set a A, a finite set B, and a set F of attributes each of which is a function from A to B. This information system is called finite if F is a finite set and infinite otherwise. We define the notion of a problem with many-valued decisions over U. Each such problem is described by a finite number of attributes from F which divide the set A into domains and a mapping that corresponds to each domain a finite set of decisions.

We consider two approaches to the study of decision and inhibitory trees and rule systems for problems with many-valued decisions over U: local and global. Local approach assumes that trees and rules can use only attributes from the problem description. Global approach allows us to use arbitrary attributes from the set F in trees and rules.

For each kind of information systems (finite and infinite) and for each approach (local and global), we study the behavior of four Shannon functions which characterize the growth in the worst case of (i) minimum depth of decision trees, (ii) minimum length of decision rule systems, (iii) minimum depth of inhibitory trees, and (iv) minimum length of inhibitory rule systems with the growth of the number of attributes in problem description.

Chapter 15
Local and Global Approaches to Study of Decision and Inhibitory Trees and Rule Systems

This chapter is devoted to the study of time complexity of decision and inhibitory trees and rule systems over arbitrary sets of attributes represented by information systems.

An information system $\mathscr{U} = (A, B, F)$ consists of a set A of elements and a set of attributes F which are functions that are defined on A and have values from a finite set B. We will assume that F does not contain constant attributes. This information system is called finite if F is a finite set, and infinite, otherwise. The notion of information system considered here has much in common with the notion of information system proposed by Zdzislaw Pawlak [8].

The notion of a problem with many-valued decisions over the information system $\mathscr{U}$ is defined as follows. Take finite number of attributes $f_1, \ldots, f_n$ from F. These attributes create a partition of the set A into domains. For each domain, values of the attributes $f_1, \ldots, f_n$ are constant on elements from the domain. A nonempty finite set of decisions is attached to each domain. The number n is called the dimension of the considered problem.

We study also problems with single-valued decisions for which a decision is attached to each domain. A problem with single-valued decisions can be considered as a special case of problems with many-valued decisions where each domain is labeled with a set of decisions containing only one decision.

There are two interpretations of each problem with many-valued decisions: decision and inhibitory. For the decision interpretation, for a given element a from A, it is required to recognize a decision from the set attached to the domain which contains a. For the inhibitory interpretation, for a given element a from A, it is required to recognize a decision for this problem which does not belong to the set attached to the domain which contains a. For problems with single-valued decisions, we consider only decision interpretation.

As algorithms for solving of problems with many-valued decisions, decision trees and decision rule systems are considered for decision interpretation, and inhibitory

© Springer Nature Switzerland AG 2020
F. Alsolami et al., *Decision and Inhibitory Trees and Rules for Decision Tables with Many-valued Decisions*, Intelligent Systems Reference Library 156,
https://doi.org/10.1007/978-3-030-12854-8_15

trees and inhibitory rule systems are considered for inhibitory interpretation. The depth of a tree (the maximum length of a path from the root to a terminal node) and the length of a rule system (the maximum length of a rule from the system) are considered as time complexity measures. For problems with single-valued decisions, we study only decision trees and decision rule systems.

We consider two approaches to the study of decision and inhibitory trees and rule systems: local and global. Local approach assumes that trees and rules can use only attributes $f_1, \ldots, f_n$ from the problem description. Global approach allows us to use arbitrary attributes from the set F in trees and rules. We use indexes l and g to distinguish local and global approaches, respectively.

For problems with single-valued decisions, for each information system and for each approach (local and global), we define two Shannon functions which characterize the growth in the worst case of (i) minimum depth of decision trees and (ii) minimum length of decision rule systems with the growth of problem dimension.

As an example, consider the notion of Shannon function $H_{\mathcal{U}}^g$ for problems with single-valued decisions over an information system $\mathcal{U} = (A, B, F)$, decision trees, and the global approach. Let z be a problem with single-valued decisions over the information system $\mathcal{U}$ and $h_{\mathcal{U}}^g (z)$ be the minimum depth of a decision tree over $\mathcal{U}$ which solves the problem z in the decision interpretation and uses arbitrary attributes from the set F. Then, for any natural n, Shannon function $H_{\mathcal{U}}^g (n)$ is equal to the maximum value of $h_{\mathcal{U}}^g (z)$ among all problems z with single-valued decisions over the information system $\mathcal{U}$ which dimension (the number of attributes in the description of z) is at most n.

For problems with many-valued decisions, for each information system and for each approach (local and global), we define four Shannon functions which characterize the growth in the worst case of (i) minimum depth of decision trees, (ii) minimum length of decision rule systems, (iii) minimum depth of inhibitory trees, and (iv) minimum length of inhibitory rule systems with the growth of problem dimension.

In Sect. 15.1, we define all these Shannon functions and study relationships between Shannon functions for problems with single-valued and many-valued decisions (they are equal), and relationships between Shannon functions for decision and inhibitory approaches (they are also equal).

Our aim is to describe the behavior of Shannon functions for problems with many-valued decisions. To this end, we should either study Shannon functions for problems with many-valued decisions directly or extend results for Shannon functions for problems with single-valued decisions to problems with many-valued decisions using statements from Sect. 15.1.

For the study of the most part of cases, it is enough to consider results published years ago in [2–7]. The only exception is the case of global approach and finite information systems. In Sect. 15.2, we study this case for problems with single-valued decisions.

In Sect. 15.3, we describe behavior of Shannon functions for local and global approaches, for decision and inhibitory trees and rule systems for problems with many-valued decisions over finite and infinite information systems.

15.1   Various Types of Shannon Functions

In this section, we define different kinds of Shannon functions and study relationships between Shannon functions for problems with single-valued and many-valued
decisions, and relationships between Shannon functions for decision and inhibitory
approaches.

Let A be a nonempty set of elements, B be a finite set with at least two elements,
and F be a nonempty set of functions from A to B. Functions from F are called
attributes and the triple $\mathcal{U} = (A, B, F)$ is called an *information system*. This information system is called *finite* if F is a finite set, and infinite otherwise. We assume
that $f \not\equiv \mathrm{const}$ for any $f \in F$.

An *equation system over* $\mathcal{U}$ is an arbitrary system of the kind

$$\{f_1(x) = \delta_1, \ldots, f_m(x) = \delta_m\}$$

where $f_1, \ldots, f_m \in F$ and $\delta_1, \ldots, \delta_m \in B$. It is possible that the considered system
does not have equations. Such system is called *empty*. The set of solutions of the
empty system coincides with the set A. There is one-to-one correspondence between
equation systems over $\mathcal{U}$ and words from the set $\Omega_{F,B} = \{(f, \delta) : f \in F, \delta \in B\}^*$:
the word $(f_1, \delta_1) \ldots (f_m, \delta_m)$ corresponds to the considered equation system, the
empty word λ corresponds to the empty equation system. For any $\alpha \in \Omega_{F,B}$, denote
by $\mathrm{Sol}_{\mathcal{U}}(\alpha)$ the set of solutions over A of the equation system corresponding to the
word α.

15.1.1   Shannon Functions for Decision Trees and Rule
Systems. Problems with Single-valued Decisions

Let $\omega = \{0, 1, 2, \ldots\}$ be the set of nonnegative integers including all natural numbers
and the number 0. A *problem over* $\mathcal{U}$ with single-valued decisions is an arbitrary
$(n + 1)$-tuple $z = (\nu, f_1, \ldots, f_n)$ where $\nu : B^n \to \omega$ and $f_1, \ldots, f_n \in F$. The number $\dim z = n$ is called the *dimension* of the problem z. Denote $\mathrm{At}(z) = \{f_1, \ldots, f_n\}$.
The problem z may be interpreted as a problem of searching for the value $z(a) =
\nu(f_1(a), \ldots, f_n(a))$ for an arbitrary $a \in A$. We say about this interpretation as about
decision one. Note that one can interpret $f_1, \ldots, f_n$ as conditional attributes and z as
a decision attribute. Denote $\mathrm{Probl}_{\mathcal{U}}$ the set of problems with single-valued decisions
over $\mathcal{U}$.

Decision trees and decision rule systems are considered as algorithms for problem
solving when we study its decision interpretation.

A *decision tree over* $\mathcal{U}$ is a labeled finite rooted directed tree in which each
terminal node is labeled with a number from ω; each node which is not terminal
(such nodes are called *working*) is labeled with an attribute from F; each edge is

labeled with an element from B, and edges starting in a working node are labeled with pairwise different elements. Denote $\text{Tree}_{\mathcal{U}}$ the set of decision trees over $\mathcal{U}$.

Let Γ be a decision tree over $\mathcal{U}$. Denote $\text{At}(\Gamma)$ the set of attributes assigned to working nodes of Γ. A *complete path* in Γ is an arbitrary path from the root to a terminal node. Denote $\text{Path}(\Gamma)$ the set of all complete paths in Γ. Let $\xi \in \text{Path}(\Gamma)$. Define a word $\pi(\xi)$ from the set $\Omega_{F,B}(\Gamma) = \{(f,\delta) : f \in \text{At}(\Gamma), \delta \in B\}^*$ associated with ξ. If there are no working nodes in ξ then $\pi(\xi) = \lambda$. Note that in this case the set $\text{Sol}_{\mathcal{U}}(\pi(\xi))$ coincides with the set A. Let $\xi = v_1, d_1, \ldots, v_m, d_m, v_{m+1}$ where $m > 0$, v_1 is the root, v_{m+1} is a terminal node, and v_i is the initial and v_{i+1} is the terminal node of the edge d_i for $i = 1, \ldots, m$. Let the node v_i be labeled with the attribute f_i, and the edge d_i be labeled with the element δ_i from B, $i = 1, \ldots, m$. Then $\pi(\xi) = (f_1, \delta_1) \ldots (f_m, \delta_m)$. Note that in this case the set $\text{Sol}_{\mathcal{U}}(\pi(\xi))$ coincides with the set of solutions over A of the equation system $\{f_1(x) = \delta_1, \ldots, f_m(x) = \delta_m\}$.

A decision tree Γ over $\mathcal{U}$ *solves* a problem z over $\mathcal{U}$ if, for each $a \in A$, there exists a complete path ξ in Γ such that $a \in \text{Sol}_{\mathcal{U}}(\pi(\xi))$, and the terminal node of the path ξ is labeled with the number $z(a)$.

For decision trees, as time complexity measure the *depth* of a decision tree is considered which is the maximum number of working nodes in a complete path in the tree. Denote $h(\Gamma)$ the depth of a decision tree Γ. For a problem z over $\mathcal{U}$, denote $h_{\mathcal{U}}^g(z)$ the minimum depth of a decision tree over $\mathcal{U}$ which solves z. For a problem z over $\mathcal{U}$, denote $h_{\mathcal{U}}^l(z)$ the minimum depth of a decision tree Γ over $\mathcal{U}$ which solves z and for which $\text{At}(\Gamma) \subseteq \text{At}(z)$. The considered parameters correspond to global and local approaches to the study of decision trees, respectively. One can show that $h_{\mathcal{U}}^g(z) \leq h_{\mathcal{U}}^l(z) \leq \dim z$ for each problem z over $\mathcal{U}$.

Define two Shannon functions $H_{\mathcal{U}}^g(n)$ and $H_{\mathcal{U}}^l(n)$. Let $n \in \omega \setminus \{0\}$. Then

$$H_{\mathcal{U}}^g(n) = \max\{h_{\mathcal{U}}^g(z) : z \in \text{Probl}_{\mathcal{U}}, \dim z \leq n\},$$
$$H_{\mathcal{U}}^l(n) = \max\{h_{\mathcal{U}}^l(z) : z \in \text{Probl}_{\mathcal{U}}, \dim z \leq n\}.$$

It is clear that $H_{\mathcal{U}}^g(n) \leq H_{\mathcal{U}}^l(n) \leq n$ for any $n \in \omega \setminus \{0\}$.

A *decision rule over* $\mathcal{U}$ is an arbitrary expression ρ of the kind

$$(f_1 = \delta_1) \wedge \ldots \wedge (f_m = \delta_m) \Rightarrow t$$

where $f_1, \ldots, f_m \in F$, $\delta_1, \ldots, \delta_m \in B$ and $t \in \omega$. The number m is called the *length* of the rule ρ. Denote $\text{At}(\rho) = \{f_1, \ldots, f_m\}$. Let us define a word $\pi(\rho)$ from the set $\Omega_{F,B}(\rho) = \{(f,\delta) : f \in \text{At}(\rho), \delta \in B\}^*$ associated with ρ. If $m = 0$ then $\pi(\xi) = \lambda$. Note that in this case the set $\text{Sol}_{\mathcal{U}}(\pi(\rho))$ coincides with the set A. Let $m > 0$. Then $\pi(\rho) = (f_1, \delta_1) \ldots (f_m, \delta_m)$. Note that in this case the set $\text{Sol}_{\mathcal{U}}(\pi(\rho))$ coincides with the set of solutions over A of the equation system $\{f_1(x) = \delta_1, \ldots, f_m(x) = \delta_m\}$. The number t is called the *right-hand side of the rule* ρ.

A *decision rule system over* $\mathcal{U}$ is a nonempty finite set of decision rules over $\mathcal{U}$. Let Δ be a decision rule system over $\mathcal{U}$ and z be a problem over $\mathcal{U}$. Denote $\text{At}(\Delta) = \bigcup_{\rho \in \Delta} \text{At}(\rho)$. The decision rule system Δ is *complete for the problem z if,* for any $a \in A$, there exists a rule $\rho \in \Delta$ such that $a \in \mathrm{Sol}_{\mathscr{U}}(\pi(\rho))$ and, for each rule $\rho \in \Delta$ such that $a \in \mathrm{Sol}_{\mathscr{U}}(\pi(\rho))$, the right-hand side of ρ is equal to $z(a)$.

For decision rule systems, as time complexity measure the *length* of a decision rule system is considered which is the maximum length of a rule from the system. Denote $l(\Delta)$ the length of a decision rule system Δ. For a problem z over $\mathscr{U}$, denote $l_{\mathscr{U}}^g(z)$ the minimum length of a decision rule system over $\mathscr{U}$ which is complete for the problem z. For a problem z over $\mathscr{U}$, denote $l_{\mathscr{U}}^l(z)$ the minimum length of a decision rule system Δ over $\mathscr{U}$ which is complete for the problem z and for which $\mathrm{At}(\Delta) \subseteq \mathrm{At}(z)$. The considered parameters correspond to global and local approaches to the study of decision rule systems, respectively. One can show that $l_{\mathscr{U}}^g(z) \le l_{\mathscr{U}}^l(z) \le \dim z$ for each problem z over $\mathscr{U}$.

Define two Shannon functions $L_{\mathscr{U}}^g(n)$ and $L_{\mathscr{U}}^l(n)$. Let $n \in \omega \setminus \{0\}$. Then

$$L_{\mathscr{U}}^g(n) = \max\{l_{\mathscr{U}}^g(z) : z \in \mathrm{Probl}_{\mathscr{U}}, \dim z \le n\},$$
$$L_{\mathscr{U}}^l(n) = \max\{l_{\mathscr{U}}^l(z) : z \in \mathrm{Probl}_{\mathscr{U}}, \dim z \le n\}.$$

It is clear that $L_{\mathscr{U}}^g(n) \le L_{\mathscr{U}}^l(n) \le n$ for any $n \in \omega \setminus \{0\}$.

Let $\mathscr{U}$ be a finite information system. A problem z over $\mathscr{U}$ is *tree-stable* if $h_{\mathscr{U}}^g(z) = \dim z$. Denote $\mathrm{ts}(\mathscr{U})$ the maximum dimension of a tree-stable problem from $\mathrm{Probl}_{\mathscr{U}}$. A problem z over $\mathscr{U}$ is *rule-stable* if $l_{\mathscr{U}}^g(z) = \dim z$. Denote $\mathrm{rs}(\mathscr{U})$ the maximum dimension of a rule-stable problem from $\mathrm{Probl}_{\mathscr{U}}$.

15.1.2 Shannon Functions for Decision Trees and Rule Systems. Problems with Many-valued Decisions

Let us consider an information system $\mathscr{U} = (A, B, F)$. A *problem with many-valued decisions over the information system* $\mathscr{U}$ is an arbitrary $(n + 1)$-tuple $z = (\nu, f_1, \ldots, f_n)$ where $\nu : B^n \to \mathrm{Fin}(\omega)$, $\mathrm{Fin}(\omega)$ is the set of all nonempty finite subsets of ω, and $f_1, \ldots, f_n \in F$. The number $\dim z = n$ is called the *dimension* of the problem z. Denote $\mathrm{At}(z) = \{f_1, \ldots, f_n\}$. The problem z may be interpreted as a problem of searching for a value from the set $z(a) = \nu(f_1(a), \ldots, f_n(a))$ for an arbitrary $a \in A$. We say about this interpretation as about decision one. Denote $\mathrm{Probl}_{\mathscr{U}}^{\infty}$ the set of problems with many-valued decisions over $\mathscr{U}$.

Decision trees and decision rule systems are considered as algorithms for problem solving when we study its decision interpretation.

A decision tree Γ over $\mathscr{U}$ *solves* a problem $z \in \mathrm{Probl}_{\mathscr{U}}^{\infty}$ if, for each $a \in A$, there exists a complete path ξ in Γ such that $a \in \mathrm{Sol}_{\mathscr{U}}(\pi(\xi))$, and the terminal node of the path ξ is labeled with a number belonging to $z(a)$. Denote $h_{\mathscr{U}}^g(z)$ the minimum depth of a decision tree over $\mathscr{U}$ which solves z. Denote $h_{\mathscr{U}}^l(z)$ the minimum depth of a decision tree Γ over $\mathscr{U}$ which solves z and for which $\mathrm{At}(\Gamma) \subseteq \mathrm{At}(z)$. The considered two parameters correspond to global and local approaches, respectively. One can show that $h_{\mathscr{U}}^g(z) \le h_{\mathscr{U}}^l(z) \le \dim z$.

Define two Shannon functions $H^g_{\mathcal{U},\infty}(n)$ and $H^l_{\mathcal{U},\infty}(n)$. Let $n \in \omega \setminus \{0\}$. Then

$$H^g_{\mathcal{U},\infty}(n) = \max\{h^g_{\mathcal{U}}(z) : z \in \mathrm{Probl}^\infty_{\mathcal{U}}, \dim z \le n\},$$
$$H^l_{\mathcal{U},\infty}(n) = \max\{h^l_{\mathcal{U}}(z) : z \in \mathrm{Probl}^\infty_{\mathcal{U}}, \dim z \le n\}.$$

It is clear that $H^g_{\mathcal{U},\infty}(n) \le H^l_{\mathcal{U},\infty}(n) \le n$ for any $n \in \omega \setminus \{0\}$.

A decision rule system Δ over $\mathcal{U}$ is *complete for the problem* $z \in \mathrm{Probl}^\infty_{\mathcal{U}}$ if, for any $a \in A$, there exists a rule $\rho \in \Delta$ such that $a \in \mathrm{Sol}_{\mathcal{U}}(\pi(\rho))$ and, for each rule $\rho \in \Delta$ such that $a \in \mathrm{Sol}_{\mathcal{U}}(\pi(\rho))$, the right-hand side of ρ belongs to $z(a)$. Denote $l^g_{\mathcal{U}}(z)$ the minimum length of a decision rule system Δ over $\mathcal{U}$ which is complete for the problem z. Denote $l^l_{\mathcal{U}}(z)$ the minimum length of a decision rule system Δ over $\mathcal{U}$ which is complete for the problem z and for which $\mathrm{At}(\Delta) \subseteq \mathrm{At}(z)$. The considered two parameters correspond to global and local approaches, respectively. One can show that $l^g_{\mathcal{U}}(z) \le l^l_{\mathcal{U}}(z) \le \dim z$.

Define two Shannon functions $L^g_{\mathcal{U},\infty}(n)$ and $L^l_{\mathcal{U},\infty}(n)$. Let $n \in \omega \setminus \{0\}$. Then

$$L^g_{\mathcal{U},\infty}(n) = \max\{l^g_{\mathcal{U}}(z) : z \in \mathrm{Probl}^\infty_{\mathcal{U}}, \dim z \le n\},$$
$$L^l_{\mathcal{U},\infty}(n) = \max\{l^l_{\mathcal{U}}(z) : z \in \mathrm{Probl}^\infty_{\mathcal{U}}, \dim z \le n\}.$$

It is clear that $L^g_{\mathcal{U},\infty}(n) \le L^l_{\mathcal{U},\infty}(n) \le n$ for any $n \in \omega \setminus \{0\}$.

Let $\mathcal{U}$ be a finite information system. A problem $z \in \mathrm{Probl}^\infty_{\mathcal{U}}$ is *tree-stable* if $h^g_{\mathcal{U}}(z) = \dim z$. Denote $\mathrm{ts}^\infty(\mathcal{U})$ the maximum dimension of a tree-stable problem from $\mathrm{Probl}^\infty_{\mathcal{U}}$. A problem $z \in \mathrm{Probl}^\infty_{\mathcal{U}}$ is *rule-stable* if $l^g_{\mathcal{U}}(z) = \dim z$. Denote $\mathrm{rs}^\infty(\mathcal{U})$ the maximum dimension of a rule-stable problem from $\mathrm{Probl}^\infty_{\mathcal{U}}$.

Proposition 15.1 *Let $\mathcal{U} = (A, B, F)$ be an information system. Then, for any $n \in \omega \setminus \{0\}$, $H^g_{\mathcal{U},\infty}(n) = H^g_{\mathcal{U}}(n)$, $H^l_{\mathcal{U},\infty}(n) = H^l_{\mathcal{U}}(n)$, $L^g_{\mathcal{U},\infty}(n) = L^g_{\mathcal{U}}(n)$, and $L^l_{\mathcal{U},\infty}(n) = L^l_{\mathcal{U}}(n)$. If $\mathcal{U}$ is a finite information system then $\mathrm{ts}^\infty(\mathcal{U}) = \mathrm{ts}(\mathcal{U})$ and $\mathrm{rs}^\infty(\mathcal{U}) = \mathrm{rs}(\mathcal{U})$.*

Proof Let $f_1, \ldots, f_n \in F$. Denote by v_n a mapping of the set B^n into the set ω such that $v_n(\bar\delta_1) \ne v_n(\bar\delta_2)$ for any $\bar\delta_1, \bar\delta_2 \in B^n, \bar\delta_1 \ne \bar\delta_2$. Let $z_n = (v_n, f_1, \ldots, f_n)$ and z be an arbitrary problem of the kind $(v, f_1, \ldots, f_n)$ where $v : B^n \to \omega$. One can show that $h^g_{\mathcal{U}}(z) \le h^g_{\mathcal{U}}(z_n)$, $h^l_{\mathcal{U}}(z) \le h^l_{\mathcal{U}}(z_n)$, $l^g_{\mathcal{U}}(z) \le l^g_{\mathcal{U}}(z_n)$, and $l^l_{\mathcal{U}}(z) \le l^l_{\mathcal{U}}(z_n)$.

Denote by v^∞_n a mapping of the set B^n into the set $\mathrm{Fin}(\omega)$ such that $v^\infty_n(\bar\delta_1) \cap v^\infty_n(\bar\delta_2) = \emptyset$ for any $\bar\delta_1, \bar\delta_2 \in B^n, \bar\delta_1 \ne \bar\delta_2$. Let $z^\infty_n = (v^\infty_n, f_1, \ldots, f_n)$ and z^∞ be an arbitrary problem of the kind $(v^\infty, f_1, \ldots, f_n)$ where $v^\infty : B^n \to \mathrm{Fin}(\omega)$. One can show that $h^g_{\mathcal{U}}(z^\infty) \le h^g_{\mathcal{U}}(z^\infty_n)$, $h^l_{\mathcal{U}}(z^\infty) \le h^l_{\mathcal{U}}(z^\infty_n)$, $l^g_{\mathcal{U}}(z^\infty) \le l^g_{\mathcal{U}}(z^\infty_n)$, and $l^l_{\mathcal{U}}(z^\infty) \le l^l_{\mathcal{U}}(z^\infty_n)$.

It is easy to see that $h^g_{\mathcal{U}}(z_n) = h^g_{\mathcal{U}}(z^\infty_n)$, $h^l_{\mathcal{U}}(z_n) = h^l_{\mathcal{U}}(z^\infty_n)$, $l^g_{\mathcal{U}}(z_n) = l^g_{\mathcal{U}}(z^\infty_n)$, and $l^l_{\mathcal{U}}(z_n) = l^l_{\mathcal{U}}(z^\infty_n)$. Using these facts it is not difficult to show that, for any $n \in \omega \setminus \{0\}$, $H^g_{\mathcal{U},\infty}(n) = H^g_{\mathcal{U}}(n)$, $H^l_{\mathcal{U},\infty}(n) = H^l_{\mathcal{U}}(n)$, $L^g_{\mathcal{U},\infty}(n) = L^g_{\mathcal{U}}(n)$, and $L^l_{\mathcal{U},\infty}(n) = L^l_{\mathcal{U}}(n)$, and if $\mathcal{U}$ is a finite information system, then $\mathrm{ts}^\infty(\mathcal{U}) = \mathrm{ts}(\mathcal{U})$ and $\mathrm{rs}^\infty(\mathcal{U}) = \mathrm{rs}(\mathcal{U})$. $\qquad\square$

15.1.3 Shannon Functions for Inhibitory Trees and Rule Systems. Problems with Many-valued Decisions

Let $\mathcal{U} = (A, B, F)$ be an information system and $z = (\nu, f_1, \ldots, f_n) \in \mathrm{Probl}_{\mathcal{U}}^{\infty}$. Denote $\mathrm{Row}_{\mathcal{U}}(z)$ the set of n-tuples $(\delta_1, \ldots, \delta_n) \in B^n$ such that the equation system $\{f_1(x) = \delta_1, \ldots, f_n(x) = \delta_n\}$ has a solution over A, and

$$D_{\mathcal{U}}(z) = \bigcup_{(\delta_1, \ldots, \delta_n) \in \mathrm{Row}_{\mathcal{U}}(z)} \nu(\delta_1, \ldots, \delta_n) .$$

We will say that the problem z is *inhibitory-correct* if $D_{\mathcal{U}}(z) \setminus \nu(\delta_1, \ldots, \delta_n) \neq \emptyset$ for any tuple $(\delta_1, \ldots, \delta_n) \in \mathrm{Row}_{\mathcal{U}}(z)$. Denote $\mathrm{IProbl}_{\mathcal{U}}^{\infty}$ the set of inhibitory-correct problems from $\mathrm{Probl}_{\mathcal{U}}^{\infty}$.

Let $z = (\nu, f_1, \ldots, f_n) \in \mathrm{IProbl}_{\mathcal{U}}^{\infty}$. The problem z may be interpreted as a problem of searching for a value from the set $D_{\mathcal{U}}(z) \setminus z(a)$ for an arbitrary $a \in A$ where $z(a) = \nu(f_1(a), \ldots, f_n(a))$. We say about this interpretation as about inhibitory one. Inhibitory trees and rule systems are considered as algorithms for problem solving when we study its inhibitory interpretation.

An inhibitory tree over $\mathcal{U}$ is a labeled finite rooted directed tree in which each terminal node is labeled with an expression $\neq t$ where $t \in \omega$; each node which is not terminal (such nodes are called *working*) is labeled with an attribute from F; each edge is labeled with an element from B, and edges starting in a working node are labeled with pairwise different elements.

Let Γ be an inhibitory tree over $\mathcal{U}$. Denote $\mathrm{At}(\Gamma)$ the set of attributes assigned to working nodes of Γ. A *complete path* in Γ is an arbitrary path from the root to a terminal node. Denote $\mathrm{Path}(\Gamma)$ the set of all complete paths in Γ. Let $\xi \in \mathrm{Path}(\Gamma)$. Define a word $\pi(\xi)$ from the set $\Omega_{F,B}(\Gamma) = \{(f, \delta) : f \in \mathrm{At}(\Gamma), \delta \in B\}^*$ associated with ξ. If there are no working nodes in ξ then $\pi(\xi) = \lambda$. Note that in this case the set $\mathrm{Sol}_{\mathcal{U}}(\pi(\xi))$ coincides with the set A. Let $\xi = v_1, d_1, \ldots, v_m, d_m, v_{m+1}$ where $m > 0$, v_1 is the root, v_{m+1} is a terminal node, and v_i is the initial and v_{i+1} is the terminal node of the edge d_i for $i = 1, \ldots, m$. Let the node v_i be labeled with the attribute f_i, and the edge d_i be labeled with the element δ_i from $B, i = 1, \ldots, m$. Then $\pi(\xi) = (f_1, \delta_1) \ldots (f_m, \delta_m)$. Note that in this case the set $\mathrm{Sol}_{\mathcal{U}}(\pi(\xi))$ coincides with the set of solutions over A of the equation system $\{f_1(x) = \delta_1, \ldots, f_m(x) = \delta_m\}$.

An inhibitory tree Γ over $\mathcal{U}$ *solves* a problem z from $\mathrm{IProbl}_{\mathcal{U}}^{\infty}$ if, for each $a \in A$, there exists a complete path ξ in Γ such that $a \in \mathrm{Sol}_{\mathcal{U}}(\pi(\xi))$, and the terminal node of the path ξ is labeled with an expression $\neq t$ where $t \in D_{\mathcal{U}}(z)$ and $t \notin z(a)$.

For inhibitory trees, as time complexity measure the *depth* of an inhibitory tree is considered which is the maximum number of working nodes in a complete path in the tree. Denote $h(\Gamma)$ the depth of an inhibitory tree Γ. For a problem z from $\mathrm{IProbl}_{\mathcal{U}}^{\infty}$, denote $ih_{\mathcal{U}}^g(z)$ the minimum depth of an inhibitory tree over $\mathcal{U}$ which solves z. For a problem z from $\mathrm{IProbl}_{\mathcal{U}}^{\infty}$, denote $ih_{\mathcal{U}}^l(z)$ the minimum depth of an inhibitory tree Γ over $\mathcal{U}$ which solves z and for which $\mathrm{At}(\Gamma) \subseteq \mathrm{At}(z)$. The considered two parameters correspond to global and local approaches, respectively. One can show that $ih_{\mathcal{U}}^g(z) \leq ih_{\mathcal{U}}^l(z) \leq \dim z$ for each problem z from $\mathrm{IProbl}_{\mathcal{U}}^\infty$.

Define two Shannon functions $IH_{\mathcal{U},\infty}^g(n)$ and $IH_{\mathcal{U},\infty}^l(n)$. Let $n \in \omega \setminus \{0\}$. Then

$$IH_{\mathcal{U},\infty}^g(n) = \max\{ih_{\mathcal{U}}^g(z) : z \in \mathrm{IProbl}_{\mathcal{U}}^\infty, \dim z \leq n\},$$
$$IH_{\mathcal{U},\infty}^l(n) = \max\{ih_{\mathcal{U}}^l(z) : z \in \mathrm{IProbl}_{\mathcal{U}}^\infty, \dim z \leq n\}.$$

It is clear that $IH_{\mathcal{U},\infty}^g(n) \leq IH_{\mathcal{U},\infty}^l(n) \leq n$ for any $n \in \omega \setminus \{0\}$.

An *inhibitory rule over* $\mathcal{U}$ is an arbitrary expression ρ of the kind

$$(f_1 = \delta_1) \wedge \ldots \wedge (f_m = \delta_m) \Rightarrow \neq t$$

where $f_1, \ldots, f_m \in F$, $\delta_1, \ldots, \delta_m \in B$ and $t \in \omega$. The number m is called the *length* of the rule ρ. Denote $\mathrm{At}(\rho) = \{f_1, \ldots, f_m\}$. Let us define a word $\pi(\rho)$ from the set $\Omega_{F,B}(\rho) = \{(f,\delta) : f \in \mathrm{At}(\rho), \delta \in B\}^*$ associated with ρ. If $m = 0$ then $\pi(\xi) = \lambda$. Note that in this case the set $\mathrm{Sol}_{\mathcal{U}}(\pi(\rho))$ coincides with the set A. Let $m > 0$. Then $\pi(\rho) = (f_1, \delta_1) \ldots (f_m, \delta_m)$. Note that in this case the set $\mathrm{Sol}_{\mathcal{U}}(\pi(\rho))$ coincides with the set of solutions over A of the equation system $\{f_1(x) = \delta_1, \ldots, f_m(x) = \delta_m\}$. The expression $\neq t$ is called the *right-hand side of the rule* ρ.

An *inhibitory rule system over* $\mathcal{U}$ is a nonempty finite set of inhibitory rules over $\mathcal{U}$. Let Δ be an inhibitory rule system over $\mathcal{U}$ and z be a problem from $\mathrm{IProbl}_{\mathcal{U}}^\infty$. Denote $\mathrm{At}(\Delta) = \bigcup_{\rho \in \Delta} \mathrm{At}(\rho)$. The inhibitory rule system Δ is *complete for the problem z* if, for any $a \in A$, there exists a rule $\rho \in \Delta$ such that $a \in \mathrm{Sol}_{\mathcal{U}}(\pi(\rho))$ and, for each rule $\rho \in \Delta$ such that $a \in \mathrm{Sol}_{\mathcal{U}}(\pi(\rho))$, the right-hand side $\neq t$ of ρ satisfies the conditions $t \in D_{\mathcal{U}}(z)$ and $t \notin z(a)$.

For inhibitory rule systems, as time complexity measure the *length* of an inhibitory rule system is considered which is the maximum length of a rule from the system. Denote $l(\Delta)$ the length of an inhibitory rule system Δ. For a problem z from $\mathrm{IProbl}_{\mathcal{U}}^\infty$, denote $il_{\mathcal{U}}^g(z)$ the minimum length of an inhibitory rule system over $\mathcal{U}$ which is complete for the problem z. For a problem z from $\mathrm{IProbl}_{\mathcal{U}}^\infty$, denote $il_{\mathcal{U}}^l(z)$ the minimum length of an inhibitory rule system Δ over $\mathcal{U}$ which is complete for the problem z and for which $\mathrm{At}(\Delta) \subseteq \mathrm{At}(z)$. The considered two parameters correspond to global and local approaches, respectively. One can show that $il_{\mathcal{U}}^g(z) \leq il_{\mathcal{U}}^l(z) \leq \dim z$ for each problem z from $\mathrm{IProbl}_{\mathcal{U}}^\infty$.

Define two Shannon functions $IL_{\mathcal{U},\infty}^g(n)$ and $IL_{\mathcal{U},\infty}^l(n)$. Let $n \in \omega \setminus \{0\}$. Then

$$IL_{\mathcal{U},\infty}^g(n) = \max\{il_{\mathcal{U}}^g(z) : z \in \mathrm{IProbl}_{\mathcal{U}}^\infty, \dim z \leq n\},$$
$$IL_{\mathcal{U},\infty}^l(n) = \max\{il_{\mathcal{U}}^l(z) : z \in \mathrm{IProbl}_{\mathcal{U}}^\infty, \dim z \leq n\}.$$

It is clear that $IL_{\mathcal{U},\infty}^g(n) \leq IL_{\mathcal{U},\infty}^l(n) \leq n$ for any $n \in \omega \setminus \{0\}$.

Proposition 15.2 *Let* $\mathcal{U} = (A, B, F)$ *be an information system. Then, for any* $n \in \omega \setminus \{0\}$, $IH_{\mathcal{U},\infty}^g(n) = H_{\mathcal{U},\infty}^g(n)$, $IH_{\mathcal{U},\infty}^l(n) = H_{\mathcal{U},\infty}^l(n)$, $IL_{\mathcal{U},\infty}^g(n) = L_{\mathcal{U},\infty}^g(n)$, *and* $IL_{\mathcal{U},\infty}^l(n) = L_{\mathcal{U},\infty}^l(n)$.

Proof Let $z = (v, f_1, \ldots, f_n) \in \mathrm{IProbl}_{\mathscr{U}}^{\infty}$. The problem $z^C = (v^C, f_1, \ldots, f_n)$ where, for any $(\delta_1, \ldots, \delta_n) \in B^n$, $v^C(\delta_1, \ldots, \delta_n) = D_{\mathscr{U}}(z) \setminus v(\delta_1, \ldots, \delta_n)$ is called the *complementary problem for z*.

Let Γ be a decision tree over $\mathscr{U}$ and Δ be a decision rule system over $\mathscr{U}$. We denote by Γ^- an inhibitory tree over $\mathscr{U}$ obtained from Γ by changing labels of terminal nodes: if a terminal node of Γ is labeled with t then the corresponding terminal node of Γ^- is labeled with $\neq t$. We denote by Δ^- an inhibitory rule system over $\mathscr{U}$ obtained from Δ by changing right-hand sides of rules: if a right-hand side of a rule from Δ is equal to t then the right-hand side of the corresponding rule in Δ^- is equal to $\neq t$.

One can show that the decision tree Γ solves the problem z^C if and only if the inhibitory tree Γ^- solves the problem z. One can show also that the decision rule system Δ is complete for the problem z^C if and only if the inhibitory rule system Δ^- is complete for the problem z. From here it follows that $ih_{\mathscr{U}}^g(z) = h_{\mathscr{U}}^g(z^C)$, $ih_{\mathscr{U}}^l(z) = h_{\mathscr{U}}^l(z^C)$, $il_{\mathscr{U}}^g(z) = l_{\mathscr{U}}^g(z^C)$, and $il_{\mathscr{U}}^l(z) = l_{\mathscr{U}}^l(z^C)$.

Let $n \in \omega \setminus \{0\}$. We now show that $IH_{\mathscr{U},\infty}^g(n) = H_{\mathscr{U},\infty}^g(n)$. Let $z \in \mathrm{IProbl}_{\mathscr{U}}^{\infty}$ and $\dim z \leq n$. We know that $\dim z = \dim z^C$ and $ih_{\mathscr{U}}^g(z) = h_{\mathscr{U}}^g(z^C)$. Therefore $IH_{\mathscr{U},\infty}^g(n) \leq H_{\mathscr{U},\infty}^g(n)$.

Let $z \in \mathrm{Probl}_{\mathscr{U}}^{\infty}$, $\dim z \leq n$, and $h_{\mathscr{U}}^g(z) = H_{\mathscr{U},\infty}^g(n)$. From the proof of Proposition 15.1 it follows that the problem z can be chosen such that $z = (v, f_1, \ldots, f_m)$ where $m \leq n$ and $v(\bar{\delta}_1) \cap v(\bar{\delta}_2) = \emptyset$ for any $\bar{\delta}_1, \bar{\delta}_2 \in \mathrm{Row}_{\mathscr{U}}(z)$, $\bar{\delta}_1 \neq \bar{\delta}_2$. Since F does not contain constant attributes, $|\mathrm{Row}_{\mathscr{U}}(z)| \geq 2$. From here it follows that $z^C \in \mathrm{IProbl}_{\mathscr{U}}^{\infty}$ and $h_{\mathscr{U}}^g(z) = h_{\mathscr{U}}^g(z^{CC})$. Denote $q = z^C$. We have $ih_{\mathscr{U}}^g(q) = h_{\mathscr{U}}^g(q^C) = h_{\mathscr{U}}^g(z) = H_{\mathscr{U},\infty}^g(n)$. Therefore $IH_{\mathscr{U},\infty}^g(n) \geq H_{\mathscr{U},\infty}^g(n)$ and $IH_{\mathscr{U},\infty}^g(n) = H_{\mathscr{U},\infty}^g(n)$.

We can prove in similar way that, for any $n \in \omega \setminus \{0\}$, $IH_{\mathscr{U},\infty}^l(n) = H_{\mathscr{U},\infty}^l(n)$, $IL_{\mathscr{U},\infty}^g(n) = L_{\mathscr{U},\infty}^g(n)$, and $IL_{\mathscr{U},\infty}^l(n) = L_{\mathscr{U},\infty}^l(n)$. $\qquad \square$

15.2 Problems with Single-valued Decisions Over Finite Information Systems. Global Approach

In this section, we study the behavior of Shannon functions for decision trees and rule systems for problems with single-valued decisions over finite information systems in the frameworks of global approach.

Let $\mathscr{U} = (A, B, F)$ be a finite information system. Define the notion of *independence dimension* (or, in short, *I-dimension*) of information system $\mathscr{U}$. A subset $\{f_1, \ldots, f_p\}$ of the set F is called an *independent set* if there exist two-element subsets $B_1, \ldots, B_p$ of the set B such that, for any $\delta_1 \in B_1, \ldots, \delta_p \in B_p$, the set $\mathrm{Sol}_{\mathscr{U}}((f_1, \delta_1) \ldots (f_p, \delta_p))$ is nonempty, i.e., the system of equations

$$\{f_1(x) = \delta_1, \ldots, f_p(x) = \delta_p\}$$

has a solution over A. If the subset $\{f_1, \ldots, f_p\}$ is not independent then it is called *dependent*. The I-dimension of $\mathscr{U}$ is the maximum cardinality of a subset of F which is an independent set. Denote it by $I(\mathscr{U})$.

A subset $\{f_1, \ldots, f_p\}$ of the set F is called *redundant* if $p \geq 2$ and there exist $i \in \{1, \ldots, p\}$ and mapping $\mu : B^{p-1} \to B$ such that

$$f_i(a) = \mu(f_1(a), \ldots, f_{i-1}(a), f_{i+1}(a), \ldots, f_p(a))$$

for any $a \in A$. If $\{f_1, \ldots, f_p\}$ is not redundant it is called *irredundant*. Denote $\mathrm{ir}(\mathscr{U})$ the maximum cardinality of an irredundant subset of F. It is clear that each redundant subset of F is a dependent set. Therefore

$$I(\mathscr{U}) \leq \mathrm{ir}(\mathscr{U}) . \tag{15.1}$$

Let us remind the notions of tree-stable and rule-stable problems. A problem z over $\mathscr{U}$ is *tree-stable* if $h^g_{\mathscr{U}}(z) = \dim z$. We denoted by $\mathrm{ts}(\mathscr{U})$ the maximum dimension of a tree-stable problem over $\mathscr{U}$. A problem z over $\mathscr{U}$ is *rule-stable* if $l^g_{\mathscr{U}}(z) = \dim z$. We denoted by $\mathrm{rs}(\mathscr{U})$ the maximum dimension of a rule-stable problem over $\mathscr{U}$. Let z be a rule-stable problem, i.e., $l^g_{\mathscr{U}}(z) = \dim z$. Then this problem is tree-stable. Let us assume the contrary: $h^g_{\mathscr{U}}(z) < \dim z$. Then there exists a decision tree Γ over $\mathscr{U}$ which solves z and for which $h(\Gamma) < \dim z$. It is easy to derive from Γ a decision rule system Δ which is complete for the problem z and for which $l(\Delta) < \dim z$ (rules in this system correspond to complete paths in Γ) but this is impossible. Thus

$$\mathrm{rs}(\mathscr{U}) \leq \mathrm{ts}(\mathscr{U}) .$$

It is clear also that, if a problem $z = (\nu, f_1, \ldots, f_m)$ is a rule-stable problem or a tree-stable problem, then the set $\{f_1, \ldots, f_m\}$ is irredundant. Therefore

$$\mathrm{rs}(\mathscr{U}) \leq \mathrm{ts}(\mathscr{U}) \leq \mathrm{ir}(\mathscr{U}) . \tag{15.2}$$

Consider now three examples which show that $I(\mathscr{U})$ can be less than $\mathrm{rs}(\mathscr{U})$, can lie between $\mathrm{rs}(\mathscr{U})$ and $\mathrm{ts}(T)$, and can be greater than $\mathrm{ts}(T)$. Note that, for the case when B contains exactly two elements, $I(T)$ is at most $\mathrm{ts}(T)$ (see proof of Theorem 5.3 in [6]).

Example 15.1 Let n be a natural number and $n \geq 3$. Consider a finite information system $\mathscr{U}_1 = (A_1, \{0, 1\}, F_1)$ such that A_1 is the two-dimensional Euclidean plane and F_1 is the set of attributes corresponding to n straight lines in the plane that form a convex polygon P with n sides. Each straight line divides the plane into open and closed half-planes on which the corresponding attribute has values 0 and 1, respectively. Two nonparallel straight lines divide the plane into four parts. There are no three straight lines that divide the plane into eight parts. Therefore $I(\mathscr{U}_1) = 2$. The membership problem for the polygon P can be stated as a problem over $\mathscr{U}_1$ with dimension n. One can show that this problem is rule-stable and tree-stable. Therefore $rs(\mathcal{U}_1) \geq n$ and $ts(\mathcal{U}_1) \geq n$. It is clear that $ir(\mathcal{U}_1) \leq |F_1| = n$. Thus $rs(\mathcal{U}_1) = ts(\mathcal{U}_1) = ir(\mathcal{U}_1) = n$. As a result, $I(\mathcal{U}_1) = 2$ and $rs(\mathcal{U}_1) = ts(\mathcal{U}_1) = ir(\mathcal{U}_1) = n$.

Example 15.2 Let n be a natural number. Consider a finite information system $\mathcal{U}_2 = (A_2, \{0, 1\}, F_2)$ where the set A_2 contains 2^n elements and F_2 contains all possible nonconstant attributes defined on A_2 and with values from $\{0, 1\}$. One can show that there exist attributes $f_1, \ldots, f_n \in F_2$ such that, for any $b_1, \ldots, b_n \in \{0, 1\}$, the system of equations $\{f_1(x) = b_1, \ldots, f_n(x) = b_n\}$ has a solution over A_2. Therefore $I(\mathcal{U}_2) \geq n$. Since $|A_2| = 2^n$, $I(\mathcal{U}_2) \leq n$ and $I(\mathcal{U}_2) = n$. Let $v' : \{0, 1\}^n \to \omega$ and, for any $\bar{a}, \bar{b} \in \{0, 1\}^n$, $v'(\bar{a}) \neq v'(\bar{b})$. It is not difficult to show that the problem $z' = (v', f_1, \ldots, f_n)$ is a tree-stable problem. Therefore $ts(\mathcal{U}_2) \geq n$. One can show that, for any problem z over $\mathcal{U}_2$, $h^g_{\mathcal{U}_2}(z) \leq n$. Therefore $ts(\mathcal{U}_2) = n$. It is easy to see that, for any problem z over $\mathcal{U}$, $l^g_{\mathcal{U}_2}(z) \leq 1$, and $l^g_{\mathcal{U}_2}(z') \geq 1$. Therefore $rs(\mathcal{U}_2) = 1$. Let $A_2 = \{a_1, \ldots, a_{2^n}\}$ and, for $i = 1, \ldots, 2^n - 1$, let g_i be an attribute from F_2 which is equal to 1 only on the element a_i. One can show that $\{g_1, \ldots, g_{2^n-1}\}$ is an irredundant set. Therefore $ir(\mathcal{U}_2) \geq 2^n - 1$. As a result, $rs(\mathcal{U}_2) = 1$, $I(\mathcal{U}_2) = ts(\mathcal{U}_2) = n$, and $ir(\mathcal{U}_2) \geq 2^n - 1$.

Example 15.3 Let n be an even natural number. Consider a finite information system $\mathcal{U}_3 = (A_3, \{0, 1, 2, 3\}, F_3)$ where the set A_3 contains 2^n elements and F_3 contains all possible nonconstant attributes defined on A_3 and with values from $\{0, 1, 2, 3\}$. As for the case of information system $\mathcal{U}_2$, one can show that $rs(\mathcal{U}_3) = 1$, $I(\mathcal{U}_3) = n$, and $ir(\mathcal{U}_3) \geq 2^n - 1$. One can show that there exist attributes $f_1, \ldots, f_{n/2} \in F_3$ such that, for any $b_1, \ldots, b_{n/2} \in \{0, 1, 2, 3\}$, the system of equations $\{f_1(x) = b_1, \ldots, f_{n/2}(x) = b_{n/2}\}$ has a solution over A_3. Let $v' : \{0, 1, 2, 3\}^{n/2} \to \omega$ and, for any $\bar{a}, \bar{b} \in \{0, 1, 2, 3\}^{n/2}$, $v'(\bar{a}) \neq v'(\bar{b})$. It is not difficult to show that the problem $z' = (v', f_1, \ldots, f_{n/2})$ is a tree-stable problem. Therefore $ts(\mathcal{U}_3) \geq n/2$. One can show that, for any problem z over $\mathcal{U}_3$, $h^g_{\mathcal{U}_3}(z) \leq n/2$. Therefore $ts(\mathcal{U}_3) = n/2$. As a result, $rs(\mathcal{U}_3) = 1$, $ts(\mathcal{U}_3) = n/2$, $I(\mathcal{U}_3) = n$, and $ir(\mathcal{U}_3) \geq 2^n - 1$.

Let us remind the definitions of two Shannon functions $H^g_{\mathcal{U}}(n)$ and $L^g_{\mathcal{U}}(n)$. Let $n \in \omega \setminus \{0\}$. Then

$$H^g_{\mathcal{U}}(n) = \max\{h^g_{\mathcal{U}}(z) : z \in \mathrm{Probl}_{\mathcal{U}}, \dim z \leq n\},$$
$$L^g_{\mathcal{U}}(n) = \max\{l^g_{\mathcal{U}}(z) : z \in \mathrm{Probl}_{\mathcal{U}}, \dim z \leq n\}.$$

Our aim is to study the behavior of functions $H^g_{\mathcal{U}}(n)$ and $L^g_{\mathcal{U}}(n)$ for an arbitrary finite information system $\mathcal{U}$. We will prove the following two theorems.

Theorem 15.1 *Let* $\mathcal{U} = (A, B, F)$ *be a finite information system,* $k = |B|$, *and* $m = \max(ts(\mathcal{U}), I(\mathcal{U}))$. *Then, for any* $n \in \omega \setminus \{0\}$, *the following statements hold:*
(a) *if* $n \leq m$ *then* $\frac{n}{\log_2 k} \leq H^g_{\mathcal{U}}(n) \leq n$;
(b) *if* $m < n \leq ir(\mathcal{U})$ *then*

$$\max\left\{ts(\mathcal{U}), \frac{I(\mathcal{U})}{\log_2 k}, \log_k(n+1)\right\} \leq H^g_{\mathcal{U}}(n)$$
$$\leq \min\left\{n - 1, 4(m+1)^4(\log_2 n)^2 + 4(m+1)^5(\log_2 n)(\log_2 k)\right\};$$

(c) if $n \geq \mathrm{ir}(\mathcal{U})$ then $H_{\mathcal{U}}^g(n) = H_{\mathcal{U}}^g(\mathrm{ir}(\mathcal{U}))$.

Theorem 15.2 *Let $\mathcal{U} = (A, B, F)$ be a finite information system. Then, for any $n \in \omega \setminus \{0\}$, the following statements hold:*
(a) if $n \leq \mathrm{rs}(\mathcal{U})$ then $L_{\mathcal{U}}^g(n) = n$;
(b) if $n \geq \mathrm{rs}(\mathcal{U})$ then $L_{\mathcal{U}}^g(n) = \mathrm{rs}(\mathcal{U})$.

The behavior of the function $H_{\mathcal{U}}^g(n)$ was studied in [6] (see Theorem 5.3) for the case of finite two-valued information systems where $|B| = 2$. In Theorem 15.1, the considered result is adapted for the case of arbitrary finite information system. The proof from [6] cannot be extended directly to this case since the behavior of independence dimension for many-valued information systems can be different in comparison with two-valued ones.

15.2.1 Proofs of Theorems 15.1 and 15.2

First, we consider some auxiliary statements. For a problem $z = (\nu, f_1, \ldots, f_n)$ over $\mathcal{U}$ with single-valued decisions, denote $\mathrm{Row}_{\mathcal{U}}(z)$ the set of n-tuples $(\delta_1, \ldots, \delta_n) \in B^n$ such that the equation system $\{f_1(x) = \delta_1, \ldots, f_n(x) = \delta_n\}$ has a solution over A, $N_{\mathcal{U}}(z)$ the cardinality of the set $\mathrm{Row}_{\mathcal{U}}(z)$, and $S_{\mathcal{U}}(z)$ the cardinality of the set $\{\nu(\delta_1, \ldots, \delta_n) : (\delta_1, \ldots, \delta_n) \in \mathrm{Row}_{\mathcal{U}}(z)\}$. The next statement follows from Theorem 4.6 from [6].

Lemma 15.1 *Let z be a problem over finite information system $\mathcal{U} = (A, B, F)$ and $k = |B|$. Then*

$$N_{\mathcal{U}}(z) \leq (k^2 \dim z)^{I(\mathcal{U})}.$$

The next statement follows from Theorems 3.2 and 4.2 from [6].

Lemma 15.2 *Let z be a problem over information system $\mathcal{U} = (A, B, F)$ and $k = |B|$. Then*

$$h_{\mathcal{U}}^g(z) \geq \lceil \log_k S_{\mathcal{U}}(z) \rceil.$$

For any word $\alpha \in \Omega_{F,B}$, we denote by $\mathrm{Alph}(\alpha)$ the set of letters from the alphabet $\{(f, \delta) : f \in F, \delta \in B\}$ contained in α. Define the parameter $M_{\mathcal{U}}(z)$ for the problem $z = (\nu, f_1, \ldots, f_n)$. For $\bar{\delta} = (\delta_1, \ldots, \delta_n) \in B^n$, denote $M_{\mathcal{U}}(z, \bar{\delta})$ the minimum length of a word γ such that $\mathrm{Alph}(\gamma) \subseteq \mathrm{Alph}((f_1, \delta_1) \ldots (f_n, \delta_n))$ and either $\mathrm{Sol}_{\mathcal{U}}(\gamma) = \emptyset$ or the function $\nu(f_1(x), \ldots, f_n(x))$ is constant on the set $\mathrm{Sol}_{\mathcal{U}}(\gamma)$. Then $M_{\mathcal{U}}(z) = \max\{M_{\mathcal{U}}(z, \bar{\delta}) : \bar{\delta} \in B^n\}$. The next statement follows from Theorems 3.5 and 4.1 from [6].

Lemma 15.3 *Let z be a problem over information system $\mathcal{U} = (A, B, F)$. Then*

$$h_{\mathcal{U}}^g(z) \leq M_{\mathcal{U}}(z) \log_2 N_{\mathcal{U}}(z).$$

For $\Gamma \in \text{Tree}_{\mathscr{U}}$, denote $L_w(\Gamma)$ the number of working nodes in Γ. The next statement follows from Lemma 3.4 from [6].

Lemma 15.4 *Let* $\Gamma \in \text{Tree}_{\mathscr{U}}$ *and* $k = |B|$. *Then* $L_w(\Gamma) \leq (k^{h(\Gamma)} - 1)/(k - 1)$.

We now adapt Lemma 5.24 from [6] for the case of arbitrary finite information system.

Lemma 15.5 *Let* $\mathscr{U} = (A, B, F)$ *be a finite information system,* $k = |B|$, *and* $m = \max(\text{ts}(\mathscr{U}), I(\mathscr{U}))$. *Then, for any natural* $n \geq 2$, *the following inequality holds:*

$$H_{\mathscr{U}}^g(n) \leq 4(m+1)^4(\log_2 n)^2 + 4(m+1)^5(\log_2 n)(\log_2 k) .$$

Proof We begin with the overview of the proof. Let $z = (\nu, f_1, \ldots, f_r)$ be an arbitrary problem over $\mathscr{U}$ with $\dim z = r \leq n$. We define a subset $J(z)$ of the set F that is an extension of the set $\{f_1, \ldots, f_r\}$. Let $J(z) = \{f_1, \ldots, f_p\}$. We consider a problem $z' = (\nu', f_1, \ldots, f_p)$ such that $z'(a) = z(a)$ for any $a \in A$. The choice of $J(z)$ allows us to prove that

$$N_{\mathscr{U}}(z') \leq 2^{2(m+1)^3(\log_2 n)^2 + 2(m+1)^4(\log_2 n)(\log_2 k)}$$

and

$$M_{\mathscr{U}}(z') \leq 2(m+1) .$$

By Lemma 15.3,

$$h_{\mathscr{U}}(z') \leq M_{\mathscr{U}}(z') \log_2 N_{\mathscr{U}}(z') \leq 4(m+1)^4(\log_2 n)^2 + 4(m+1)^5(\log_2 n)(\log_2 k) .$$

It is clear that $h_{\mathscr{U}}(z) = h_{\mathscr{U}}(z')$. Since z is an arbitrary problem over $\mathscr{U}$ with $\dim z \leq n$, $H_{\mathscr{U}}^g(n) \leq 4(m+1)^4(\log_2 n)^2 + 4(m+1)^5(\log_2 n)(\log_2 k)$. Note that both the description and the study of the set $J(z)$ as well as the study of the problem z' require some effort.

It is clear that $H_{\mathscr{U}}^g(m+1) \leq m$. For an arbitrary $p \in \omega \setminus \{0\}$, denote ν_p the mapping of the set B^p into the set ω such that $\nu_p(\bar{\delta}_1) \neq \nu_p(\bar{\delta}_2)$ for any $\bar{\delta}_1, \bar{\delta}_2 \in B^p$, $\bar{\delta}_1 \neq \bar{\delta}_2$.

By Lemma 15.1, for any problem z over $\mathscr{U}$, the following inequality holds:

$$N_{\mathscr{U}}(z) \leq k^{2m}(\dim z)^m . \tag{15.3}$$

Let $f_1, \ldots, f_{m+1}$ be pairwise distinct attributes from F (if in the set F there are no $m+1$ pairwise distinct attributes, then $H_{\mathscr{U}}^g(n) \leq m$ for any natural $n \geq 2$, and the statement of the lemma holds). Denote

$$z(f_1, \ldots, f_{m+1}) = (\nu_{m+1}, f_1, \ldots, f_{m+1}) .$$

From $H_{\mathcal{U}}^g(m+1) \le m$ it follows the existence of a decision tree $\Gamma(f_1, \ldots, f_{m+1})$ over $\mathcal{U}$ such that $h(\Gamma(f_1, \ldots, f_{m+1})) \le m$ and the decision tree $\Gamma(f_1, \ldots, f_{m+1})$ solves the problem $z(f_1, \ldots, f_{m+1})$. Evidently, for any $\delta_1, \ldots, \delta_{m+1} \in B$ and for any complete path ξ in the tree $\Gamma(f_1, \ldots, f_{m+1})$, either

$$\mathrm{Sol}_{\mathcal{U}}(\pi(\xi)) \cap \mathrm{Sol}_{\mathcal{U}}((f_1, \delta_1) \ldots (f_{m+1}, \delta_{m+1})) = \emptyset$$

or

$$\mathrm{Sol}_{\mathcal{U}}(\pi(\xi)) \subseteq \mathrm{Sol}_{\mathcal{U}}((f_1, \delta_1) \ldots (f_{m+1}, \delta_{m+1})) .$$

By Lemma 15.4,

$$L_w(\Gamma(f_1, \ldots, f_{m+1})) \le k^m . \tag{15.4}$$

Let $f_1, \ldots, f_q$ be pairwise distinct attributes from F and let $q \le m$. Define a decision tree $\Gamma(f_1, \ldots, f_q)$ over $\mathcal{U}$ in the following way. Let every working node of the tree $\Gamma(f_1, \ldots, f_q)$ have exactly k edges issuing from it, and let every complete path in the tree $\Gamma(f_1, \ldots, f_q)$ contain exactly q working nodes. Let $\xi = v_1, d_1, \ldots, v_q, d_q, v_{q+1}$ be an arbitrary complete path in the tree $\Gamma(f_1, \ldots, f_q)$. Then, for $i = 1, \ldots, q$, the node v_i is labeled with the attribute f_i, and the node v_{q+1} is labeled with the number 0.

For every problem z over $\mathcal{U}$, we define by induction a subset $J(z)$ of the set F. If $\dim z \le m$ then $J(z) = \mathrm{At}(z)$. Assume that for some $n, n \ge m+1$, for any problem z' over $\mathcal{U}$ with $\dim z' < n$ the set $J(z')$ has already been defined. Define the set $J(z)$ for a problem $z = (\nu, f_1, \ldots, f_n)$ over $\mathcal{U}$. Let $n = t(m+1) + q$, where $t \in \omega \setminus \{0\}$ and $0 \le q \le m$. For $i = 1, \ldots, t$, denote $\Gamma_i = \Gamma(f_{(m+1)(i-1)+1}, \ldots, f_{(m+1)(i-1)+m+1})$. Define a decision tree Γ_{t+1} over $\mathcal{U}$. If $q = 0$ then the tree Γ_{t+1} contains only the node labeled with the number 0. If $q > 0$ then $\Gamma_{t+1} = \Gamma(f_{(m+1)t+1}, \ldots, f_{(m+1)t+q})$. Define decision trees $G_1, \ldots, G_{t+1}$ from $\mathrm{Tree}_{\mathcal{U}}$ in the following way: $G_1 = \Gamma_1$ and, for $i = 1, \ldots, t$, the tree G_{i+1} is obtained from the tree G_i by replacing of every terminal node v in the tree G_i with the tree Γ_{i+1} (the edge which had entered the node v will be entered the root of the tree Γ_{i+1}). Denote by $\Gamma(z)$ the decision tree that consists of all nodes and edges of the tree G_{t+1} for each of which there exists a complete path ξ containing it and satisfying the condition $\mathrm{Sol}_{\mathcal{U}}(\pi(\xi)) \ne \emptyset$. One can show that $\bigcup_{\xi \in \mathrm{Path}(\Gamma(z))} \mathrm{Sol}_{\mathcal{U}}(\pi(\xi)) = A$. Denote $c = 2m/(2m+1)$. One can easily show $h(G_{t+1}) \le mt + q \le cn$. Therefore

$$h(\Gamma(z)) \le cn . \tag{15.5}$$

From (15.4) and from the description of the tree Γ_{t+1} it follows that $|\mathrm{At}(G_{t+1})| \le tk^m + q \le nk^m$. Using these inequalities and the inequality (15.3) we conclude that the tree G_{t+1} contains at most $k^{2m}(nk^m)^m = n^m k^{m^2+2m}$ complete paths ξ such that $\mathrm{Sol}_{\mathcal{U}}(\pi(\xi)) \ne \emptyset$. Therefore

$$|\mathrm{Path}(\Gamma(z))| \le n^m k^{m^2+2m} . \tag{15.6}$$

We correspond to every complete path ξ in the tree $\Gamma(z)$ a problem z_ξ over $\mathcal{U}$. Let $\{f_{i_1}, \ldots, f_{i_p}\}$ be the set of attributes from F attached to working nodes of the path ξ. Then $z_\xi = (v_p, f_{i_1}, \ldots, f_{i_p})$. From (15.5) it follows that, for any $\xi \in \mathrm{Path}(\Gamma(z))$, the inequality

$$\dim z_\xi \le cn \tag{15.7}$$

holds. Hence, by assumption, the set $J(z_\xi)$ has already been determined for any $\xi \in \mathrm{Path}(\Gamma(z))$. Set

$$J(z) = \mathrm{At}(z) \cup \left(\bigcup_{\xi \in \mathrm{Path}(\Gamma(z))} J(z_\xi) \right) . \tag{15.8}$$

For $n \in \omega \setminus \{0\}$, denote

$$J_{\mathcal{U}}(n) = \max\{|J(z)| : z \in \mathrm{Probl}_{\mathcal{U}}, \dim z \le n\} .$$

The inequality

$$J_{\mathcal{U}}(n) \le n^{2(m+1)^2 \ln n} k^{2(m+1)^3 \ln n} \tag{15.9}$$

will be proven by induction on $n \ge 1$. It is clear that if $n \le m$ then $J_{\mathcal{U}}(n) \le n$. Hence for $n \le m$ the inequality (15.9) holds. Let for some n, $n \ge m + 1$, for any n', $1 \le n' < n$, the inequality (15.9) hold. Let us show that it holds also for n. Let $z \in \mathrm{Probl}_{\mathcal{U}}$ and $\dim z \le n$. If $\dim z < n$, then using induction hypothesis one can show $|J(z)| \le n^{2(m+1)^2 \ln n} k^{2(m+1)^3 \ln n}$. Let $\dim z = n$. Evidently, $1 \le \lfloor cn \rfloor < n$ and $J_{\mathcal{U}}(\lfloor cn \rfloor) \ge 1$. Using (15.6)–(15.8) obtain $|J(z)| \le n + n^m k^{m^2+2m} J_{\mathcal{U}}(\lfloor cn \rfloor) \le n^m k^{m^2+2m+1} J_{\mathcal{U}}(\lfloor cn \rfloor)$. Using the induction hypothesis obtain

$$|J(z)| \le n^m k^{(m+1)^2} (\lfloor cn \rfloor)^{2(m+1)^2 \ln \lfloor cn \rfloor} k^{2(m+1)^3 \ln \lfloor cn \rfloor}$$
$$\le n^{m+2(m+1)^2 \ln(cn)} k^{(m+1)^2+2(m+1)^3 \ln(cn)} .$$

From the inequality $\ln(1 + 1/r) > 1/(r + 1)$ which is true for any natural r it follows that $\ln c < -1/(2m + 1) < -1/2(m + 1)$. Hence

$$|J(z)| \le n^{2(m+1)^2 \ln n} k^{2(m+1)^3 \ln n} .$$

Since z is an arbitrary problem over $\mathcal{U}$ such that $\dim z \le n$, the inequality (15.9) holds.

Prove the following statement by induction on n. Let $z = (v, f_1, \ldots, f_n) \in \mathrm{Probl}_{\mathcal{U}}$, $J(z) = \{f_1, \ldots, f_p\}$, $\bar{\delta} = (\delta_1, \ldots, \delta_p) \in B^p$, let

$$\alpha(z, \bar{\delta}) = (f_1, \delta_1) \ldots (f_p, \delta_p)$$

and $\beta(z, \bar{\delta}) = (f_1, \delta_1) \ldots (f_n, \delta_n)$. Then there exists a word $\gamma(z, \bar{\delta})$ from the set $\Omega_{F,B}$ such that $\mathrm{Alph}(\gamma(z, \bar{\delta})) \subseteq \mathrm{Alph}(\alpha(z, \bar{\delta}))$, $\mathrm{Sol}_{\mathscr{U}}(\gamma(z, \bar{\delta})) \subseteq \mathrm{Sol}_{\mathscr{U}}(\beta(z, \bar{\delta}))$, and the length of the word $\gamma(z, \bar{\delta})$ is at most $2(m + 1)$.

For $n \leq 2(m + 1)$, this statement is true since we can take the word $\beta(z, \bar{\delta})$ as the word $\gamma(z, \bar{\delta})$. Suppose that for certain n, $n \geq 2(m + 1) + 1$, the statement is true for any problem z over $\mathscr{U}$ with $\dim z < n$. Let us show that the considered statement holds for an arbitrary problem $z = (\nu, f_1, \ldots, f_n)$ over $\mathscr{U}$. Let $J(z) = \{f_1, \ldots, f_p\}$ and $\bar{\delta} = (\delta_1, \ldots, \delta_p) \in B^p$. One can show that $\mathrm{At}(\Gamma(z)) \subseteq J(z)$. Consider a directed path $\kappa = v_1, d_1, \ldots, v_r, d_r, v_{r+1}$ in the tree $\Gamma(z)$ starting in the root and possessing the following properties:

(1) if the node v_i, $i \in \{1, \ldots, r\}$, is labeled with an attribute f_l, then the edge d_i is labeled with δ_l;

(2) if v_{r+1} is a working node in the tree $\Gamma(z)$ which is labeled with the attribute f_l then an edge labeled with δ_l does not issue from v_{r+1}.

First, assume that κ is a complete path in the tree $\Gamma(z)$. Let $n = t(m + 1) + q$ where $t \geq 1$ and $0 \leq q \leq m$. For $i = 1, \ldots, t$, denote

$$\Gamma_i = \Gamma(f_{(m+1)(i-1)+1}, \ldots, f_{(m+1)(i-1)+m+1}) \,.$$

Define a decision tree Γ_{t+1} over $\mathscr{U}$. If $q = 0$, then Γ_{t+1} consists of the root labeled with 0. If $q > 0$, then $\Gamma_{t+1} = \Gamma(f_{(m+1)t+1}, \ldots, f_{(m+1)t+q})$. Define words $\beta_1, \ldots, \beta_{t+1}$. For $i = 1, \ldots, t$, let

$$\beta_i = (f_{(m+1)(i-1)+1}, \delta_{(m+1)(i-1)+1}) \ldots (f_{(m+1)(i-1)+m+1}, \delta_{(m+1)(i-1)+m+1}) \,.$$

If $q = 0$, then $\beta_{t+1} = \lambda$. If $q > 0$, then

$$\beta_{t+1} = (f_{(m+1)t+1}, \delta_{(m+1)t+1}) \ldots (f_{(m+1)t+q}, \delta_{(m+1)t+q}) \,.$$

Evidently, $\beta(z, \bar{\delta}) = \beta_1 \ldots \beta_{t+1}$. One can show that the word $\pi(\kappa)$ can be represented in the form $\pi(\kappa) = \pi(\xi_1) \ldots \pi(\xi_{t+1})$ where ξ_i is a complete path in the tree Γ_i, $i = 1, \ldots, t + 1$.

Let there exist $i \in \{1, \ldots, t\}$ such that $\mathrm{Sol}_{\mathscr{U}}(\beta_i) \cap \mathrm{Sol}_{\mathscr{U}}(\pi(\xi_i)) = \emptyset$. Denote $\gamma = \beta_i \pi(\xi_i)$. It is clear that $\mathrm{Alph}(\gamma) \subseteq \mathrm{Alph}(\alpha(z, \bar{\delta}))$ and $\mathrm{Sol}_{\mathscr{U}}(\gamma) = \emptyset$. Hence $\mathrm{Sol}_{\mathscr{U}}(\gamma) \subseteq \mathrm{Sol}_{\mathscr{U}}(\beta(z, \bar{\delta}))$ and the length of the word γ is at most $m + 1 + m < 2(m + 1)$. Thus, in the considered case the word γ can be taken as the word $\gamma(z, \bar{\delta})$.

Let $\mathrm{Sol}_{\mathscr{U}}(\beta_i) \cap \mathrm{Sol}_{\mathscr{U}}(\pi(\xi_i)) \neq \emptyset$ for $i = 1, \ldots, t$. Then, as mentioned above, $\mathrm{Sol}_{\mathscr{U}}(\pi(\xi_i)) \subseteq \mathrm{Sol}_{\mathscr{U}}(\beta_i)$ for $i = 1, \ldots, t$. Evidently, $\mathrm{Sol}_{\mathscr{U}}(\pi(\xi_{t+1})) = \mathrm{Sol}_{\mathscr{U}}(\beta_{t+1})$ and hence

$$\mathrm{Sol}_{\mathscr{U}}(\pi(\kappa)) \subseteq \mathrm{Sol}_{\mathscr{U}}(\beta(z, \bar{\delta})) \,. \tag{15.10}$$

Consider the problem z_κ. Let $z_\kappa = (\nu_l, f_{j_1}, \ldots, f_{j_t})$ and $J(z_\kappa) = \{f_{j_1}, \ldots, f_{j_u}\}$. From (15.8) it follows that $J(z_\kappa) \subseteq J(z)$. Denote $\bar{\delta}' = (\delta_{j_1}, \ldots, \delta_{j_u})$. Using (15.7) obtain $\dim z_\kappa < n$. From this inequality and from the induction hypothesis it follows that there exists a word $\gamma(z_\kappa, \bar{\delta}') \in \Omega_{F,B}$ such that $\mathrm{Alph}(\gamma(z_\kappa, \bar{\delta}')) \subseteq$

Alph($\alpha(z_\kappa, \bar{\delta}')$), Sol$_\mathcal{U}$ ($\gamma(z_\kappa, \bar{\delta}')$) $\subseteq$ Sol$_\mathcal{U}$ ($\beta(z_\kappa, \bar{\delta}')$), and the length of the word $\gamma(z_\kappa, \bar{\delta}')$ is at most $2(m + 1)$. It is clear that Alph($\alpha(z_\kappa, \bar{\delta}')$) $\subseteq$ Alph($\alpha(z, \bar{\delta})$) and Alph($\gamma(z_\kappa, \bar{\delta}')$) $\subseteq$ Alph($\alpha(z, \bar{\delta})$). One can easily show

$$\text{Sol}_\mathcal{U}\ (\pi(\kappa)) = \text{Sol}_\mathcal{U}\ (\beta(z_\kappa, \bar{\delta}')) \ .$$

Using (15.10) obtain Sol$_\mathcal{U}$ ($\gamma(z_\kappa, \bar{\delta}')$) $\subseteq$ Sol$_\mathcal{U}$ ($\beta(z, \bar{\delta})$). Hence in this case the word $\gamma(z_\kappa, \bar{\delta}')$ can be taken as the word $\gamma(z, \bar{\delta})$.

Suppose now that the path κ is not a complete path in the tree $\Gamma(z)$. Evidently, there exists a complete path ξ in the tree $\Gamma(z)$ containing the node v_{r+1}. Consider the problem z_ξ. Let $z_\xi = (v_l, f_{j_1}, \ldots, f_{j_l})$ and $J(z_\xi) = \{f_{j_1}, \ldots, f_{j_u}\}$. From (15.8) it follows that $J(z_\xi) \subseteq J(z)$. Denote $\bar{\delta}' = (\delta_{j_1}, \ldots, \delta_{j_u})$. Recalling that the path κ is not a complete path in the tree $\Gamma(z)$ one can show Sol$_\mathcal{U}$ ($\beta(z_\xi, \bar{\delta}')$) $= \emptyset$. Using (15.7) obtain dim $z_\xi < n$. From this inequality and from the induction hypothesis it follows that there exists a word $\gamma(z_\xi, \bar{\delta}') \in \Omega_{F,B}$ such that Alph($\gamma(z_\xi, \bar{\delta}')$) $\subseteq$ Alph($\alpha(z_\xi, \bar{\delta}')$), Sol$_\mathcal{U}$ ($\gamma(z_\xi, \bar{\delta}')$) $\subseteq$ Sol$_\mathcal{U}$ ($\beta(z_\xi, \bar{\delta}')$), and the length of the word $\gamma(z_\xi, \bar{\delta}')$ is at most $2(m + 1)$. It is clear that Alph($\alpha(z_\xi, \bar{\delta}')$) $\subseteq$ Alph($\alpha(z, \bar{\delta})$). Therefore Alph($\gamma(z_\xi, \bar{\delta}')$) $\subseteq$ Alph($\alpha(z, \bar{\delta})$). From the relations Sol$_\mathcal{U}$ ($\gamma(z_\xi, \bar{\delta}')$) $\subseteq$ Sol$_\mathcal{U}$ ($\beta(z_\xi, \bar{\delta}')$) $= \emptyset$ it follows that

$$\text{Sol}_\mathcal{U}\ (\gamma(z_\xi, \bar{\delta}')) \subseteq \text{Sol}_\mathcal{U}\ (\beta(z, \bar{\delta})) \ .$$

Thus, in the considered case the word $\gamma(z_\xi, \bar{\delta}')$ can be taken as the word $\gamma(z, \bar{\delta})$.

Let $n \geq 2$. Consider an arbitrary problem z over $\mathcal{U}$ with dim $z \leq n$. Let $z = (v, f_1, \ldots, f_r)$ and $J(z) = \{f_1, \ldots, f_p\}$. Consider also the problem $z' = (v', f_1, \ldots, f_p)$ where $v' : B^p \to \omega$ and the equality $v'(\bar{\delta}) = v(\delta_1, \ldots, \delta_r)$ holds for any tuple $\bar{\delta} = (\delta_1, \ldots, \delta_p) \in B^p$. Using (15.3) and (15.9) obtain

$$\begin{aligned} N_\mathcal{U}\ (z') &\leq k^{2m}\,(n^{2(m+1)^2 \ln n} k^{2(m+1)^3 \ln n})^m \\ &= n^{2m(m+1)^2 \ln n} k^{2m+2m(m+1)^3 \ln n} \\ &\leq 2^{2(m+1)^3 (\log_2 n)^2 + 2(m+1)^4 (\log_2 n)(\log_2 k)} \ . \end{aligned} \tag{15.11}$$

Let us show that

$$M_\mathcal{U}\ (z') \leq 2(m + 1) \ . \tag{15.12}$$

Let $\bar{\delta} = (\delta_1, \ldots, \delta_p) \in B^p$. Then, by proved above, there exists a word $\gamma(z, \bar{\delta})$ from the set $\Omega_{F,B}$ such that Alph($\gamma(z, \bar{\delta})$) $\subseteq$ Alph($\alpha(z, \bar{\delta})$),

$$\text{Sol}_\mathcal{U}\ (\gamma(z, \bar{\delta})) \subseteq \text{Sol}_\mathcal{U}\ (\beta(z, \bar{\delta})) \ ,$$

and the length of the word $\gamma(z, \bar{\delta})$ is at most $2(m + 1)$. It is clear that Alph($\gamma(z, \bar{\delta})$) $\subseteq \{(f_1, \delta_1), \ldots, (f_p, \delta_p)\}$. Taking into account Sol$_\mathcal{U}$ ($\gamma(z, \bar{\delta})$) $\subseteq$ Sol$_\mathcal{U}$ ($\beta(z, \bar{\delta})$) we can easily show that either Sol$_\mathcal{U}$ ($\gamma(z, \bar{\delta})$) $= \emptyset$ or the function $v'(f_1(x), \ldots, f_p(x))$ is constant on the set Sol$_\mathcal{U}$ ($\gamma(z, \bar{\delta})$). Therefore $M_\mathcal{U}$ ($T, \bar{\delta}$) $\leq 2(m + 1)$. Recalling that

$\bar{\delta}$ is an arbitrary tuple from B^p we conclude that the inequality (15.12) holds. From Lemma 15.3 and from inequalities (15.11) and (15.12) it follows that $h^g_{\mathcal{U}}(z') \leq M_{\mathcal{U}}(z') \log_2 N_{\mathcal{U}}(z') \leq 4(m+1)^4 (\log_2 n)^2 + 4(m+1)^5 (\log_2 n)(\log_2 k)$. Evidently, for any element $a \in A$, the equality $z(a) = z'(a)$ holds. Therefore $h^g_{\mathcal{U}}(z) = h^g_{\mathcal{U}}(z')$. Since z is an arbitrary problem over $\mathcal{U}$ with $\dim z \leq n$, $H^g_{\mathcal{U}}(n) \leq 4(m+1)^4 (\log_2 n)^2 + 4(m+1)^5 (\log_2 n)(\log_2 k)$. □

Proof (*of Theorem* 15.1) (a) Let $n \leq \text{ts}(\mathcal{U})$, and $z = (v, f_1, \ldots, f_{\text{ts}(\mathcal{U})})$ be a problem over $\mathcal{U}$ such that $h^g_{\mathcal{U}}(z) = \text{ts}(\mathcal{U})$. Consider the problem $z' = (v_n, f_1, \ldots, f_n)$ over $\mathcal{U}$ (the mapping v_p for any natural p is defined at the beginning of the proof of Lemma 15.5). Assume that $h^g_{\mathcal{U}}(z') < n$. One can show that in this case there exists a decision tree Γ' over $\mathcal{U}$ such that Γ' solves the problem z', $h(\Gamma') < n$, and $\text{Sol}_{\mathcal{U}}(\pi(\xi)) \neq \emptyset$ for any complete path ξ of Γ'. It is clear that, for any terminal node of Γ', the solution of z' attached to this node allows one to restore the values of the attributes $f_1, \ldots, f_n$. Therefore, by computing values of additional attributes $f_{n+1}, \ldots, f_{\text{ts}(\mathcal{U})}$, one can transform Γ' into a decision tree Γ over $\mathcal{U}$ which solves z and for which $h(\Gamma) < n + \text{ts}(\mathcal{U}) - n = \text{ts}(\mathcal{U})$. In this case $h^g_{\mathcal{U}}(z) < \text{ts}(\mathcal{U})$ which is impossible. Therefore $h^g_{\mathcal{U}}(z') = n$ and $H^g_{\mathcal{U}}(n) \geq n$. It is clear that $H^g_{\mathcal{U}}(n) \leq n$. Thus, $H^g_{\mathcal{U}}(n) = n$.

Let $n \leq I(\mathcal{U})$. Then there exist attributes $f_1, \ldots, f_n \in F$ and two-element subsets $B_1, \ldots, B_n$ of the set B such that, for any $\delta_1 \in B_1, \ldots, \delta_n \in B_n$, the set

$$\text{Sol}_{\mathcal{U}}((f_1, \delta_1) \ldots (f_n, \delta_n))$$

is nonempty. Consider the problem $z = (v_n, f_1, \ldots, f_n)$ over $\mathcal{U}$. It is clear that $S_{\mathcal{U}}(z) \geq 2^n$. By Lemma 15.2, $h^g_{\mathcal{U}}(z) \geq \log_k 2^n = \frac{n}{\log_2 k}$ where $k = |B|$. Using the equality $\dim z_n = n$ obtain $\frac{n}{\log_2 k} \leq H^g_{\mathcal{U}}(n)$. It is clear that $H^g_{\mathcal{U}}(n) \leq n$.

(b) Let $m < n \leq \text{ir}(\mathcal{U})$. From the part (a) of the proof it follows that $\text{ts}(\mathcal{U}) \leq H^g_{\mathcal{U}}(n)$ and $\frac{I(\mathcal{U})}{\log_2 k} \leq H^g_{\mathcal{U}}(n)$. Let us show that $\log_k(n+1) \leq H^g_{\mathcal{U}}(n)$. From the inequality $n \leq \text{ir}(\mathcal{U})$ it follows that there exists an independent subset $\{f_1, \ldots, f_n\}$ of the set F. It is clear that $n \geq 2$. For $i = 1, \ldots, n$, denote $z_i = (v_i, f_1, \ldots, f_i)$. Since $f_1 \not\equiv \text{const}$, $N_{\mathcal{U}}(z_1) \geq 2$. Let us show that $N_{\mathcal{U}}(z_i) < N_{\mathcal{U}}(z_{i+1})$ for $i = 1, \ldots, n - 1$. Assume the contrary: let $N_{\mathcal{U}}(z_i) = N_{\mathcal{U}}(z_{i+1})$ for some $i \in \{1, \ldots, n-1\}$. One can show that in this case there exists a mapping $\mu : B^i \to B$ such that $f_{i+1}(a) = \mu(f_1(a), \ldots, f_i(a))$ for any $a \in A$ which is impossible. Thus, $N_{\mathcal{U}}(z_1) \geq 2$ and $N_{\mathcal{U}}(z_i) < N_{\mathcal{U}}(z_{i+1})$ for $i = 1, \ldots, n - 1$. Therefore $N_{\mathcal{U}}(z_n) \geq n + 1$ and $S_{\mathcal{U}}(z_n) \geq n + 1$. By Lemma 15.2, $h^g_{\mathcal{U}}(z_n) \geq \log_k(n+1)$ where $k = |B|$. Using the equality $\dim z_n = n$ obtain $\log_k(n+1) \leq H^g_{\mathcal{U}}(n)$.

It is clear that $H^g_{\mathcal{U}}(n) \leq n - 1$. The inequality $H^g_{\mathcal{U}}(n) \leq 4(m+1)^4 (\log_2 n)^2 + 4(m+1)^5 (\log_2 n)(\log_2 k)$ follows from Lemma 15.5.

(c) Let $n \geq \text{ir}(\mathcal{U})$. Consider an arbitrary problem z over $\mathcal{U}$ such that $\dim z \leq n$. Let $\{f_1, \ldots, f_t\}$ be an independent subset of the set $\text{At}(z)$ with maximum cardinality. It is clear that $t \leq \text{ir}(\mathcal{U})$. One can show that there exists a mapping $v : B^t \to \omega$ such that, for the problem $z' = (v, f_1, \ldots, f_t)$, the equality $z(a) = z'(a)$ holds for any $a \in A$. It is clear that $h^g_{\mathcal{U}}(z) = h^g_{\mathcal{U}}(z')$. Therefore $h^g_{\mathcal{U}}(z) \leq H^g_{\mathcal{U}}(t) \leq H^g_{\mathcal{U}}(\text{ir}(\mathcal{U}))$.

Since z is an arbitrary problem over $\mathcal{U}$ such that $\dim z \leq n$, $H_{\mathcal{U}}^g(n) \leq H_{\mathcal{U}}^g(\mathrm{ir}(\mathcal{U}))$. It is clear that $H_{\mathcal{U}}^g(n) \geq H_{\mathcal{U}}^g(\mathrm{ir}(\mathcal{U}))$. Thus, $H_{\mathcal{U}}^g(n) = H_{\mathcal{U}}^g(\mathrm{ir}(\mathcal{U}))$. □

Proof (of Theorem 15.2) (a) Let $n \leq \mathrm{rs}(\mathcal{U})$, and $z = (\nu, f_1, \ldots, f_{\mathrm{rs}(\mathcal{U})})$ be a problem over $\mathcal{U}$ such that $l_{\mathcal{U}}^g(z) = \mathrm{rs}(\mathcal{U})$. Consider the problem $z' = (\nu_n, f_1, \ldots, f_n)$ over $\mathcal{U}$ (the mapping ν_p for any natural p is defined at the beginning of the proof of Lemma 15.5). Assume that $l_{\mathcal{U}}^g(z') < n$. One can show that in this case there exists a decision rule system Δ' over $\mathcal{U}$ such that Δ' is complete for the problem z', $l(\Delta') < n$, and $\mathrm{Sol}_{\mathcal{U}}(\pi(\rho)) \neq \emptyset$ for any rule $\rho \in \Delta'$. It is clear that, for any rule ρ from Δ', the solution of z' from the right-hand side of ρ allows one to restore the values of the attributes $f_1, \ldots, f_n$. Therefore, by adding different groups of conditions of the kind $f_{n+1} = \delta_{n+1}, \ldots, f_{\mathrm{rs}(\mathcal{U})} = \delta_{\mathrm{rs}(\mathcal{U})}$, where $\delta_{n+1}, \ldots, \delta_{\mathrm{rs}(\mathcal{U})} \in B$, to the left-hand sides of rules and changing the right-hand sides, we can transform Δ' into a decision rule system Δ over $\mathcal{U}$ which is complete for the problem z and for which $l(\Delta) < n + \mathrm{rs}(\mathcal{U}) - n = \mathrm{rs}(\mathcal{U})$. In this case $l_{\mathcal{U}}^g(z) < \mathrm{rs}(\mathcal{U})$ which is impossible. Therefore $l_{\mathcal{U}}^g(z') = n$ and $L_{\mathcal{U}}^g(n) \geq n$. It is clear that $L_{\mathcal{U}}^g(n) \leq n$. Thus, $L_{\mathcal{U}}^g(n) = n$.

(b) Let us prove by induction on n that, for any $n \geq \mathrm{rs}(\mathcal{U})$, $L_{\mathcal{U}}^g(n) = \mathrm{rs}(\mathcal{U})$. It is clear that $L_{\mathcal{U}}^g(\mathrm{rs}(\mathcal{U})) = \mathrm{rs}(\mathcal{U})$. Let us assume that, for some $n \geq \mathrm{rs}(\mathcal{U})$, $L_{\mathcal{U}}^g(m) = \mathrm{rs}(\mathcal{U})$ for any m, $\mathrm{rs}(\mathcal{U}) \leq m \leq n$. Let us show that $L_{\mathcal{U}}^g(n + 1) = \mathrm{rs}(\mathcal{U})$. Since $n + 1 > \mathrm{rs}(\mathcal{U})$, $L_{\mathcal{U}}^g(n + 1) < n + 1$. Let z be a problem over $\mathcal{U}$ such that $\dim z \leq n + 1$. One can show that there exists a decision rule system Δ' over $\mathcal{U}$ such that Δ' is complete for the problem z, $l(\Delta') \leq n$, and $\mathrm{Sol}_{\mathcal{U}}(\pi(\rho)) \neq \emptyset$ for any rule $\rho \in \Delta'$. Treat each rule $\rho \in \Delta'$ in the following way. If $l(\rho) \leq \mathrm{rs}(\mathcal{U})$, keep the rule ρ untouched. Let $\mathrm{rs}(\mathcal{U}) < l(\rho) \leq n$ and ρ be equal to $(f_1 = \delta_1) \wedge \ldots \wedge (f_{l(\rho)} = \delta_{l(\rho)}) \Rightarrow \sigma$. Consider the problem $z_\rho = (\nu_\rho, f_1, \ldots, f_{l(\rho)})$ where, for any $(b_1, \ldots, b_{l(\rho)}) \in B^{l(\rho)}$, $\nu_\rho(b_1, \ldots, b_{l(\rho)}) = \sigma$ if $(b_1, \ldots, b_{l(\rho)}) = (\delta_1, \ldots, \delta_{l(\rho)})$ and $\nu_\rho(b_1, \ldots, b_{l(\rho)}) = \sigma + 1$, otherwise. According to the induction hypothesis, there exists a decision rule system Δ_ρ over $\mathcal{U}$ such that Δ_ρ is complete for the problem z_ρ and $l(\Delta_\rho) \leq \mathrm{rs}(\mathcal{U})$. Remove ρ from the rule system Δ' and add to Δ' all rules from Δ_ρ that have σ on the right-hand side. As a result, obtain a rule system Δ over $\mathcal{U}$. One can prove that Δ is complete for the problem z and $l(\Delta) \leq \mathrm{rs}(\mathcal{U})$. Therefore, $l_{\mathcal{U}}^g(z) \leq \mathrm{rs}(\mathcal{U})$. Since z is an arbitrary problem over $\mathcal{U}$ such that $\dim z \leq n + 1$, $L_{\mathcal{U}}^g(n + 1) \leq \mathrm{rs}(\mathcal{U})$. It is clear that $L_{\mathcal{U}}^g(n + 1) \geq \mathrm{rs}(\mathcal{U})$. Therefore $L_{\mathcal{U}}^g(n + 1) \geq \mathrm{rs}(\mathcal{U})$. Thus $L_{\mathcal{U}}^g(n) = \mathrm{rs}(\mathcal{U})$ for any $n \geq \mathrm{rs}(\mathcal{U})$. □

15.3 Behavior of Shannon Functions for Problems with Many-valued Decisions

In this section, we consider behavior of Shannon functions for local and global approaches, for decision and inhibitory trees and rule systems for problems with many-valued decisions over finite and infinite information systems.

15.3.1 Local Approach. Infinite Information Systems

We will say that an information system $\mathcal{U} = (A, B, F)$ is *restricted* if there exists a number $r \in \omega \setminus \{0\}$ such that, for each consistent (having a solution over the set A) system of equations of the kind

$$\{f_1(x) = \delta_1, \ldots, f_m(x) = \delta_m\}$$

where $m \in \omega \setminus \{0\}$, $f_1, \ldots, f_m \in F$ and $\delta_1, \ldots, \delta_m \in B$, there exists a subsystem of this system which has the same set of solutions over the set A and contains at most r equations.

In [2] (see also [4, 6]), the behavior of Shannon function $H_{\mathcal{U}}^l(n)$ was studied for infinite information systems. The next theorem follows from the obtained results (see Theorem 4.3 from [6]) and Propositions 15.1 and 15.2.

Theorem 15.3 *Let $\mathcal{U}$ be an infinite information system. Then the following statements hold:*
(a) if $\mathcal{U}$ is restricted, then $IH_{\mathcal{U},\infty}^l(n) = H_{\mathcal{U},\infty}^l(n) = \Theta(\log n)$;
(b) if $\mathcal{U}$ is not restricted, then $IH_{\mathcal{U},\infty}^l(n) = H_{\mathcal{U},\infty}^l(n) = n$ for each $n \in \omega \setminus \{0\}$.

In [7], the behavior of Shannon function $L_{\mathcal{U}}^l(n)$ was studied for infinite information systems. The next theorem follows from the obtained results (see Theorem 8.2 from [7]) and Propositions 15.1 and 15.2.

Theorem 15.4 *Let $\mathcal{U} = (A, B, F)$ be an infinite information system. Then the following statements hold:*
(a) if $\mathcal{U}$ is restricted, then $IL_{\mathcal{U},\infty}^l(n) = L_{\mathcal{U},\infty}^l(n) = O(1)$;
(b) if $\mathcal{U}$ is not restricted, then $IL_{\mathcal{U},\infty}^l(n) = L_{\mathcal{U},\infty}^l(n) = n$ for each $n \in \omega \setminus \{0\}$.

Now we extend Example 8.3 from [7] to the case of problems with many-valued decisions.

Example 15.4 Let $m, t \in \omega \setminus \{0\}$. We denote by $Pol(m)$ the set of all polynomials which have integer coefficients and depend on variables $x_1, \ldots, x_m$. We denote by $Pol(m, t)$ the set of all polynomials from $Pol(m)$ such that the degree of each polynomial is at most t. We define information systems $\mathcal{U}(m)$ and $\mathcal{U}(m, t)$ as follows: $\mathcal{U}(m) = (\mathbb{R}^m, E, F(m))$ and $\mathcal{U}(m, t) = (\mathbb{R}^m, E, F(m, t))$ where $E = \{-1, 0, +1\}$, $F(m) = \{\text{sign}(p) : p \in Pol(m), \text{sign}(p) \not\equiv \text{const}\}$ and $F(m, t) = \{\text{sign}(p) : p \in Pol(m, t), \text{sign}(p) \not\equiv \text{const}\}$. Here $\text{sign}(x) = -1$ if $x < 0$, $\text{sign}(x) = 0$ if $x = 0$, and $\text{sign}(x) = +1$ if $x > 0$. One can prove that $IH_{\mathcal{U}(m),\infty}^l(n) = H_{\mathcal{U}(m),\infty}^l(n) = IL_{\mathcal{U}(m),\infty}^l(n) = L_{\mathcal{U}(m),\infty}^l(n) = n$ for each $n \in \omega \setminus \{0\}$, $IH_{\mathcal{U}(1,1),\infty}^l(n) = H_{\mathcal{U}(1,1),\infty}^l(n) = \Theta(\log n)$, $IL_{\mathcal{U}(1,1),\infty}^l(n) = L_{\mathcal{U}(1,1),\infty}^l(n) = O(1)$, and if $m > 1$ or $t > 1$ then

$$IH_{\mathcal{U}(m,t),\infty}^l(n) = H_{\mathcal{U}(m,t),\infty}^l(n) = IL_{\mathcal{U}(m,t),\infty}^l(n) = L_{\mathcal{U}(m,t),\infty}^l(n) = n$$

for each $n \in \omega \setminus \{0\}$.

15.3.2 Local Approach. Finite Information Systems

Let $\mathcal{U} = (A, B, F)$ be a finite information system.

Let us remind that a set of attributes $\{f_1, \ldots, f_n\} \subseteq F$ is called *redundant* if $n \geq 2$ and there exist $i \in \{1, \ldots, n\}$ and $\mu : B^{n-1} \to B$ such that

$$f_i(a) = \mu(f_1(a), \ldots, f_{i-1}(a), f_{i+1}(a), \ldots, f_n(a))$$

for each $a \in A$. If the set $\{f_1, \ldots, f_n\}$ is not redundant then it is called *irredundant*. We denoted by $\mathrm{ir}(\mathcal{U})$ the maximum number of attributes in an irredundant subset of the set F.

A systems of equations over $\mathcal{U}$

$$\{f_1(x) = \delta_1, \ldots, f_n(x) = \delta_n\} \tag{15.13}$$

is called *cancelable* if $n \geq 2$ and there exists a number $i \in \{1, \ldots, n\}$ such that the system of equations

$$\{f_1(x) = \delta_1, \ldots, f_{i-1}(x) = \delta_{i-1}, f_{i+1}(x) = \delta_{i+1}, \ldots, f_n(x) = \delta_n\}$$

has the same set of solutions over A just as the system (15.13). If the system (15.13) is not cancelable then it is called *uncancelable*. We denote by $\mathrm{un}(\mathcal{U})$ the maximum number of equations in an uncancelable consistent system of equations over $\mathcal{U}$.

Let $S = \{f_1(x) = \delta_1, \ldots, f_n(x) = \delta_n\}$ be an uncancelable system of equations over $\mathcal{U}$. It is clear that the set of attributes $\{f_1, \ldots, f_n\}$ is irredundant. Therefore $\mathrm{un}(\mathcal{U}) \leq \mathrm{ir}(\mathcal{U})$. Evidently, $1 \leq \mathrm{un}(\mathcal{U})$. Thus

$$1 \leq \mathrm{un}(\mathcal{U}) \leq \mathrm{ir}(\mathcal{U}) .$$

In [3], the behavior of Shannon functions $H^l_{\mathcal{U},\infty}(n)$ and $L^l_{\mathcal{U},\infty}(n)$ for finite information systems was studied. The next two theorems follow from the obtained results (see Theorems 3 and 4 from [3]) and Proposition 15.2.

Theorem 15.5 *Let $\mathcal{U} = (A, B, F)$ be a finite information system, and $n \in \omega \setminus \{0\}$. Then the following statements hold:*
(a) if $n \leq \mathrm{un}(\mathcal{U})$ then $IH^l_{\mathcal{U},\infty}(n) = H^l_{\mathcal{U},\infty}(n) = n;$
(b) if $\mathrm{un}(\mathcal{U}) \leq n \leq \mathrm{ir}(\mathcal{U})$ then

$$\max\{\mathrm{un}(\mathcal{U}), \log_k(n+1)\} \leq IH^l_{\mathcal{U},\infty}(n) = H^l_{\mathcal{U},\infty}(n)$$

$$\leq \min\{n, 2(\mathrm{un}(\mathcal{U}))^2 \log_2 2(kn+1)\}$$

where $k = |B|$;
(c) if $n \geq \mathrm{ir}(\mathcal{U})$ then $IH^l_{\mathcal{U},\infty}(n) = H^l_{\mathcal{U},\infty}(n) = IH^l_{\mathcal{U},\infty}(\mathrm{ir}(\mathcal{U})) = H^l_{\mathcal{U},\infty}(\mathrm{ir}(\mathcal{U})).$

Theorem 15.6 *Let $\mathscr{U} = (A, B, F)$ be a finite information system, and $n \in \omega \setminus \{0\}$.*
Then the following statements hold:
(a) if $n \leq \mathrm{un}(\mathscr{U})$ then $IL_{\mathscr{U},\infty}^{l}(n) = L_{\mathscr{U},\infty}^{l}(n) = n$;
(b) if $n \geq \mathrm{un}(\mathscr{U})$ then $IL_{\mathscr{U},\infty}^{l}(n) = L_{\mathscr{U},\infty}^{l}(n) = \mathrm{un}(\mathscr{U})$.

We now consider Example 8.7 from [7].

Example 15.5 Denote by P the set of all points in the two-dimensional Euclidean plane. Consider an arbitrary straight line l, which divides the plane into positive and negative open half-planes, and the line l itself. Assign a function $f : P \to \{0, 1\}$ to the line l. The function f takes the value 1 if a point is situated on the positive half-plane, and f takes the value 0 if a point is situated on the negative half-plane or on the line l. Denote by F the set of functions which correspond to certain r mutually disjoint finite classes of parallel straight lines. Consider a finite information system $\mathscr{U} = (P, \{0, 1\}, F)$. One can show that $\mathrm{ir}(\mathscr{U}) = |F|$ and $\mathrm{un}(\mathscr{U}) \leq 2r$.

15.3.3 Global Approach. Infinite Information Systems

Extend the notion of *independence dimension* (or, in short, *I-dimension*) to the case of infinite information system $\mathscr{U} = (A, B, F)$. A finite subset $\{f_1, \ldots, f_p\}$ of the set F is called an *independent set* if there exist two-element subsets $B_1, \ldots, B_p$ of the set B such that, for any $\delta_1 \in B_1, \ldots, \delta_p \in B_p$, the system of equations

$$\{f_1(x) = \delta_1, \ldots, f_p(x) = \delta_p\} \tag{15.14}$$

is consistent (has a solution over the set A). If for any natural p there exists a subset of the set F, which cardinality is equal to p and which is an independent set, then we will say that the information system $\mathscr{U}$ has infinite I-dimension. Otherwise, I-dimension of $\mathscr{U}$ is the maximum cardinality of a subset of F, which is an independent set.

The notion of I-dimension is closely connected with well known notion of Vapnik–Chervonenkis dimension [9]. In particular, an information system

$$(A, \{0, 1\}, F)$$

has finite I-dimension if and only if it has finite VC-dimension [1].

Now we consider the condition of decomposition for the information system $\mathscr{U}$. Let $p \in \omega \setminus \{0\}$. A nonempty subset D of the set A will be called a $(p, \mathscr{U})$-set if D coincides with the set of solutions over A of a system of the kind (15.14) where $f_1, \ldots, f_p \in F$ and $\delta_1, \ldots, \delta_p \in B$ (we admit that among the attributes $f_1, \ldots, f_p$ there are identical ones).

We will say that the information system $\mathscr{U}$ satisfies the *condition of decomposition* if there exist numbers $m, t \in \omega \setminus \{0\}$ such that every $(m + 1, \mathscr{U})$-set is a union of t sets each of which is an $(m, \mathscr{U})$-set (we admit that among the considered t sets there are identical ones).

We now consider Example 8.8 from [7].

Example 15.6 Let P be the set of all points in the two-dimensional Euclidean plane and l be a straight line in the plane. This line divides the plane into two open half-planes H_1 and H_2, and the line l. We correspond one attribute to the line l. This attribute takes value 0 on points from H_1, and value 1 on points from H_2 and l. Denote by F the set of all attributes corresponding to lines in the plane. Let us consider the information system $\mathcal{U} = (P, \{0, 1\}, F)$. The information system $\mathcal{U}$ has finite I-dimension: there are no three lines which divide the plane into eight domains. The information system $\mathcal{U}$ satisfies the condition of decomposition: each $(4, \mathcal{U})$-set is a union of two $(3, \mathcal{U})$-sets.

In [5], the behavior of Shannon functions $H_{\mathcal{U}}^g(n)$ and $L_{\mathcal{U}}^g(n)$ for infinite information systems was studied. The next two theorems follows from the obtained results (see Theorem 2.1 and Lemma 3.7 from [5]) and Propositions 15.1 and 15.2.

Theorem 15.7 *Let $\mathcal{U} = (A, B, F)$ be an infinite information system. Then the following statements hold:*

(a) if $\mathcal{U}$ has finite I-dimension and satisfies the condition of decomposition then, for any ε, $0 < \varepsilon < 1$, $IH_{\mathcal{U},\infty}^g(n) = H_{\mathcal{U},\infty}^g(n) = \Omega(\log n)$ and $IH_{\mathcal{U},\infty}^g(n) = H_{\mathcal{U},\infty}^g(n) = O((\log n)^{1+\varepsilon})$.

(b) if $\mathcal{U}$ has infinite I-dimension or does not satisfy the condition of decomposition then, for any $n \in \omega \setminus \{0\}$, $IH_{\mathcal{U},\infty}^g(n) = H_{\mathcal{U},\infty}^g(n) = n$.

Theorem 15.8 *Let $\mathcal{U} = (A, B, F)$ be an infinite information system. Then either $IL_{\mathcal{U},\infty}^g(n) = L_{\mathcal{U},\infty}^g(n) = O(1)$ or $IL_{\mathcal{U},\infty}^g(n) = L_{\mathcal{U},\infty}^g(n) = n$ for   any $n \in \omega \setminus \{0\}$.*

Let us consider Example 8.12 from [7].

Example 15.7 Let $m, t \in \omega \setminus \{0\}$. We consider the same information systems $\mathcal{U}(m)$ and $\mathcal{U}(m, t)$ as in Example 15.4. One can prove that $\mathcal{U}(m, t)$ has finite I-dimension and satisfies the condition of decomposition, and $\mathcal{U}(m)$ has infinite I-dimension.

15.3.4   Global Approach. Finite Information Systems

Let $\mathcal{U} = (A, B, F)$ be a finite information system. In this section, we consider the following previously defined parameters of the information system $\mathcal{U}$:

- $I(\mathcal{U})$ (I-dimension of $\mathcal{U}$) – the maximum cardinality of a subset of F which is an independent set (Sect. 15.2);
- $ir(\mathcal{U})$ – the maximum cardinality of an irredundant subset of F (Sect. 15.2);
- $ts(\mathcal{U})$ – the maximum dimension of a tree-stable problem from $Probl_{\mathcal{U}}$ (Sect. 15.1.1);

- $\mathrm{rs}(\mathscr{U})$ – the maximum dimension of a rule-stable problem from $\mathrm{Probl}_{\mathscr{U}}$ (Sect. 15.1.1);
- $\mathrm{ts}^{\infty}(\mathscr{U})$ – the maximum dimension of a tree-stable problem from $\mathrm{Probl}_{\mathscr{U}}^{\infty}$ (Sect. 15.1.2);
- $\mathrm{rs}^{\infty}(\mathscr{U})$ – the maximum dimension of a rule-stable problem from $\mathrm{Probl}_{\mathscr{U}}^{\infty}$ (Sect. 15.1.2).

From (15.1) it follows that

$$I(\mathscr{U}) \leq \mathrm{ir}(\mathscr{U}) .$$

According to Proposition 15.1, $\mathrm{rs}^{\infty}(\mathscr{U}) = \mathrm{rs}(\mathscr{U})$ and $\mathrm{ts}^{\infty}(\mathscr{U}) = \mathrm{ts}(\mathscr{U})$. From here and from (15.2) it follows that

$$\mathrm{rs}^{\infty}(\mathscr{U}) \leq \mathrm{ts}^{\infty}(\mathscr{U}) \leq \mathrm{ir}(\mathscr{U}) .$$

Examples 15.1–15.3 show different behavior of parameters $\mathrm{rs}(\mathscr{U}) = \mathrm{rs}^{\infty}(\mathscr{U})$, $\mathrm{ts}(\mathscr{U}) = \mathrm{ts}^{\infty}(\mathscr{U})$, $I(\mathscr{U})$, and $\mathrm{ir}(\mathscr{U})$.

The next two theorems follows from Theorems 15.1 and 15.2, and Propositions 15.1 and 15.2.

Theorem 15.9 *Let* $\mathscr{U} = (A, B, F)$ *be a finite information system,* $k = |B|$, *and* $m = \max(\mathrm{ts}^{\infty}(\mathscr{U}), I(\mathscr{U}))$. *Then, for any* $n \in \omega \setminus \{0\}$, *the following statements hold:*
(a) if $n \leq m$ *then* $\frac{n}{\log_2 k} \leq I H_{\mathscr{U},\infty}^{g}(n) = H_{\mathscr{U},\infty}^{g}(n) \leq n$;
(b) if $m < n \leq \mathrm{ir}(\mathscr{U})$ *then*

$$\max\left\{\mathrm{ts}^{\infty}(\mathscr{U}), \tfrac{I(\mathscr{U})}{\log_2 k}, \log_k(n+1)\right\} \leq I H_{\mathscr{U},\infty}^{g}(n) = H_{\mathscr{U},\infty}^{g}(n)$$
$$\leq \min\left\{n-1, 4(m+1)^4 (\log_2 n)^2 + 4(m+1)^5 (\log_2 n)(\log_2 k)\right\} ;$$

(c) if $n \geq \mathrm{ir}(\mathscr{U})$ *then* $I H_{\mathscr{U},\infty}^{g}(n) = H_{\mathscr{U},\infty}^{g}(n) = I H_{\mathscr{U},\infty}^{g}(\mathrm{ir}(\mathscr{U})) = H_{\mathscr{U},\infty}^{g}(\mathrm{ir}(\mathscr{U}))$.

Theorem 15.10 *Let* $\mathscr{U} = (A, B, F)$ *be a finite information system. Then, for any* $n \in \omega \setminus \{0\}$, *the following statements hold:*
(a) if $n \leq \mathrm{rs}^{\infty}(\mathscr{U})$ *then* $I L_{\mathscr{U},\infty}^{g}(n) = L_{\mathscr{U},\infty}^{g}(n) = n$;
(b) if $n \geq \mathrm{rs}^{\infty}(\mathscr{U})$ *then* $I L_{\mathscr{U},\infty}^{g}(n) = L_{\mathscr{U},\infty}^{g}(n) = \mathrm{rs}^{\infty}(\mathscr{U})$.

References

1. Laskowski, M.: Vapnik-Chervonenkis classes of definable sets. J. Lond. Math. Soc. **45**, 377–384 (1992)
2. Moshkov, M.: Decision Trees. Theory and Applications (in Russian). Nizhny Novgorod University Publishers, Nizhny Novgorod (1994)

3. Moshkov, M.: Unimprovable upper bounds on complexity of decision trees over information systems. Found. Comput. Decis. Sci. **21**, 219–231 (1996)
4. Moshkov, M.: On time complexity of decision trees. In: Polkowski, L., Skowron, A. (eds.) Rough Sets in Knowledge Discovery 1: Methodology and Applications. Studies in Fuzziness and Soft Computing, vol. 18, pp. 160–191. Physica-Verlag (1998)
5. Moshkov, M.: Classification of infinite information systems depending on complexity of decision trees and decision rule systems. Fundam. Inf. **54**, 345–368 (2003)
6. Moshkov, M.: Time complexity of decision trees. In: Peters, J.F., Skowron, A. (eds.) Trans. Rough Sets III. Lecture Notes in Computer Science, vol. 3400, pp. 244–459. Springer (2005)
7. Moshkov, M., Zielosko, B.: Combinatorial Machine Learning – A Rough Set Approach. Studies in Computational Intelligence, vol. 360. Springer, Heidelberg (2011)
8. Pawlak, Z.: Information systems theoretical foundations. Inf. Syst. **6**(3), 205–218 (1981)
9. Vapnik, V., Chervonenkis, A.: On the uniform convergence of relative frequencies of events to their probabilities. Theory Probab. Appl. **16**, 264–280 (1971)

Final Remarks

The aim of this book is to study the decision and inhibitory trees, rules, and rule systems for the decision tables with many-valued decisions.

First, we discussed various examples which show that the considered decision and inhibitory interpretations of decision tables with many-valued decisions make sense. We mentioned without proofs some relatively simple results for decision trees, tests, rules, and rule systems obtained earlier for binary decision tables with many-valued decisions, and extended these results to the case of inhibitory trees, tests, rules, and rule systems.

Next, we extended multi-stage and bi-criteria optimization approaches to the case of decision trees, rules, and rule systems for decision tables with many-valued decisions, and generalized the obtained results to the case of inhibitory trees, rules, and rule systems. We proved the correctness of the considered algorithms and analyzed their time complexity. We described classes of decision tables for which these algorithms have polynomial time complexity depending on the number of conditional attributes in the input decision tables. We considered different applications of the created techniques related to the study of problem complexity, data mining, knowledge representation, and machine learning.

Finally, we studied the time complexity of decision and inhibitory trees and rule systems over arbitrary sets of attributes (finite and infinite) represented by information systems.

Future study will be devoted to the extension of multi-stage and bi-criteria optimization approaches to various combinatorial optimization problems represented by special circuits.

© Springer Nature Switzerland AG 2020
F. Alsolami et al., *Decision and Inhibitory Trees and Rules for Decision
Tables with Many-valued Decisions*, Intelligent Systems Reference Library 156,
https://doi.org/10.1007/978-3-030-12854-8

Index

© Springer Nature Switzerland AG 2020
F. Alsolami et al., *Decision and Inhibitory Trees and Rules for Decision Tables with Many-valued Decisions*, Intelligent Systems Reference Library 156,
https://doi.org/10.1007/978-3-030-12854-8

Printed in the United States
By Bookmasters